SYSTEMS ANALYSIS AND SIMULATION IN WILDLIFE AND FISHERIES SCIENCES

SYSTEMS ANALYSIS AND SIMULATION IN WILDLIFE AND FISHERIES SCIENCES

WILLIAM E. GRANT

Department of Wildlife and Fisheries Sciences
Texas A&M University
College Station, Texas

A Wiley-Interscience Publication

JOHN WILEY & SONS

New York • Chichester • Brisbane • Toronto • Singapore

Library of Congress Cataloging-in-Publication Data

Grant, William E. (William Edward), 1947–
Systems analysis and simulation in wildlife and
fisheries sciences.

"A Wiley-Interscience publication."
Bibliography: p.
Includes index.
1. Wildlife management—Mathematical models.
2. Fishery management—Mathematical models. 3. Systems
analysis. I. Title.
SK355.G73 1986 639.9 85-32311
ISBN 0-471-89236-X

Printed in the United States of America

10 9 8 7 6 5 4 3 2 1

To

Linda, Jennifer, Stephanie and David

PREFACE

This book presents both the philosophy and techniques of systems analysis and simulation, or modeling, as applied in wildlife and fisheries sciences. Several books on systems ecology and chapters in several books on either wildlife or fisheries provide excellent overviews of the value of modeling. However, texts providing a hands-on approach to learning modeling skills, similar to those found in engineering and management fields, have not been written for students in wildlife and fisheries sciences. My objectives are to describe the systems approach to problem solving, demonstrate its applicability to wildlife and fisheries problems, and develop a general understanding and basic vocabulary in selected areas of mathematics, statistics, and computer science as they relate to ecological modeling.

The book is targeted at graduate and upperclass undergraduate students in wildlife and fisheries sciences and is intended to serve as the basis for a one-semester course. The material should also be of interest to practicing professionals in wildlife and fisheries sciences and to students and professionals in other renewable natural resource fields. I assume that readers have little background in mathematics, statistics, or computer science.

I have attempted to achieve an appropriate balance between the presentation of conceptual and quantitative aspects of modeling. Although there is a natural tendency to become preoccupied with descriptions of quantitative techniques associated with modeling, the most significant intellectual contribution that a wildlife or fisheries professional makes to a modeling project usually involves initial development and subsequent improvement of the conceptual model. One must master a relatively small set of quantitative techniques, at least at the computational level, to gain the working knowledge of modeling that comes only from first-hand experience. But acquisition of additional techniques is most efficient if postponed until the modeling process itself is well understood.

The particular quantitative techniques included in this book are those that I have found most useful in modeling wildlife and fisheries systems. Calculations associated with each technique are described in detail; however, no derivations or rationales justifying the use of particular techniques are presented. Assumptions of specific techniques are discussed only if they apply in a special way to a particular situation presented. Derivations, rationales for use, and assumptions of the quantitative techniques used are available in standard mathematics or statistics texts.

Chapter 1 traces the development of systems ecology, introduces basic concepts of systems analysis and simulation, and presents two brief case histories of modeling in wildlife and fisheries sciences. Chapters 2 through 5 present the four phases of systems analysis: conceptual model formulation (Chapter 2), quantitative specification of the model (Chapter 3), model validation (Chapter 4), and model use (Chapter 5). Within each chapter the various steps associated with the given phase of systems analysis are identified and discussed. First the theory or concept associated with each step is presented along with brief examples. Then application of the steps is demonstrated through a detailed example. The same example, which deals with modeling the growth and harvest of fish in a hypothetical pond system, is followed through each chapter, thus giving readers an opportunity to "participate" in the entire modeling process. At the ends of Chapters 2 through 5 are exercises that deal with modeling the dynamics of a plant-herbivore system. As with the detailed example, the same problem is followed through each chapter. Chapter 6 presents a few words of caution regarding development and use of models within a management framework. The first three appendixes define terminology and present basic calculations associated with selected topics in matrix algebra (Appendix 1), differential and difference equations (Appendix 2), and regression analysis (Appendix 3). Appendix 4 presents a brief overview of the general nature of computers and programming. Appendix 5 contains an interactive, user-friendly microcomputer program with which to quantify, validate, and use the pond model from Chapters 2 through 5. Appendix 6 contains answers to the exercises in Chapters 2 through 5.

Several people helped me significantly in preparing this book. Jim Matis served as my mentor regarding use of statistical and mathematical procedures throughout development of the book and reviewed the final manuscript. Norm French, Bob Stickney, Carl Walters, and Milt Weller also provided most helpful reviews of the manuscript. Paul Carothers, Paul Ettestad, Danny Lee, and John Mellen provided many helpful suggestions from a student perspective, particularly regarding format of presentation. John Mellen also helped with many details associated with assembling the final draft. Lee George spent countless hours developing and redesigning (in response to my ever-changing goals) the computer program for the pond model example. Carol Masser prepared most of the figures. Janice Crenshaw typed the final manuscript and Connie Norton, Alice Campo, and Carla Cole typed earlier drafts. The book is much improved as a result of the effort and concern of these people. Any shortcomings that remain are mine. Finally, I owe a great deal to the students who have participated in my graduate class over the past 10 years. Collectively, they have provided me with the incentive and enthusiasm necessary to write this book and have influenced significantly my views on the use of systems analysis and simulation in wildlife and fisheries sciences.

WILLIAM E. GRANT

College Station, Texas
January 1986

CONTENTS

1. USE OF SYSTEMS ANALYSIS AND SIMULATION IN WILDLIFE AND FISHERIES SCIENCES **1**

1.1 Introduction, 1
1.2 Some basic concepts and definitions, 5
1.3 Four phases of systems analysis, 14
1.4 Two brief case histories, 22
1.5 Additional reading, 34

2. CONCEPTUAL MODEL FORMULATION **37**

2.1 Introduction, 37
2.2 Bounding the system of interest, 38
2.3 Categorizing components within the system of interest, 42
2.4 Identifying relationships between components that are of interest, 45
2.5 Formally representing the conceptual model, 48
2.6 Example of conceptual model formulation: Hypothetical pond system, 52
2.7 Exercises, 58

3. QUANTITATIVE SPECIFICATION OF THE MODEL **60**

3.1 Introduction, 60
3.2 Choosing the general quantitative structure for the model, 61
3.3 Choosing functional forms of model equations, 65
3.4 Choosing the basic time unit for simulations and parameterizing model equations, 69
3.5 Formally presenting and computer coding model equations and executing the baseline simulation, 80
3.6 Example of quantitative specification of the model: Hypothetical pond system, 88
3.7 Exercises, 123

4. MODEL VALIDATION **129**

4.1 Introduction, 129
4.2 Examining capability of the model to address the problem of interest, 130
4.3 Examining reasonableness of model structure and individual model mechanisms, 132
4.4 Examining qualitative reasonableness of overall model behavior, 133
4.5 Examining quantitative correspondence between overall model behavior and real system behavior, 137
4.6 Sensitivity analysis of the model, 149
4.7 Example of model validation: Hypothetical pond model, 151
4.8 Exercises, 166

5. MODEL USE **168**

5.1 Introduction, 168
5.2 Identifying management policies or environmental situations to be evaluated and representing them in the model, 169
5.3 Developing the experimental design for simulations, 170
5.4 Analyzing and interpreting simulation results, 172
5.5 Further examining selected types of management policies or environmental situations, 175
5.6 Example of model use: Hypothetical pond model, 178
5.7 Exercises, 199

6. A FEW WORDS OF CAUTION **200**

6.1 Introduction, 200
6.2 Example 1: Fishery model, 200
6.3 Example 2: Wildlife model, 208
6.4 Concluding remarks, 221

APPENDIX 1. MATRIX ALGEBRA **223**

A1.1 Introduction, 223
A1.2 Matrix terminology and basic matrix calculations, 224
A1.3 Examples of use of matrix algebra in wildlife and fisheries sciences, 231

APPENDIX 2. DIFFERENTIAL AND DIFFERENCE EQUATIONS **241**

A2.1 Introduction, 241
A2.2 Differential and difference equation terminology and basic calculus concepts, 242

A2.3 Examples of use of differential and difference equations in wildlife and fisheries sciences, 248

APPENDIX 3. REGRESSION ANALYSIS 252
A3.1 Introduction, 252
A3.2 Regression terminology and simple linear regression, 253
A3.3 Multiple linear regression, 259
A3.4 Linear transformations, 263
A3.5 Examples of use of regression analysis in wildlife and fisheries sciences, 265

APPENDIX 4. COMPUTERS AND PROGRAMMING 266
A4.1 Introduction, 266
A4.2 Analog computers, 267
A4.3 Digital computers, 267

APPENDIX 5. COMPUTER PROGRAM FOR THE POND SIMULATION MODEL 271
A5.1 Introduction, 271
A5.2 User's guide, 271
A5.3 Program listing, 278

APPENDIX 6. ANSWERS TO EXERCISES 304
A6.1 Answers to Chapter 2 exercises, 304
A6.2 Answers to Chapter 3 exercises, 306
A6.3 Answers to Chapter 4 exercises, 314
A6.4 Answers to Chapter 5 exercises, 322

References, 327

Index, 333

CHAPTER 1

USE OF SYSTEMS ANALYSIS AND SIMULATION IN WILDLIFE AND FISHERIES SCIENCES

1.1. INTRODUCTION

The most interesting and critical problems that we face as wildlife and fisheries scientists and managers are inherently complex. We may need to consider many interacting biological and ecological, as well as economic, social, and legal factors when addressing a given wildlife or fisheries issue. Attempts to deal with such complex problems in a narrow or fragmentary way often lead to poor research design and ultimately to poor management decisions. Management plans based on sound biological information fail when unforeseen economic factors shift hunting or fishing pressure. Economic policies fail when unforeseen biological limits are exceeded.

Systems analysis is both a philosophical approach and a collection of techniques (including simulation) developed explicitly to address complex problems. The origins of systems analysis can be traced to the military's attempts to deal with complex logistical problems during World War II. It has been applied successfully in a variety of fields in engineering, industrial dynamics, business management, and economics, and more recently has found increasing application in the fields of biology, ecology, and renewable natural resource management.

Systems analysis emphasizes a holistic approach to problem solving and the use of mathematical models to identify and simulate important characteristics of complex systems. We will define terms in greater detail shortly, but in the simplest sense a system is any set of objects that interact. A mathematical model is a set of equations which describes the interrelationships among system components. By solving the equations comprising a mathematical model we can mimic, or simulate, the dynamic (time-varying) behavior of the system.

Application of systems analysis in wildlife and fisheries sciences, as in other ecological fields, is a natural outgrowth of our growing need to quantitatively address increasingly complex problems and our search for appropriate methodologies in other disciplines that historically have had a quantitative problem-solving orientation. This is not to say that the transfer of methodology across disciplines has been easy or is complete. Indeed, as with any such transfer of technology, translation of ideas into a more familiar framework and custom-fitting of techniques to specific problems have not happened spontaneously and have been accompanied by legitimate apprehension.

Kitching (1983, pp. 1–4) provides an excellent account of the gradual integration of mathematics and computing into ecology that has given birth to the relatively new field of "systems ecology," which subsequently has provided the main impetus for recent interest in use of systems analysis in wildlife and fisheries. In the 1960s C. S. Holling and K. E. F. Watt were among the first to recognize the value of applying available techniques of computing and systems analysis to modeling ecological systems. Watt's book *Systems Analysis in Ecology* (1966) was the first attempt at a synthesis in this area. A second book by Watt, *Ecology and Resource Management* (1968), was the first to present the philosophy, techniques, and application of ecological modeling. Also in the 1960s, others were developing modeling techniques to examine the fates of radiotracers in ecosystems (e.g., Olson 1960). These techniques, which legitimately can be viewed as a specialized part of systems analysis, have become known as "compartment modeling" (see review by O'Neill 1979). During the late 1960s and early 1970s the ideas of Holling and Watt and the concepts of compartment modeling served as the basis for development of several large ecosystem models under the aegis of the International Biological Program (IBP) (e.g., Innis 1978). Results from the IBP studies, which included a tremendous volume of diverse field data, have been synthesized using the techniques of systems analysis (e.g., Breymeyer and Van Dyne 1980, Reichle 1981, Bliss et al. 1981). Through the IBP a relatively large cohort of "systems ecologists" were educated and many other ecologists were exposed in one way or another to the application of systems analysis in ecology.

Systems ecology has grown rapidly during the 1970s and early 1980s and a wide range of ecological systems have been modeled using a diversity of mathematical and computing techniques. B. C. Patten's landmark four volume series *Systems Analysis and Simulation in Ecology* (1971a, 1972, 1975, 1976) exemplifies the breadth of the field (see also Innis and O'Neill 1979, Matis et al. 1979, Shugart and O'Neill 1979). This breadth also is reflected in the very successful international scientific journal *Ecological Modelling*, which was established in

1975. The rapid growth of systems ecology has been accompanied, albeit with a slight time lag, with maturation in the sense that practitioners are beginning to reflect more critically on the potential dangers, as well as benefits, of modeling in ecology (e.g., Pielou 1981).

The use of mathematical models in fisheries work was established in the late 1950's by Beverton and Holt (1957). Building on this cornerstone, and drawing upon the work of early systems ecologists, fisheries scientists were among the first to use simulation models in an applied renewable natural resource field. In 1969 the *Transactions of the American Fisheries Society* devoted part of an issue (vol. 98, no. 3) to the use of simulation models which contained important articles by Walters, Silliman, Patten, and Paulik. Subsequently a number of models have been applied to specific fisheries [e.g., Larkin (1971) on sockeye salmon, Tillman and Stadelman (1976) on anchovy, Grant and Griffin (1979) on Gulf of Mexico shrimp] and a number of specific modeling techniques have been used [e.g., dynamic optimization (Walters and Hilborn 1976), stochastic dynamic programming (Hilborn 1976), catastrophy theory (Jones and Walters 1976), Leslie matrices (Horst 1977)]. Walters (1980), in a chapter for a recent fisheries text, provides an excellent overview of the use of systems analysis in fisheries management.

The use of mathematical models in wildlife work has a shorter history, although one could argue that the basis for much of the more recent population dynamics modeling was established in the 1940s by Leslie (1945, 1948). Walters and Brandy (1972) were among the first to demonstrate the usefulness of simulation in evaluating alternative harvesting schemes. Subsequently a number of models have been applied to specific game species [e.g., Sinclair (1973) on African buffalo, Walters et al. (1975) on barren-ground caribou, Brown et al. (1976) on North American mallard, Roseberry (1979) on bobwhite quail]. However, in general, mathematical modeling has been used relatively less in wildlife than in fisheries work. This is surprising given the many commonalities between wildlife and fisheries problems and is not easily explained. Tipton (1980) provides an excellent overview of the principles of mathematical modeling as applied in wildlife management.

The benefits of a ''systems'' approach in wildlife and fisheries work, such as the emphasis that it places on clear delineation of program objectives and its usefulness in identifying areas where critical data are lacking, now are well accepted. But other aspects of systems analysis, such as the use of simulation models in evaluating management policy alternatives, are more controversial. This controversy results from a lack of knowledge regarding philosophy and techniques involved in systems analysis and simulation, and from a concern about adequacy of available data bases in wildlife and fisheries to support application of these techniques. These two areas—(1) description of the philosophy and techniques of systems analysis and simulation and (2) application of these techniques to various types of information bases—form the primary subject matter of this book.

We will focus primarily on the use of systems analysis and simulation within a management-oriented framework and on the role of models in providing information for the decision-making process. Such usage often involves development of a quantitative model predicting effects of a management policy or environmental change on some aspect of a wildlife or fisheries system. The model is used to simulate behavior of the real system under the particular circumstances in which we are interested. To the extent that we have confidence in the model as a reliable representation of the real system, simulation results become input into the decision-making process.

But the role of models in wildlife and fisheries research and education also should be clear. Just as models allow managers to evaluate alternative management policies via computer simulation, models also allow researchers to explore alternative hypotheses about relationships within a system and to simulate a variety of experiments before choosing experiments that will be conducted in the real system. Information obtained from real-world experiments then can be used to further refine the model in an iterative process within the research or management program. In an educational setting models can be used to demonstrate basic principles of ecology and management via "gaming" simulations in which students interact with the model as they would with the real system.

Before discussing some basic concepts in systems analysis, a few words of caution are in order. Although systems analysis and simulation hold great promise as useful tools in wildlife and fisheries sciences, there are at least five general areas of potential danger that both beginning and experienced modelers should always keep in mind. All are concerned in one way or another with becoming enamored of models themselves, rather than viewing models simply as a means to an end.

First, many valuable resources (both in terms of money and brainpower) have been wasted building models that cannot be used because the purpose for which they were built was never clearly identified. It is not sufficient justification to begin a modeling effort simply because we have the capability to do so. Questions that the model is to address must be clearly stated, relevant, and answerable. The formulation of specific objectives is the most important, and often most neglected, aspect of almost any undertaking, and modeling is no exception. In fact, a clear statement of project objectives may indicate that mathematical modeling is not appropriate.

A second common pitfall is to become preoccupied with the "automated" aspects of model formulation and evaluation, at the expense of good common sense. It is tempting to look for statistical or mathematical procedures that when applied in a completely objective manner will guarantee development of a proper model. Likewise it is tempting to search for completely objective criteria, often in the form of statistical tests, by which a model may be certified safe for use. We must be masters rather than slaves of such automated procedures and resist the tendency to use them as a substitute for critical thought. There is no substitute for first-hand familiarity with the systems being modeled. It is all too easy to build nonsense into a model if we are naive about subtleties of the real-world, as the oft-quoted modeling maxim "garbage in, garbage out" suggests.

A third common danger involves the perception that a model, once completed, is *the* model of the system. This implied permanence of a model can inhibit our ability to gain new insights into system behavior, which should be one of the main benefits of modeling. A model should evolve as our understanding of the system changes as the result of newly available information, and as our questions about the system change.

Yet a fourth potential pitfall is to take a "shotgun" approach when using a model to address project objectives. After a model finally is fully developed, validated, and ready to use, it usually is very easy to run a great many simulations quickly and inexpensively. However, since each simulation is analogous to an experiment conducted in the real system, the experimental design for simulations should be as carefully thought out as would be the case if we were designing a set of field or laboratory experiments. Although computers provide the capability to simulate many experiments with relative ease, it still is incumbent upon the modeler to design experiments well.

A final danger is the failure to distinguish between simulation results and results from experimentation with the real system. As we use a given model to address our initial questions about a system, initial results often raise new questions that lead to more simulations and, in turn, more new questions. Such exploratory simulations, if done in a systematic fashion, can yield much insight into model behavior, but the procedure can become hypnotic. At some point, if at all possible, important simulation results must be corroborated by field or laboratory observations. Answers to real-world questions ultimately must be found in the real world.

1.2. SOME BASIC CONCEPTS AND DEFINITIONS

System

Like many words, "system" is difficult to define precisely. In relation to the physical or biological sciences, a system is an organized collection of interrelated physical components characterized by a boundary and functional unity.

A system is any collection of "communicating" materials and processes that together perform some set of functions. A system is an interlocking complex of processes characterized by many reciprocal cause–effect pathways.

Clearly any collection of interacting objects can be viewed as a system. The principal attribute of a system is that we can understand it only by viewing it as a whole. Another important attribute of a system is that it is chosen for a particular purpose: to answer a question, to demonstrate a theory, or to classify part of the natural world. Common examples of ecological systems include ecosystems, communities, populations, and individuals. But the rumen of a deer may be viewed as a system, as can the planet earth.

We may view, or conceptualize, the same real-world system in different ways depending upon our particular interests. A film that has appeared on television demonstrates this idea nicely. The film begins with the camera focused on a 10 square meter area from a height of 10 meters. From this perspective we see that the camera is centered on a man and a woman having a picnic lunch. In addition to the couple we see a picnic basket and the beach blanket upon which they are sitting. As the film proceeds, the camera backs off and increases its coverage by an order of magnitude every 5 seconds. After 5 seconds we are viewing an area of 100 m^2 from a height of 100 m, after 10 seconds an area of 1000 m^2 from a height of 1000 m, and so on. As the perspective changes we see that the couple is on a beach with many other people, that the beach is near a big city with tall buildings, and that the lake also is quite large. As the perspective changes further we see that the couple must be on a beach near Chicago or Milwaukee as the outline of Lake Michigan becomes clear. Soon we are viewing all of North America, then the entire earth, and so it goes until we behold the Milky Way!

Of course none of these views of the real world is any more valid or real than any other view. But we would all agree that for any particular set of interests some views would be more appropriate and enlightening than others. If our primary interest is in the movements of people on the beach, we would be ill advised to worry about the stars in the Milky Way. Likewise we would be ill advised to confine our attention solely to the original couple on the beach. The system that we could most fruitfully analyze to obtain results relevant to our primary interest lies in between these two extremes.

Summarizing these ideas more formally, systems have two properties that are most useful. First, systems may be nested: an individual is part of a population, a population is part of a community, and so on. But systems at any level of resolution can be studied using a common set of principles and techniques often referred to as "general systems theory" (von Bertalanffy 1964). Second, systems at the same level of resolution may overlap: the system that we define to study population dynamics of species A will overlap with the system that we define to study population dynamics of species B if the two species are competitors.

For our purposes it is useful to think in terms of a subjective "system of interest" that is defined arbitrarily relative to a specific problem. This is not to imply that the systems we study can be defined capriciously. We must take great care to define the boundaries of the system of interest appropriately for the problem at hand. As we shall see shortly, this is the first step in systems analysis and definitely is not a trivial task.

Systems Analysis

Perhaps systems analysis can be defined most directly as application of the scientific method to problems involving complex systems. It is a body of theory and techniques for studying, describing, and making predictions about complex systems, which often is characterized by use of advanced mathematical and statistical procedures and by use of computers. However, the essence of systems analysis is not found in a collection of quantitative techniques, but rather in a broad problem-solving strategy.

The goal of systems analysis within a wildlife or fisheries management framework quite simply is to promote good decision-making in practical situations. The systems approach to problem solving is nothing more than the application of good common sense as we think through a problem. The rigor that is introduced into the procedure and the emphasis that is placed on quantification of ideas and relationships serve only to clarify our thinking in terms of what we really understand about our problem and what we really do not. Yet the introduction of this formalism into our thinking can be a powerful aid in tackling complex problems.

Model

A model is an abstraction of reality. It is a formal description of the essential elements of a problem. Since the essential elements of a problem are exactly those that we define to be in our system of interest, we can think of a model as a formal description of the system of interest. The description can be physical, mathematical, or verbal, although some modelers object to the idea of "word models" because language can be so ambiguous (Jeffers 1978).

Models can be classified in a variety of ways (e.g., see Forrester 1961, Gold 1977, Jeffers 1978). Some of the more relevant dichotomies for our purposes include:

1. physical versus abstract,
2. dynamic versus static,
3. empirical (correlative) versus mechanistic (explanatory),

4. deterministic versus stochastic, and
5. simulation versus analytical.

Physical versus Abstract

Physical models usually are physical replicas on a reduced scale of objects under study. A scaled-down architectural model used to help us visualize floor plans and space relationships is an example. Of course even physical models still are abstractions of reality in the sense of our original definition of a model. Abstract models use symbols rather than physical devices to represent the system being studied. The symbolism used can be a written language, a verbal description, or a thought process. A mathematical model is a special type of abstract model written in the language of mathematics. It is fundamentally the same as other abstract models in that it is a description of the system that it represents. But since mathematical notation is more specific than language, a mathematical model is less ambiguous than most word models. The process of converting a word model to a mathematical model is not inherently difficult. The word model "energy requirements of individual A are 100 kcal/day when the ambient temperature is 0°C and increase by 20 kcal/day with each 10°C decrease in ambient temperature" can be translated easily into the mathematical model "$Y = 100 - 2X$," where Y represents energy requirements (kcal/day) and X represents ambient temperature (°C). But difficulties arise when the word model is not an adequate description of the system, as often is the case with complex models.

Dynamic versus Static

Models may or may not represent systems that change with time. A static model describes a relationship or set of relationships that do not change with time. Common examples include regression models that do not have time as an independent variable. A dynamic model describes a time-varying relationship. Examples include simulation models, which we will discuss further in a moment, and regression models that do include time as an independent variable.

Empirical (Correlative) versus Mechanistic (Explanatory)

Empirical, or correlative, models are developed primarily to describe and summarize a set of relationships, without regard for appropriate representation of processes or mechanisms that actually are operating in the real system. The goal is prediction, not explanation. A model predicting metabolic rate of an animal solely as a function of body size is an example. Mechanistic, or explanatory, models are developed primarily to represent internal dynamics of the system of interest appropriately. The goal is explanation through representation of the causal mechanisms underlying system behavior. A model representing metabolic rate of an animal as a function of body size, level of activity, environmental temperature, wind, and length of exposure to ambient conditions is an example. Solely mechanistic models and solely empirical models form two ends of a continuum, and classification relative to this dichotomy is based more on intent of the modeler than on structure of the model. A model that we view as explanatory at one level of resolution we might view as correlative at a finer level of resolution. A model representing annual population recruitment as a function of population size appears explanatory compared to a model representing annual recruitment simply as a constant determined by averaging historical data. But this same model appears correlative compared to a model that calculates recruitment based on age-specific natality rates of individuals within the population, which, in turn, are based on the individuals' social rank and nutritional status during the breeding season.

Deterministic versus Stochastic

A model is deterministic if it contains no random variables. Deterministic model predictions under a specific set of conditions are always exactly the same. The simple model that we developed relating energy requirements of an indivudual (Y, in kcal/day) to ambient temperature (X, in °C),

$$Y = 100 - 2X$$

is a deterministic model. When ambient temperature is 0°C, the model always predicts that energy requirements are 100 kcal/day. When temperature is −10°C, requirements always are 120 kcal/day, and so on. A model is stochastic if it contains one or more random variables. Stochastic model predictions under a specified set of conditions are not always exactly the same, because random variables within the model potentially can take on different values each time the model is solved. Our deterministic energy requirements model has the general form

$$Y = a - bX$$

where a and b are constants ($a = 100$, $b = 2$). We can change this into a stochastic model by representing either a or b as a random variable. Suppose we represent b as a random variable that can take on values of 2.0 or 2.5 with equal probability. Now each time we compute a model prediction we must randomly choose a value for b from the distribution of b values that we have specified. We could flip a coin: heads means $b = 2.0$, tails means $b = 2.5$. Model predictions of energy requirements when ambient temperature is −10°C (or any other temperature except 0°C) no longer are always the same. If the coin lands heads-up we compute

$$Y = 100 - 2.0(-10) = 120$$

if the coin lands tails-up we compute

$$Y = 100 - 2.5(-10) = 125$$

Of course random variables in most stochastic models are chosen from more "sophisticated" distributions (uniform, normal, negative binomial, etc.), but the general procedure is exactly the same.

The choice between using a deterministic or a stochastic model depends on the specific objectives of the modeling project. Deterministic models generally are easier to build because they require only point estimation of constants, rather than specification of complete distributions for random variables as is the case for stochastic models. They also are easier to use because predictions for a given situation need be made only once (since they always are the same), whereas stochastic model predictions must be replicated an appropriate number of times to obtain the average response for a given situation. However if project objectives require explicit representation of variability, either the variability associated with estimates of system parameters or the inherent variability of the system itself, we must use a stochastic model. Likewise if we wish to make statistical comparisons of model predictions for different situations we must use a stochastic model.

Simulation versus Analytical

Models that can be solved in closed form mathematically are analytical models. Regression models, models of standard theoretical statistical distributions, and some simple differential equation models are examples. For such models a general solution can be obtained which is applicable to all situations the model can represent. A simple analytical model of population growth in an unlimited environment (exponential growth) is

$$N_t = N_0 e^{rt}$$

where N_t = population size at time t
N_0 = initial population size
r = intrinsic rate of population increase
t = time

We can solve this model for population size (N_t) at any time (t) by substituting the desired value for t into the model, assuming that we have defined initial population size (N_0) and intrinsic rate of population increase (r). If initial population size is 100 individuals and intrinsic rate of increase is 0.1, population size at time 5 is

$$N_5 = 100e^{0.1(5)} = 164.9$$

at time 8 is

$$N_8 = 100e^{0.1(8)} = 222.6$$

and so on. Models for which a general analytical solution is not possible must be solved numerically, using a specified set of arithmetic operations, for each particular situation the model can represent. These are simulation models. Many ecological models are of this type. For example, models representing population dynamics in response to density-dependent, competitive relationships, which, in turn, are influenced by changing environmental conditions, might be summarized in general form as

$$N_{t+1} = f(N_t, E_t)$$

where N_{t+1} = population size at time $t+1$
$f(N_t, E_t)$ = some complex function of population size and environmental conditions at time t

Often we cannot solve such models analytically because the function, or set of equations, comprising $f(N_t, E_t)$ is too complex. The model must be solved numerically for each successive time step. We will take a more detailed look at simulation (including a numerical example) in the next section.

Philosophically the choice between analytical and simulation models involves whether we sacrifice ecological realism to obtain an analytical model or sacrifice mathematical power to include more ecological realism. From a more practical standpoint these considerations are no less interesting and are influenced primarily by objectives of the modeling project. If the level of resolution at which the system of interest must be represented to meet our objectives permits use of an analytical model, then we should use an analytical model. However, if the appropriate level of resolution requires a model too complex to be represented in an analytically tractable form, then we should use a simulation model. In most cases in wildlife and fisheries management we will need to represent the system of interest in a manner too complex for analytical treatment. Thus simulation models have been and should continue to be more widely used to address management-oriented questions concerning wildlife and fisheries systems.

Simulation

Simulation is the process of using a model to mimic, or trace through step by step, the behavior of the system we are studying. Simulation models are composed of a series of arithmetic and logical operations that together represent the structure (state) and behavior (change of state) of the system of interest. These two very simple yet powerful concepts of the state of the system and the change of state of the system are fundamental to simulation (Patten 1971b). The system of interest exists in different states at different points in time and there are rules governing the manner in which the state of the system changes as time passes. The rules governing change also may change from time to time, since they are themselves a function of the state of the system. If we choose appropriate variables to describe the system and appropriately represent the rules governing change, then we should be able to trace the state of the system through time, which is to say we can simulate behavior of the system.

Suppose that we wish to simulate fluctuations in weight (in g) of an animal over time. Conceptually our system of interest includes amount of biomass contained in the individual animal, which varies over time as a function of amount of food consumed and amount of energy respired. Assume that assimilation efficiency is 100 percent. We might choose number of g in the animal at any time t (W_t) as the variable describing the state of the system. Further suppose we know that rate of consumption of food (in g/day) is a function of weight of the animal and that respiration rate (in g/day) is a function of weight of the animal and environmental temperature (Figure 1.1). (The symbols used in Figure 1.1 have specific meanings which we will define formally in Chapter 2. For now note only that the figure is a simple diagrammatic representation of the weight fluctuation model just described.) We might appropriately represent the rules governing changes in the state of the system using two equations, one representing consumption (C) as a function of animal weight (W) and one

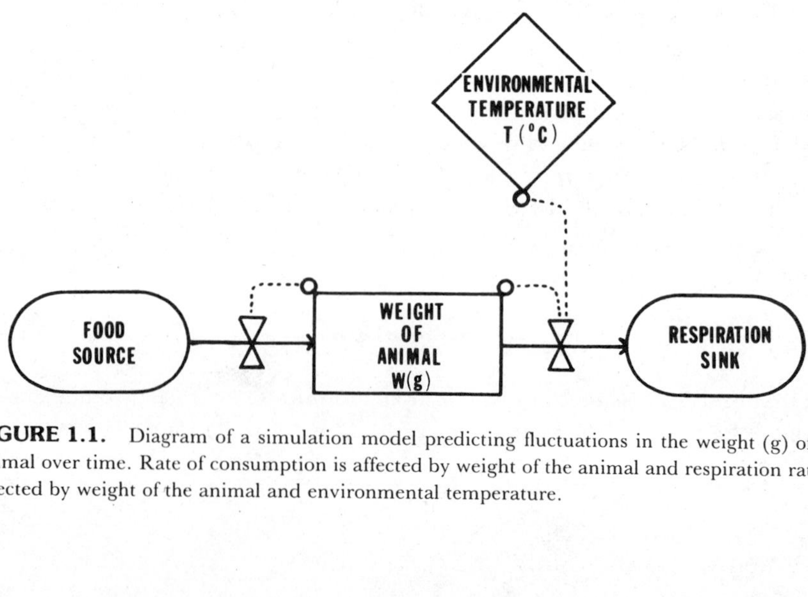

FIGURE 1.1. Diagram of a simulation model predicting fluctuations in the weight (g) of an animal over time. Rate of consumption is affected by weight of the animal and respiration rate is affected by weight of the animal and environmental temperature.

representing respiration (R) as a function of weight and environmental temperature (T).

We now can construct a simulation model predicting fluctuations in weight of the animal under a specific set of environmental temperatures, given that we know the initial weight. Mathematical representation of simulation models may take a variety of forms. One useful format involves use of difference equations. We might construct a simulation model for our example using difference equations as follows.

$$W_{t+1} = W_t + \frac{\Delta W}{\Delta t} \tag{1}$$

where

$$\frac{\Delta W}{\Delta t} = C_{t,t+1} - R_{t,t+1} \tag{2}$$

$$C_{t,t+1} = f(W_t) \tag{3}$$

$$R_{t,t+1} = f(W_t, T_t) \tag{4}$$

Equation (1) states that number of g in the animal one time unit into the future—at time $t+1$ (W_{t+1})—is equal to number of g in the animal now—at time t (W_t)—plus net change in number of g in the animal over the next time unit—from t to $t+1$ ($\Delta W/\Delta t$).

Equation (2), which is the difference equation from which this mathematical format takes its name, states that net change in number of g in the animal over the next time unit ($\Delta W/\Delta t$) is equal to number of g consumed over the next time unit ($C_{t,t+1}$) minus number of g respired over the next time unit ($R_{t,t+1}$).

Equation (3) states that number of g consumed over the next time unit ($C_{t,t+1}$) is equal to a function (f) of animal weight at the beginning of the time unit (W_t).

Equation (4) states that number of g respired over the next time unit ($R_{t,t+1}$) is equal to a function (f) of animal weight at time t (W_t) and environmental temperature at the beginning of the time unit (T_t). Of course $f(W_t)$ and $f(W_t, T_t)$ need to be made explicit. For example if C was an increasing linear function of W, then $f(W_t)$ might be the equation for a straight line:

$$C_{t,t+1} = f(W_t) = a + bW_t$$

We can simulate fluctuations in animal weight by evaluating these equations each time unit. Suppose that the time unit (Δt) is 1 day, that initial weight of the animal (W_0) is 100 g, and that environmental temperature for 2 consecutive days is $T_0 = 20°C$ and $T_1 = 19°C$. Further suppose that

$$C_{t,t+1} = f(W_t) = 0.05\,W_t$$

$$R_{t,t+1} = f(W_t, T_t) = 0.05\, W_t\left(\frac{T_t}{20}\right)$$

Thus

$$C_{0,1} = 0.05\, W_0 = 0.05(100) = 5$$

$$R_{0,1} = 0.05\, W_0 \left(\frac{T_0}{20}\right) = 0.05\,(100)\left(\frac{20}{20}\right) = 5$$

$$\frac{\Delta W}{\Delta t} = C_{0,1} - R_{0,1} = 5 - 5 = 0$$

$$W_1 = W_0 + \frac{\Delta W}{\Delta t} = 100 + 0 = 100$$

and

$$C_{1,2} = 0.05\, W_1 = 0.05(100) = 5$$

$$R_{1,2} = 0.05\, W_1\left(\frac{T_1}{20}\right) = 0.05(100)\left(\frac{19}{20}\right) = 4.75$$

$$\frac{\Delta W}{\Delta t} = C_{1,2} - R_{1,2} = 5 - 4.75 = 0.25$$

$$W_2 = W_1 + \frac{\Delta W}{\Delta t} = 100 + 0.25 = 100.25$$

Our simulation model predicts that a 100 g animal weighs 100.25 g after 2 days given that environmental temperatures are 20°C and 19°C on days 0 and 1, respectively.

1.3. FOUR PHASES OF SYSTEMS ANALYSIS

We have defined systems analysis as the application of the scientific method to problems involving complex systems. A variety of schemes summarizing the systems approach have been suggested (Gold 1977, Jeffers 1978, Innis 1979, Kitching 1983, and others). These differ with regard to details (number of steps, names of steps), but all convey the same underlying philosophy. The particular scheme presented here (Figure 1.2) reflects my personal view of the process, which has been shaped by application of the systems approach to a variety of problems in wildlife and fisheries management.

There are several important aspects of problem definition that always should be considered before application of the systems approach. These considerations are an integral part of the scientific method and should be part of any

PHASES OF SYSTEMS ANALYSIS

Phase I: Conceptual Model Formulation

1. Bounding the system of interest
2. Categorizing components within the system of interest
3. Identifying relationships between components that are of interest
4. Formally representing the conceptual model

↓

Phase II: Quantitative Specification of the Model

1. Choosing the general quantitative structure for the model
2. Choosing functional forms of model equations
3. Choosing the basic time unit for simulations and parameterizing model equations
4. Formally presenting and computer coding model equations and executing the baseline simulation

↓

Phase III: Model Validation

1. Examining capability of the model to address the problem of interest
2. Examining reasonableness of model structure and individual model mechanisms
3. Examining qualitative reasonableness of overall model behavior
4. Examining quantitative correspondence between overall model behavior and real system behavior
5. Sensitivity analysis of the model

↓

Phase IV: Model Use

1. Identifying management policies or environmental situations to be evaluated and representing them in the model
2. Developing the experimental design for simulations
3. Analyzing and interpreting simulation results
4. Further examining selected types of management policies or environmental situations

FIGURE 1.2. Four phases in systems analysis and various steps within each phase.

common sense approach to problem solving. We begin with a clear statement of our interests in terms of a problem to be solved or a question to be answered. Questions may arise from general observations of a system, as is the usual case in scientific inquiry, or may be imposed by the practical necessity of evaluating proposed management schemes. Often the initial question, such as "what are the effects of hunting on a white-tailed deer population," is too broad to address directly. We must develop a set of specific questions that can be answered directly to satisfy our interests. Questions such as "Will the number of deer increase or decrease over the next 10 years if 25 percent of the does and 90 percent of the bucks are harvested annually?" and "Will the average weight of harvested deer be increased if population size is reduced and maintained by hunting at 1 deer per 6 hectares?" are examples.

Next we must determine if the questions we have stated are important enough to justify detailed analysis and, if so, whether the systems approach is required. Our questions, although valid and interesting, simply may not be important enough to justify the manpower and money necessary to analyze thoroughly. Alternatively, we may be able to answer our questions completely using simple mathematical or statistical methods of analysis, or perhaps using qualitative descriptions. In either case, systems analysis is unnecessary. Suppose that we wanted to know whether the fish we stocked in a series of ponds would grow faster on feed A or on feed B. We could develop a complex mathematical model of the pond system, representing a multitude of ecological relationships, and simulate fish growth on each feed. Much time and money would be needed to collect the data to build such a model. We could answer the question in a much simpler fashion by feeding equal amounts of feed A and feed B in different ponds and measuring fish growth. If we had enough ponds to replicate the experiment we could analyze the data using a simple statistical model (a *t*-test).

These initial considerations of problem definition may seem trivial, but they most certainly are not. The choice to use systems analysis must be justified based on the nature of the problem we wish to address. Having chosen the systems approach, the decisions that must be made throughout the various phases of systems analysis can be made only in view of a specific set of objectives. If the problem is ill defined or unimportant, our results will be misunderstood or ignored regardless of the apparent sophistication of our techniques.

Next we provide a general overview of the four phases of systems analysis following the scheme in Figure 1.2. We will examine each of the four phases in more detail in Chapters 2 through 5.

Phase I: Conceptual Model Formulation

The goal of the first phase of systems analysis is to develop a conceptual, or qualitative, model of the system of interest (Figure 1.2). We abstract from the real system those components that must be considered to address our questions. By including these components within our model and excluding all others, we bound the system of interest. Next we categorize model components as state, driving, or auxiliary variables depending on their specific roles in describing model structure and behavior (we will describe the criteria for this categorization in Chapter 2), and identify specific relationships between components. Finally we formally represent the resulting conceptual model, often as a box and arrow diagram indicating points of accumulation of material (boxes), such as individuals, energy, biomass, nutrients, or dollars, and routes by which the material flows within the system (arrows).

In many respects conceptual model formulation is the most intellectually challenging phase of systems analysis. The best basis for the many difficult, and often highly subjective, decisions that must be made regarding choice of model components is a thorough familiarity with the real system. Prior modeling experience also is an asset. There are two general approaches to identifying model components: One is to make the initial choice of components as simple as possible and subsequently add critical components that were overlooked; the other is to include initially all components that possibly could have any importance and then delete superfluous ones. Although it can be argued that the later approach is more consistent with the holistic philosophy of systems analysis, in practice the approach taken reflects the particular preference of the individual modeler. The end product of either approach should be a conceptual model that is no more complex than is absolutely necessary to address our interests.

Phase II: Quantitative Specification of the Model

The goal of the second phase of systems analysis is to develop a quantitative, or mathematical, model of the system of interest (Figure 1.2). Using the conceptual model as a template for this quantitative development, we describe the rules governing the flow of materials in the model (the dynamics of the model) using mathematical equations. The first step is to choose a general quantitative structure for the model. Perhaps the simplest and most flexible general structure consists of modules containing two model components connected by a flow of material (in terms of our conceptual model, two boxes connected by an arrow). This general compartmental structure lends itself well to description of complex models, because it facilitates decomposition of complex interrelations into simpler cause–effect pathways.

There is a variety of specific mathematical techniques, such as matrix algebra, differential equations, and difference equations, that can be used within this general compartmental framework. Some pertinent details regarding these particular techniques are presented in Appendices 1 and 2, and Kitching (1983, p. 37) provides references to several more specialized techniques. Seldom will our system of interest have only one possible mathematical formulation, and we may use different techniques to develop alternative models of the same system. Choice of an appropriate formulation depends on objectives of the modeling project and on familiarity with capabilities of the various techniques. Models formulated using certain techniques have the potential to be solved analytically. Deterministic models comprised of algebraic equations or simple sets of linear differential equations are examples. Deterministic models comprised of more complex sets of linear differential equations, non-linear differential equations, or difference equations usually must be solved numerically by simulation. Most stochastic models must be solved numerically, however analytical solutions are possible in a few special cases. Obviously, we would be well advised to develop a sufficient mathematical vocabulary to communicate effectively with mathematicians regarding choice of appropriate techniques.

Having decided on the general mathematical form of the model, we next must develop the specific equations that collectively comprise the model. This consists of choosing the functional form (linear, exponential, sigmoid, sinusoidal, etc.) for, and estimating the parameters of, each equation. We also must choose the basic time step for solution of model equations (one day, one week, one year, etc.) if we are developing a simulation model, since this will affect our estimation of rate parameters. The best type of information that we can use to develop model equations is data from the real system being modeled. Available data often can be analyzed using standard statistical procedures to quantify various aspects of the model. The calculation of a simple mean to estimate a constant required in an equation, and the use of regression analysis to specify an equation relating rate of change of one variable to rate of change of another variable are examples. However, most commonly there will be some aspects of the model for which no data are available and for which we cannot collect new data at the present time. In such cases we may be able to use information based on theoretical or generally applicable empirical relationships. The use of the general empirical equation relating metabolic rate to body size (metabolic rate $= A \times$ body weight^B), which had been parameterized for a species closely related to the species being modeled, is an example. Alternatively, we may be able to convert qualitative information, either from the technical literature or from "expert opinion," into a quantitative form. For example, an expert may be able to tell us that his catch of fish per unit of fishing effort is roughly twice as great during the spring as it is during the fall, although he has not kept exact records to document his observations. Finally, after the model is fairly well developed, we may be able to use information generated from experimentation with the model itself to gain insight into the quantification of a certain relationship. This is educated guesswork, but based on model behavior it often is possible to greatly limit the reasonable alternatives.

The final step in quantitative specification of the model is to present model equations formally and, for simulation models, to execute the baseline simulation, which usually requires computer coding of the model. Baseline simulation refers to solving the model, or simulating behavior of the system of interest, under a specific set of conditions that usually represent the "normal" or "status quo" situation for the system. Model equations are presented formally by listing them sequentially in some logical order that unambiguously describes how to solve the model. Computer coding of the model consists of translating model equations from the language of mathematics to a language that digital computers understand. There is a variety of computer languages and also a variety of digital computers. (Analog computers seldom are used in ecological modeling, although they are quite useful in certain situations.) Choice of an appropriate computer, as with choice of an appropriate mathematical technique, depends on objectives of the modeling project. Appendix 4 describes the basic characteristics of both analog and digital computers and presents an overview of general programming strategy.

Phase III: Model Validation

The goal of the third phase of systems analysis is to ensure that the model is good enough to satisfy our needs (Figure 1.2). The process by which we define "good enough" is perhaps the most controversial aspect of modeling. Model validation has received much attention from a theoretical standpoint (e.g., Caswell 1976, Mankin et al. 1977, Holling 1978, Steinhorst 1979), but in practice it all too often is treated rather superficially. The difficulty arises primarily because it is impossible to develop a unique set of validation criteria. Models can be validated only in terms of their capability to address the problems in which we are interested.

Although it is possible to outline several steps in the validation process (Figure 1.2), the relative emphasis given to each step depends on the type of model required. If we construct an empirical, or correlative, model primarily to predict some particular aspect of system behavior well, then validation is concerned primarily with the quantitative correspondence between that aspect of model behavior and behavior of the real system. We are concerned less with examining the ecological soundness of individual mechanisms in the model, or with the qualitative reasonableness of overall model behavior. If the model is stochastic and replicated data are available from the real system, we can use standard statistical techniques to determine if the model has met our quantitative performance criteria. If we construct a mechanistic, or explanatory, model primarily to represent appropriately internal dynamics of the system of interest, then validation is concerned primarily with the correspondence between mechanisms included in the model and the details of those mechanisms in the real system. We are concerned less with the quantitative predictive capabilities of the model than with the qualitative reasonableness of its behavior.

The model in Figure 1.1 was constructed primarily to predict fluctuations in weight of an animal over time. Thus the validation procedure should emphasize comparison of model predictions of animal weight to data on a real animal growing under the specified conditions. But suppose that the model had been developed primarily to reflect appropriately the relationship between consumption and respiration of an animal over a range of environmental temperatures. Then the validation procedure should emphasize comparison of model predictions of the respiration/consumption ratio to this ratio in the real system. The ability of the model to predict absolute magnitudes of consumption and respiration would be only of secondary importance.

The final step in the validation process, assuming the model has met the performance criteria that we set for earlier validation steps, is sensitivity analysis. Sensitivity analysis consists of varying one model parameter (or set of parameters) at a time and monitoring the subsequent effects on model behavior. By identifying those parameters that most affect model behavior, sensitivity analysis provides valuable insight into the functioning of the model and also suggests the level of confidence we should have in the model. If estimates for those parameters to which behavior is most sensitive have been based on reliable data, our confidence in the model should be high. If, on the other hand, we have guessed at values of some influential parameters, we should have less confidence in the model. Thus sensitivity analysis differs from earlier validation steps in that it is not a test by which the model might be invalidated, but rather a means to measure our confidence in an already validated model. An initially troublesome implication is that we may have little confidence in a valid model. However this apparent contradiction is resolved when the entire model validation procedure is viewed within the more general framework of scientific hypothesis testing.

The analogy between model validation and hypothesis testing is a good one because, in a sense, a model is nothing more than a collection of hypotheses concerning how we think the system of interest works. The strategy that scientists use to test hypotheses is to attempt to reject, or disprove, them in as many ways as they can. The more often we try and fail to disprove a hypothesis, the more confident we are that it may indeed be true, but some doubt always remains. The strategy for validating a model is essentially the same. We try to disprove, or invalidate, the model in as many ways as we can that are consistent with objectives of the modeling project. In fact, Holling (1978) has suggested that the term "model validation" is a misnomer and that we should refer to the process as "model invalidation." Thus a valid model is one that we currently cannot refute. We may or may not have confidence that a valid model is indeed correct, depending on the rigor of our attempts to invalidate it.

Phase IV: Model Use

The goal of the final phase of systems analysis is to answer the questions that were identified at the beginning of the modeling project (Figure 1.2). Indeed, the justification for the entire modeling exercise is that it ultimately will produce a tool (the model) that is useful in pursuing our interests. Most often we are interested in using the model to simulate system behavior under alternative management policies or environmental situations. This involves developing an experimental design for simulation exercises in the same manner that we would for real-world experiments. At this point care must be taken not to abandon the strategy of a well planned experimental design in favor of the apparent expediency of a "shotgun" approach made possible by the tremendous number crunching capabilities of high speed computers. The ability to generate voluminous results does not preclude the need for a logical approach to the problem. The analysis and interpretation of simulation results also follows the same scheme that we would employ for real-world results. For stochastic models this includes the use of standard statistical procedures such as analysis of variance to compare system behavior under different circumstances. Undoubtedly, results of initial simulations, while hopefully answering our original questions, will raise new questions about the system of interest. This suggests further examination of selected management policies or environmental situations, which often can be done using the original model, and we may conduct several rounds of simulated experiments.

Model use within a management framework implies effective communication of model results to those managers and policy makers whose decisions ultimately impact wildlife and fisheries resources. As with the design, analysis, and interpretation of simulated experiments, the communication of simulation results to decision-makers is no different than communication of results of real-world experiments. We must describe the data base, the technical method used to analyze the data (i.e., to develop the model), the results of validation procedures and simulated experiments, and our conclusions. Communication with potential users of the model is facilitated greatly by their early involvement in the modeling project. Model development usually requires many subjective decisions which, when viewed en masse by users for the first time only after completion of the model, can be quite difficult to explain. User confidence in the model, as well as the overall quality of the model, almost always is higher as the result of early and continued dialogue with potential users. Communication also is easier if emphasis is placed on interpreting general trends in model behavior in ecological terms. General trends usually are of more relevance in a management context than are specific numerical values of model predictions. Users and modelers also have more confidence in the model's ability to predict trends, and feel less comfortable with any given single numerical prediction. However for most modelers there is a tendency even with relatively simple models to become preoccupied with presenting detailed results at the expense of a clear overview. This tendency is ironic considering that an underlying rationale for use of a modeling approach is to attain a holistic view of our problem.

Iteration of Phases

The four phases of systems analysis and the various steps within each phase are highly interconnected. Although we may think of the process as formally proceeding step by step in the indicated order (Figure 1.2), several steps that are considered formally during later phases of systems analysis actually are anticipated and at least partially accomplished during earlier steps. The first step in model validation, examining capability of the model to address the problem of interest, is considered first during various steps in conceptual model formulation. Likewise the second step in model validation, examining reasonableness of model structure and individual model mechanisms, is considered first during various steps in quantitative specification of the model. Finally the first step in model use, identifying management policies or environmental situations to be evaluated and representing them in the model, is considered first during conceptual model formulation and again during quantitative specification of the model.

We also may cycle through several phases and steps of systems analysis more than once. During any step we may find that we have overlooked or misrepresented an important system component or process and need to return to an earlier step in the procedure, often to a step in conceptual model formulation or quantitative specification of the model. During various steps in model validation in particular we examine the model to detect any inadequacies that may require us to cycle back to earlier steps. Discovery of such inadequacies in the model during its development usually provides additional insight into behavior of the system of interest and is an important benefit of modeling.

1.4. TWO BRIEF CASE HISTORIES

Two brief case histories demonstrate application of the basic concepts in systems analysis and simulation modeling in the wildlife and fisheries field: The first example concerns the effect of pesticides on non-target wildlife populations and the second deals with management of the Gulf of Mexico shrimp fishery. The intent is to provide a general overview of the entire modeling process before we begin a more detailed examination of each phase of systems analysis in Chapters 2 through 5.

Case History I: Effect of Vertebrate Pesticides on Non-Target Wildlife

Overview

Widespread use of pesticides in modern agriculture and concomitant concern about state of the environment have focused attention on the need to assess both benefits and potential side effects of pesticides. Evaluation of potential effects of vertebrate pesticides on non-target wildlife populations is one important aspect of this complex issue, with particular attention being focused on use of Compound 1080 in coyote control programs in the western United States. Unfortunately, data on response of most wildlife populations to application of vertebrate pesticides are lacking. Nonetheless this issue has been much debated for the past two decades and continues to be a sensitive one.

Compound 1080 (sodium monofluoroacetate) was used in impregnated meat baits for coyote control from 1946 until 1972 when it was banned. Compound 1080 was injected into eviscerated carcasses of large domestic animals which were fastened to immovable objects at "bait stations" distributed at a density of roughly one per square mile throughout the treated area. It is highly toxic to canids and rodents but is less toxic to other species. Specificity of the poison and sparce density of bait stations together were supposed to reduce risk to non-target (non-coyote) animals since most non-targets have much smaller home ranges and are much less sensitive to the poison than are coyotes.

The Animal and Plant Health Inspection Service (APHIS) of the United States Department of Agriculture is interested in developing large-scale, comprehensive models that are capable of assessing impact of pesticides on the environment. Models capable of assessing effects of vertebrate pesticides on non-target wildlife populations are among the components of this comprehensive scheme. One model designed to evaluate population-level effects of Compound 1080 on non-target wildlife in the western United States was developed as part of a Cooperative Agreement between the APHIS of the U.S. Department of Agriculture and Texas A&M University (Grant et al. 1984).

Phase I: Conceptual Model Formulation

The main objective of the modeling project was to predict dynamics of non-target wildlife populations in response to different schemes of pesticide use. First a conceptual model of population dynamics of a given non-target species was developed (Figure 1.3). Dynamics of the non-target population were represented as resulting from addition of animals to the population through births (and immigration) and loss of animals from the population through natural mortality (and emigration) and pesticide-induced mortality. Birth and natural mortality both potentially were a function of density of prey, predator, and/or competitor populations, and/or physical environmental factors. Pesticide-induced mortality was a function of non-target density and the pest management scheme. The conceptual model represented movements of individuals into and out of the non-target population based on "information" about levels of various populations, state of the physical environment, and the pest management scheme.

The modelers abstracted from the real system only what they thought were

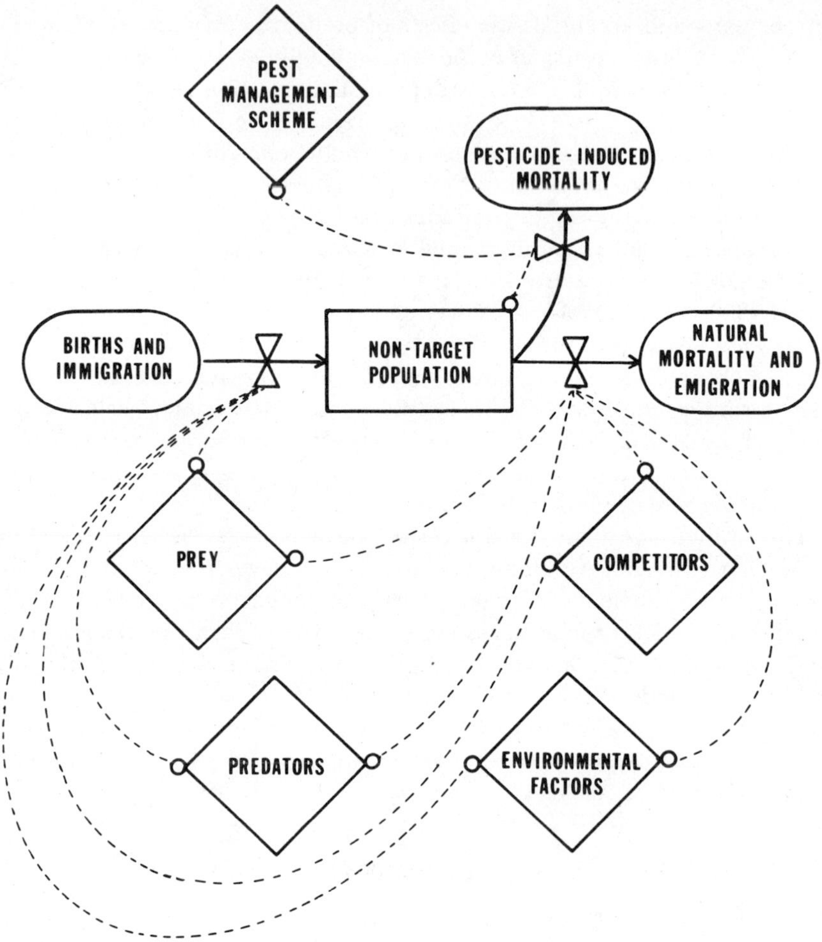

FIGURE 1.3. Conceptual model of population dynamics of a non-target wildlife population (Grant et al. 1984).

the most important components needed to represent non-target population dynamics at a level of resolution consistent with accomplishing the original objectives of the project. Population dynamics of prey, predator, and competitor populations were not represented as being affected by each other, by the non-target population, or by any other population or environmental factor. Rather, they were represented as externally specified variables (driving variables) that affected the system but were not themselves affected by the system. In fact prey, predator, or competitor populations or environmental factors were represented in the model only when investigating a non-target species that was thought to be closely linked to another population or environmental factor. Otherwise, non-target population dynamics were represented simply as a function of intraspecific density-dependent relationships. The model was a simplified abstraction of reality containing only those components thought to be most crucial for solving the problem at hand.

Phase II: Quantitative Specification of the Model

Quantitative specification of the model, or writing model equations, was based on a variety of information sources. For most non-target species modeled at least some data on natality rates were available. However, few data on natural mortality rates were available, and the density dependency of natality and mortality rates, although widely accepted for most species modelled, had not been quantified experimentally for any species. Estimation of natural mortality rates was based on the assumption that the non-target population was in equilibrium in the absence of coyote control programs. Age-specific mortality rates were specified such that when coupled with age-specific natality rates obtained from the literature the population was in equilibrium. Density-dependent relationships affecting natality and/or natural mortality rates were inferred from qualitative information in literature.

Information on level of pesticide-induced mortality resulting from particular schemes of coyote control was not available in literature. Pesticide-induced mortality rates were estimated based on expert opinion of Animal Damage Control personnel of the U.S. Fish and Wildlife Service who had had extensive field experience with the use of Compound 1080.

The model was specified as a set of difference equations and was programmed in FORTRAN for solution on a digital computer.

Phase III: Model Validation

The model was validated by assessing reasonableness of ecologically interpretable equations in the model in view of current knowledge concerning relationships that those equations represented. Since few data on population dynamics of any non-target species were available for comparison with population dynamics predicted by the model, further validation involved examination of both (1) ecological relationships within the model, and (2) model predictions, by expert authorities on the non-target species being modeled. The model was judged to be a valid representation of non-target population dynamics, relative to project objectives, based on inability of species experts to refute any relationships in the model and based on their consensus that model predictions were reasonable.

Phase IV: Model Use

The model was used to predict response of each of 11 non-target populations to each of four hypothetical schemes of coyote control using Compound 1080. The four hypothetical pesticide use schemes included:

1. no coyote control (baseline situation),
2. control every year for 10 years,
3. control every second year for 20 years, and
4. control every fifth year for 50 years.

Results of simulation of effects of Compound 1080 use on great horned owl populations are presented in Figure 1.4. Four types of information that are of particular interest to natural resource managers can be obtained from Figure 1.4:

1. general trend (up or down) in non-target density over the period of the control program,

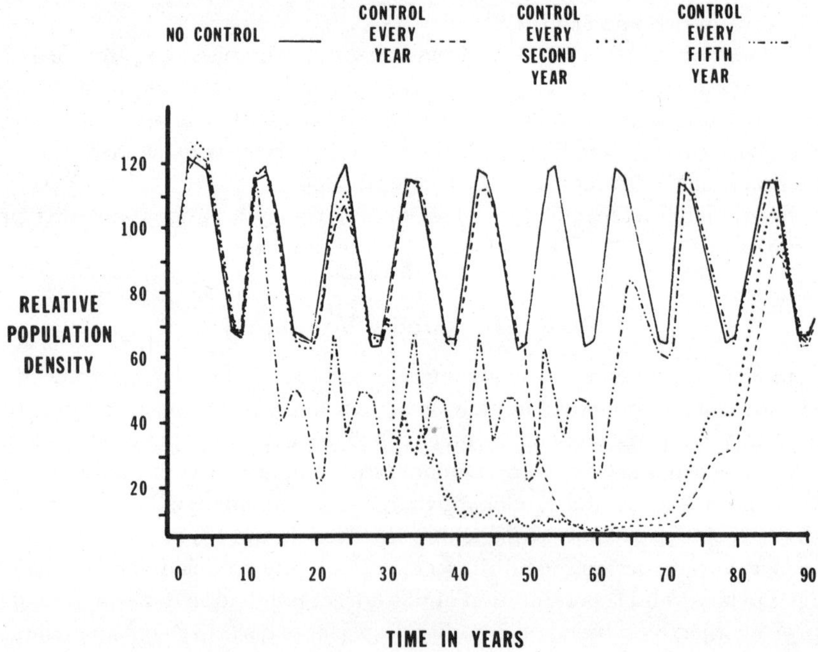

FIGURE 1.4. Great horned owl population dynamics resulting from each of four hypothetical coyote control schemes. The different schemes are initiated in different years, but in each scheme control ends at year 60.

2. lowest density reached,
3. length of time taken to recover to "normal" density after cessation of the control program, and
4. probability of extinction.

Simulation results from this model on several non-game species have been used by APHIS in legal hearings concerned with the possible reregistration of Compound 1080 for predator control.

Case History II: Gulf of Mexico Shrimp Fishery

Overview

The Gulf of Mexico shrimp fishery is one of the most valuable fisheries in the United States. It produces roughly 4 percent of total weight that the United States harvests from the sea and provides roughly 20 percent of the gross dollar value of the harvest. The fishery harvests three main species of penaeid shrimp: (1) pink shrimp (*Penaeus duorarum*) are harvested primarily off Florida and (2) brown shrimp (*P. aztecus*) and (3) white shrimp (*P. setiferus*) are harvested primarily from Alabama to Texas. Shrimp are harvested by trawlers ranging in size from 35 to 70 feet, with smaller boats fishing primarily in the bays and near shore in the Gulf, and larger vessels fishing in deeper Gulf waters.

Reproduction of penaeid shrimp occurs in species-specific areas in the Gulf. Following a period of planktonic existence as larvae, juveniles migrate into shallow estuarine nursery grounds and subsequently enter the bay fishery by moving from nursery grounds into deeper water of the bays. These shrimp then move out of bays into the offshore or Gulf fishery. Spawning completes the annual cycle.

The Fishery Conservation and Management Act passed by Congress in 1976, in addition to extending U.S. jurisdiction to 200 miles offshore, established eight regional fisheries management councils including the Gulf of Mexico Fisheries Management Council. The regional councils were given the responsibility of developing fisheries management plans for important resource species within their area. The first management plan developed by the Gulf Council was for the Gulf shrimp fishery. The Gulf Council identified a model of the Gulf shrimp resource capable of assessing the impact of alternative management policies as a critically needed management tool. Potential management policies of interest were concerned primarily with seasonal closures of the fishery in certain geographical areas, but size restrictions on harvest also were of interest. One model capable of evaluating such policies was developed as part of the Texas A&M University Sea Grant Program (Grant and Griffin 1979).

Phase I: Conceptual Model Formulation

The main objective of the modeling project was to develop a model representing biological and economic dynamics of the Gulf shrimp fishery that could predict harvest and rent (= revenue from harvest − cost of fishing) in the fishery under a variety of hypothetical environmental conditions and management policies. First a conceptual model of the Gulf shrimp fishery that represented important processes and relationships within the fishery was developed (Figure 1.5). The conceptual model was composed of a biological submodel and an economic submodel. The biological submodel represented movement of shrimp between various "compartments" within the fishery. The economic submodel represented transfer of "information" about various economic aspects of the fishery.

The biological submodel represented movement of shrimp from the Gulf via nursery grounds into the bay fishery; movement from the bay fishery to the nearshore (1–19 m depth) fishery; and movement from nearshore to the offshore (20–91 m depth) fishery. Spawning of shrimp and movement from the Gulf back into the nursery grounds was not represented explicitly in the model (1) because the modelers were interested in the fate of shrimp only after they were large enough to enter the fishery (large enough to be caught) and (2) because it was assumed that there was no stock-recruitment relationship within the fishery. Number of shrimp recruited to the fishery each year was assumed to be primarily a function of environmental factors that determined the number of the essentially limitless supply of shrimp eggs that survived to harvestable size.

Choice of the three different areas explicitly represented in the model—bay fishery, 1–19-m fishery, and 20–91-m fishery—was based on need to represent (1) different levels of fishing effort exerted on shrimp depending on their physical location and (2) areas that were under different management authorities which might be subjected to different management policies. For this particular application of the model, which examined catch of brown shrimp off the Texas coast, boundaries were chosen to represent separately:

1. the bay fishery, which is fished by a different group of fishermen than is the offshore fishery,
2. the 1–19-m fishery, which represents the Territorial Sea of the State of Texas extending from shoreline 9 miles into the Gulf, and
3. the 20–91-m fishery, which represents the Fishery Conservation Zone extending to the 200-mile limit and is under the jurisdiction of the Gulf Council.

Boundaries were defined in terms of depths rather than in terms of miles offshore because catch data available from the fishery were identified by depth.

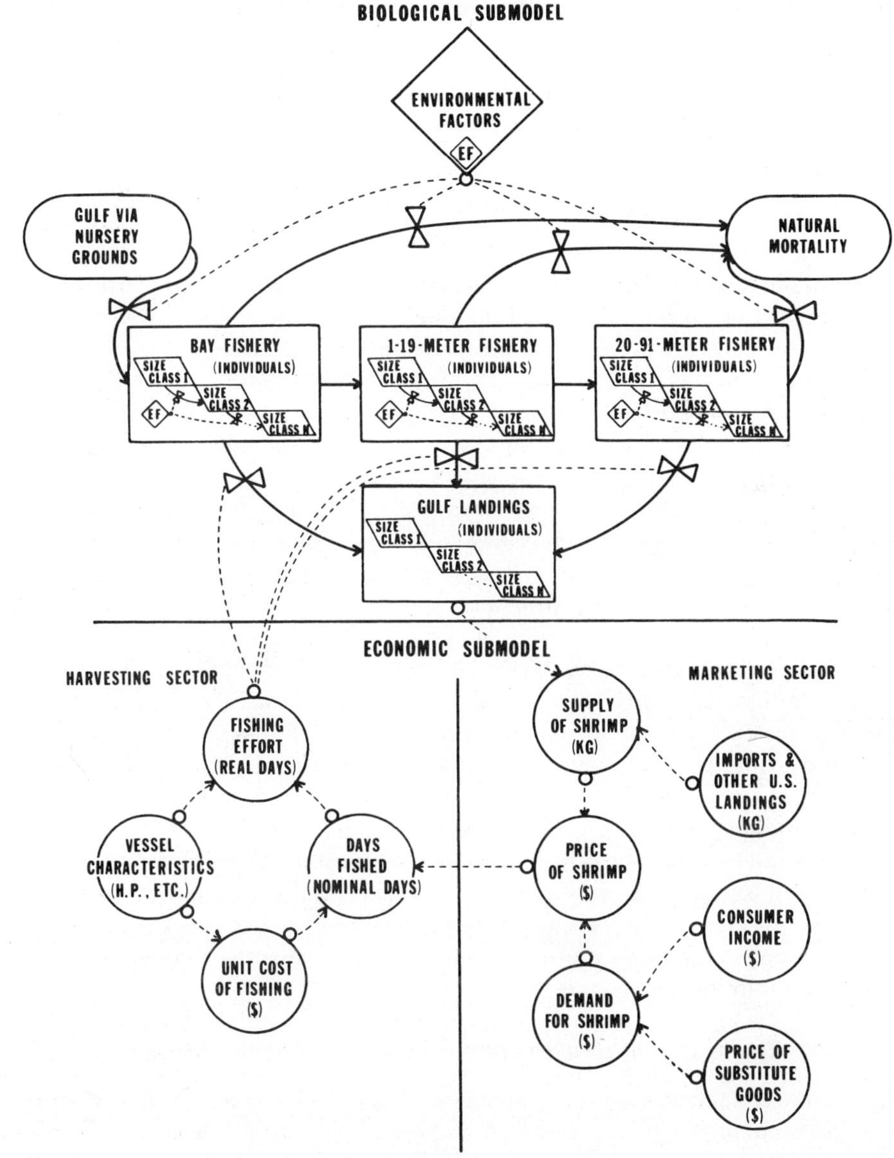

FIGURE 1.5. Conceptual model of Gulf of Mexico shrimp fishery (Grant and Griffin 1979).

Growth of shrimp was represented as a function of current size of shrimp and environmental factors to which they were exposed, which in turn were a function of their physical location in the bay, nearshore, or offshore fishery. It was particularly important to represent size class distributions of shrimp in the different areas (1) to verify that the model was representing growth and movement appropriately and (2) because price of shrimp and therefore revenue from the harvest differs for shrimp of different sizes. Natural mortality of shrimp was represented as a transfer or "movement" of shrimp from bay, nearshore, and offshore fisheries into a natural mortality sink.

Fishing mortality of shrimp was represented as a transfer or "movement" of shrimp from bay, nearshore, and offshore fisheries into the Gulf landings compartment. Fishing mortality or harvest from each area of the fishery was represented as a function of the number of shrimp in the area and level of fishing effort exerted.

The economic submodel represented both harvesting and marketing sectors of the fishery. Fishing effort in terms of nominal or unstandardized days fished was represented as a function of cost of fishing and price of shrimp. Cost of fishing was represented as a function of vessel characteristics such as size of vessel as indicated by horsepower of engine. Nominal days fished were converted to real or standardized days fished based on vessel characteristics to represent the fact that larger vessels catch more per nominal day fished than do smaller vessels. Price of shrimp was represented as a function of supply and demand, with supply of shrimp represented as a function of Gulf landings and imports and other U.S. landings, and demand for shrimp represented as a function of consumer income and price of substitute goods.

The modelers abstracted from the real Gulf shrimp fishery system those aspects of the system that were relevant to accomplishing original objectives of the project. They did not try to represent the fishery in all of its complexity. They sought to represent the system in the simplest manner that would allow them to attain their objectives.

Phase II: Quantitative Specification of the Model

Quantitative specification of the model was based on a variety of information sources including both experimental and observational data from the fishery, theoretical relationships, and in areas where both data and theory were lacking, "expert opinion." The model was specified as a set of difference equations and was programmed in FORTRAN for solution on a digital computer.

Phase III: Model Validation

The model was validated by assessing reasonableness of biologically or economically interpretable equations in the model in view of current knowledge concerning relationships that those equations represented. Modelers searched for inconsistencies between what was known about specific relationships in the fishery and the manner in which they were represented in the model. The model also was validated by comparing predictions of shrimp catch, partitioned by month, depth, and size class, to observed catches as indicated by National Marine Fisheries Service data. The model was judged to be a valid representation of the fishery relative to project objectives based on inability of the modelers to refute any relationships in the model and based on ability of the model to predict observed catches reasonably well (Figure 1.6).

Phase IV: Model Use

The model was used to predict annual harvest, revenue, and rent in the Gulf shrimp fishery under a variety of hypothetical environmental conditions and

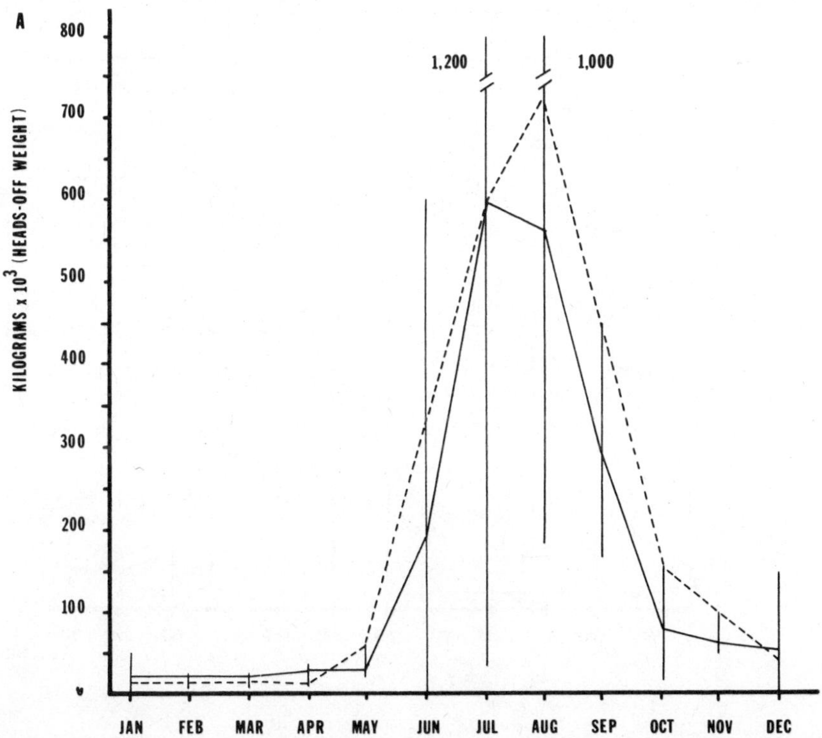

FIGURE 1.6. Comparisons of simulation model predictions of brown shrimp harvest (dashed lines) to average harvest from 1963 to 1971 (solid lines) for (A) total catch, (B) catch by depth, and (C) catch by size class. Vertical bars represent $\pm$ 1 standard deviation of the mean.

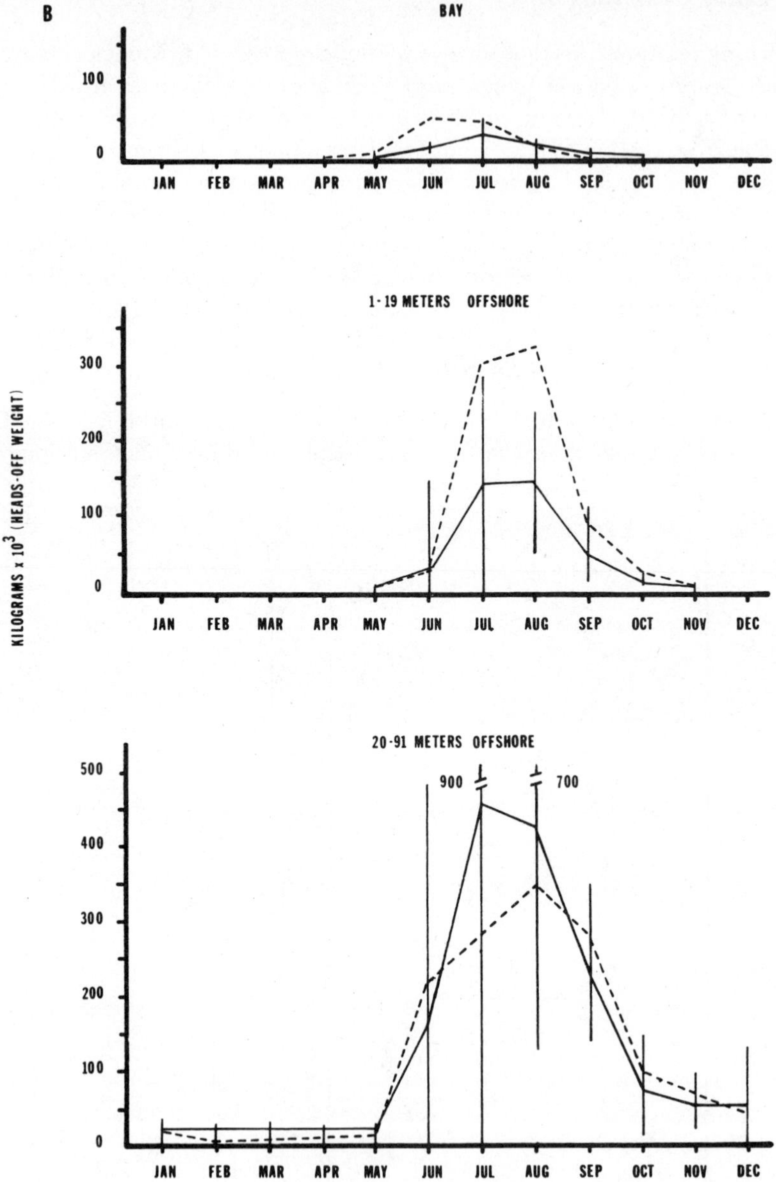

FIGURE 1.6. *(Continued)*

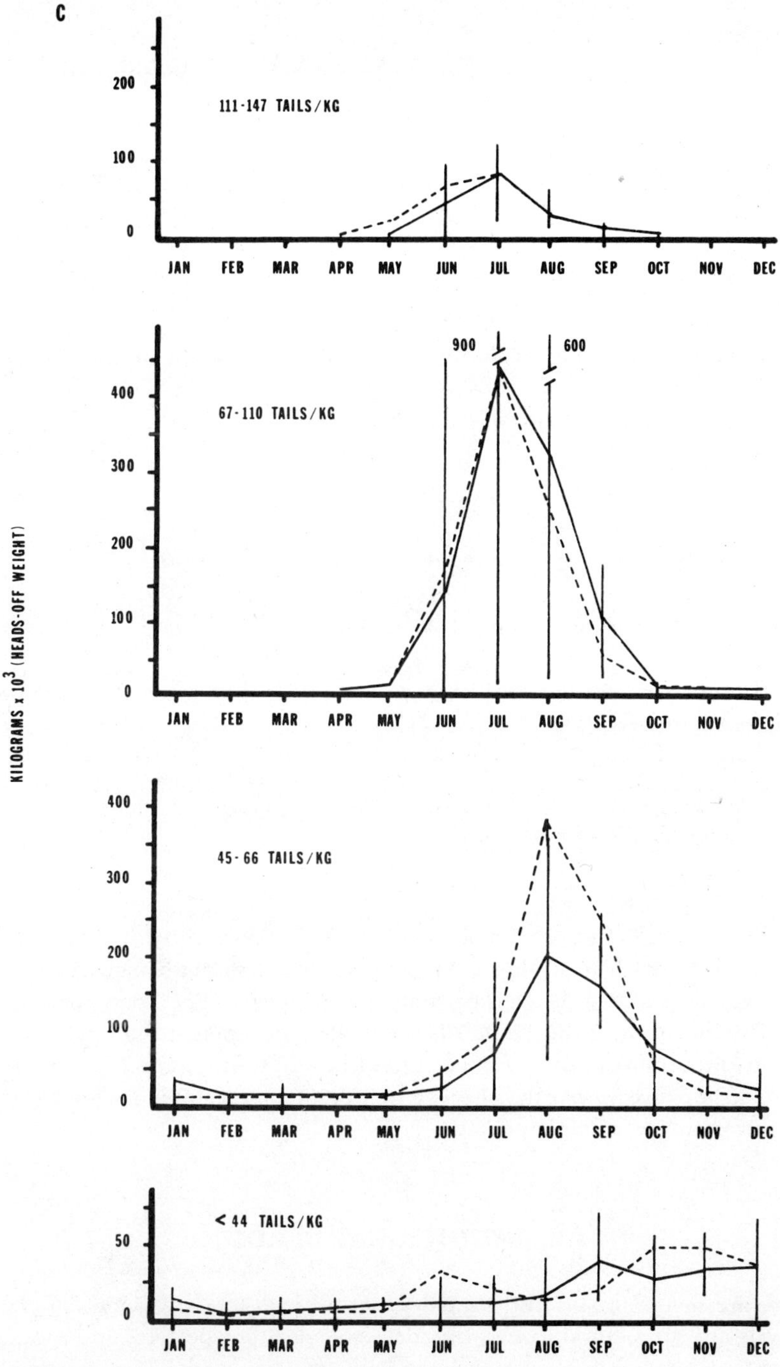

FIGURE 1.6. (*Continued*)

management schemes. Simulations of environmental conditions included (1) doubling and (2) halving recruitment of shrimp into the fishery, which represented years during which conditions in estuarine nursery grounds were excellent and poor, respectively. Simulations of management schemes included

1. doubling and
2. halving number of nominal days fished each month,
3. doubling number of nominal days fished each month only in the bay fishery,
4. completely closing the bay fishery,
5. increasing number of nominal days fished in the bay fishery during May through August, and
6. allowing harvest of shrimp of any size (shrimp smaller than 115 mm were not harvested in other simulations).

Complete output from the model for each month included

1. number of pounds of shrimp of each species harvested at each depth by each vessel class,
2. number of days fished,
3. cost of fishing,
4. revenue from the harvest, and
5. rent for each vessel class.

Simulation results from this model have been considered by the Gulf Council and the National Marine Fisheries Service, although neither group has used the model formally in developing management plans. The Food and Agriculture Organization of the United Nations also has applied the model to both shrimp and cephalopods (octopus, cuttlefish, and squid) in preliminary analyses of the northwest African fishery, although management plans for these fisheries have not been implemented.

1.5. ADDITIONAL READING

There are several books and chapters that present good overviews of systems analysis and simulation from a broader ecological perspective, and numerous others that describe applications of modeling to various ecological systems. Jeffers (1978, chapters 1, 2, and 8), Kitching (1983, chapters 1, 2, 3, and 9), Gold (1977, chapter 1), and Berryman (1981, chapter 1) provide excellent introductions to the systems approach. Chapters by Walters (1980) and Tipton (1980) also provide excellent introductions from a fisheries and a wildlife perspective, respectively. Hall and Day (1977, chapter 1) provide a good, although somewhat more complicated, general introduction, and several other early chapters on more specialized topics also should be of interest. The four volumes of Patten (1971a, 1972, 1975, 1976) describe the applications of systems analysis and simulation to a wide variety of ecological systems. Chapters by Kowal (1971) on a rationale for modeling ecological systems, Caswell (1976) on model validation, Glass (1971) on predation energetics of largemouth bass, Lassiter and Hayne (1971) on a finite difference model for simulation of dynamic processes in ecosystems, Bledsoe and Van Dyne (1971) on a compartment model simulation of secondary succession, Rykiel and Kuenzel (1971) on an analog computer model of the wolves of Isle Royale, Saila (1972) on systems analysis applied to some fisheries problems, and Paulik (1972) on digital simulation modeling in resource management and the training of applied ecologists may be of particular interest. Several chapters in Patten (1975) describe modeling in the U.S. International Biological Program. These provide a good historical perspective on a large scale effort to adapt systems analysis and simulation to ecosystem modeling. Watt (1966) contains several chapters written by early leaders in the application of systems analysis in ecology, with the introductory chapter by Watt and the chapter by Paulik and Greenough on management analysis for a salmon resource system of special interest. Jeffers (1972) contains chapters on use of a variety of mathematical models in ecology. The chapter on developments in the Leslie Matrix model by Usher is particularly relevant to those interested in modeling population dynamics. Shugart and O'Neill (1979) present a collection of benchmark papers in systems ecology that document the historical roots, major developments, and present directions in systems ecology. The chapter by Innis in Innis and O'Neill (1979) presents a primer on ecosystem simulation, while other chapters address more specialized topics in systems analysis of ecosystems. Watt (1968, chapters 8–13) describes a modeling approach to natural resource management and Holling (1978, chapters 1–10) presents a systems approach to environmental assessment and management. Odum (1983) presents a lengthy introduction to systems ecology in which he uses energy circuit language to model many ecological processes. Spain (1982) provides interesting applications of microcomputer models in biology, with several examples being of ecological relevance.

The literature on systems analysis and simulation in fields other than ecology, particularly engineering and business management, is voluminous. Many of the ecologically oriented references above provide citations to this literature. Forrester (1961), Graybeal and Pooch (1980), and Law and Kelton (1982) are fairly readable representatives of simulation texts directed toward engineers and business managers. Halfon (1979) and Zeigler (1976) are representatives of the theoretical approach to systems analysis and simulation. Weinberg (1975) provides an interesting philosophical treatment of systems analysis.

The number of articles in scientific wildlife and fisheries journals describing simulation models has increased markedly over the past decade. Examples of use (and occasionally misuse) of models to address wildlife and fisheries questions can be found in such journals as *The Journal of Wildlife Management*, *The Transactions of the American Fisheries Society*, and *The Canadian Journal of Fisheries and Aquatic Sciences* (formerly the *Journal of the Fisheries Research Board of Canada*).

CHAPTER 2

CONCEPTUAL MODEL FORMULATION

2.1. INTRODUCTION

By conceptual model formulation we mean the process of abstracting from reality, from the real-world system, those aspects that are relevant to solution of our problems. As mentioned in Chapter 1 we might abstract from the real system in many different ways, formulate many different models of the system, depending upon our specific interests. The trick is to formulate a useful model. Clearly the best guide to formulating a useful model is a clear statement of the problem that we want to solve. We need to pose specific questions relating to the problem we wish the model to address. This is sound general advice but not particularly helpful as a practical guide to model formulation. As a more practical guide to how we might proceed we will identify four steps in conceptual model formulation:

1. bounding the system of interest (Section 2.2),
2. categorizing components within the system of interest (Section 2.3),
3. identifying relationships between components that are of interest (Section 2.4), and
4. formally representing the conceptual model (Section 2.5).

Although we will examine these four steps sequentially, in practice they most commonly occur simultaneously. For purposes of formally identifying and attempting to explain the thought processes associated with conceptualizing the system of interest it is helpful to view each step as a logically distinct activity. But conceptual model formulation is such a dynamic process that experienced modelers seldom consciously classify their activities during this phase of systems analysis. Decisions made within each step influence decisions in each of the other steps. This high degree of interrelatedness will become apparent as we describe the details of each step.

2.2. BOUNDING THE SYSTEM OF INTEREST

Bounding the system of interest consists of separating those components that should be included within the system of interest from those that should be excluded. We do not want our model to be unnecessarily complex but likewise we do not want to exclude components that might be critical to solution of our problem. Recall the simple simulation model presented in Figure 1.1. Our objective was to predict fluctuations in weight of an animal over time. The only components included within the system of interest were weight of the animal and environmental temperature. Other potential system components such as rainfall, weight of available food, and number of animals in the population were excluded because they were considered unimportant in predicting fluctuations in weight of the animal.

Obviously this first step in conceptual model formulation is a highly subjective one. Often difficult decisions arise regarding the need to include certain components. Our prior modeling experience is an asset in making such decisions but ultimately decisions must be based on the best information available

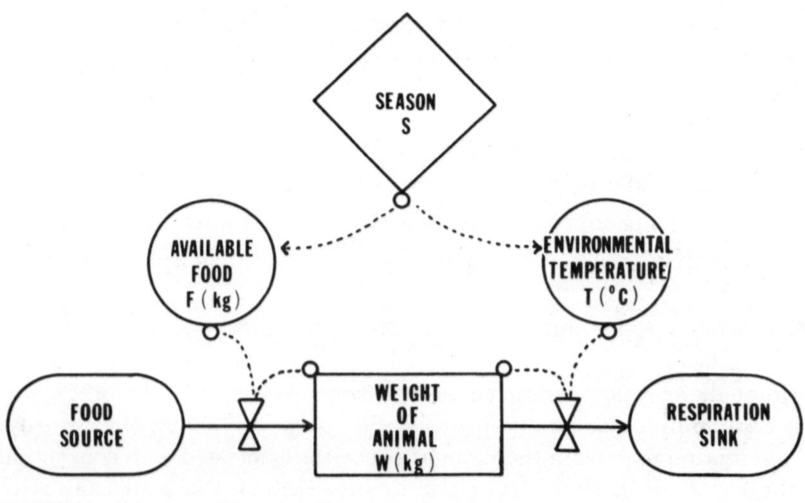

FIGURE 2.1. Diagram of the weight fluctuation model (Figure 1.1) modified to represent available food and season as system components affecting weight fluctuations.

about the system in question. Again returning to the example in Figure 1.1, it may well have been that rainfall clearly was irrelevant. But perhaps the decision to exclude weight of available food and number of animals in the population was based not on clear evidence that they were unimportant, but rather on the more tenuous assumption that their effect would be negligible over the range of conditions that we wished to simulate.

Suppose that after reconsidering the problem of predicting fluctuations in animal weight we decided that weight of available food was an important factor affecting animal consumption after all, and that variability of both weight of available food and temperature should be represented in the model as a function of season. We then would bound the system differently by including available food as a system component affecting transfer of material from the food source to the animal and by indicating the influence of season on both availability of food and environmental temperature (Figure 2.1). (As we mentioned

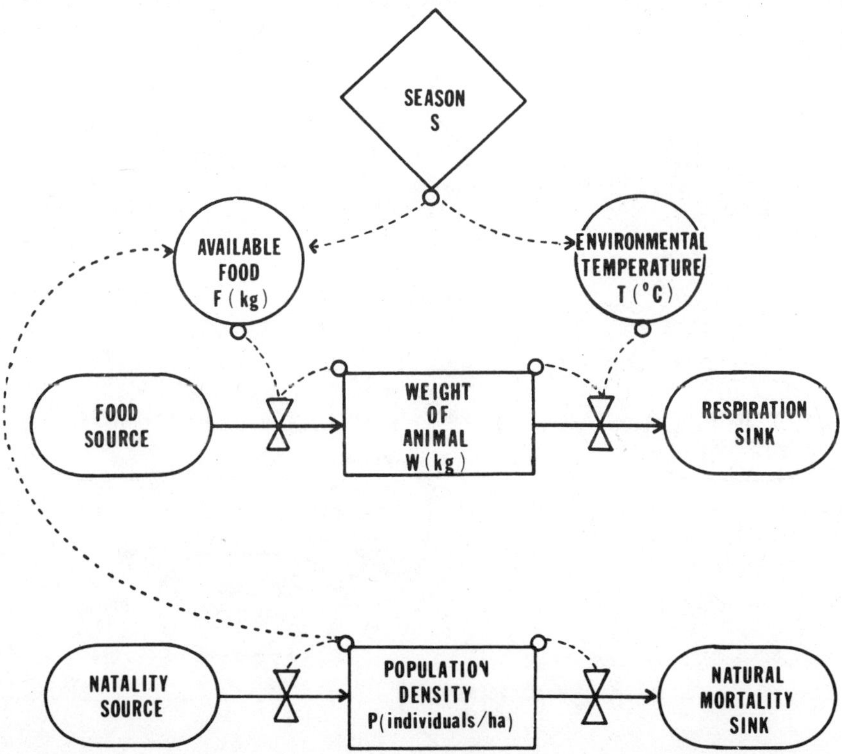

FIGURE 2.2. Diagram of the weight fluctuation model (Figure 2.1 and Figure 1.1) modified to represent number of individuals in the population as a system component affecting availability of food and affected by density-dependent natality and mortality rates.

in Chapter 1, the symbols that we are using to diagrammatically represent conceptual models have specific meanings. We will define these formally in Sections 2.3 and 2.4, but figures presented in this section (Figures 2.1, 2.2, and 2.3) can be interpretated informally without losing meaning for our present purpose.) Further suppose that we could not predict fluctuations in weight of individual animals without considering the number of animals in the population because availability of food to individuals was affected by population density. Number of individuals in the population was of course a function of natality and mortality rates which, let's suppose, were both density dependent. We would again change our bounding of the system by including the number of individuals in the population as a system component affecting availability of food, and, in turn, affected by density dependent natality and mortality rates (Figure 2.2).

Another aspect of bounding the system of interest is identifying particular attributes or units of measure of system components that are of interest. In some cases we may be interested in only a single attribute throughout the entire system.

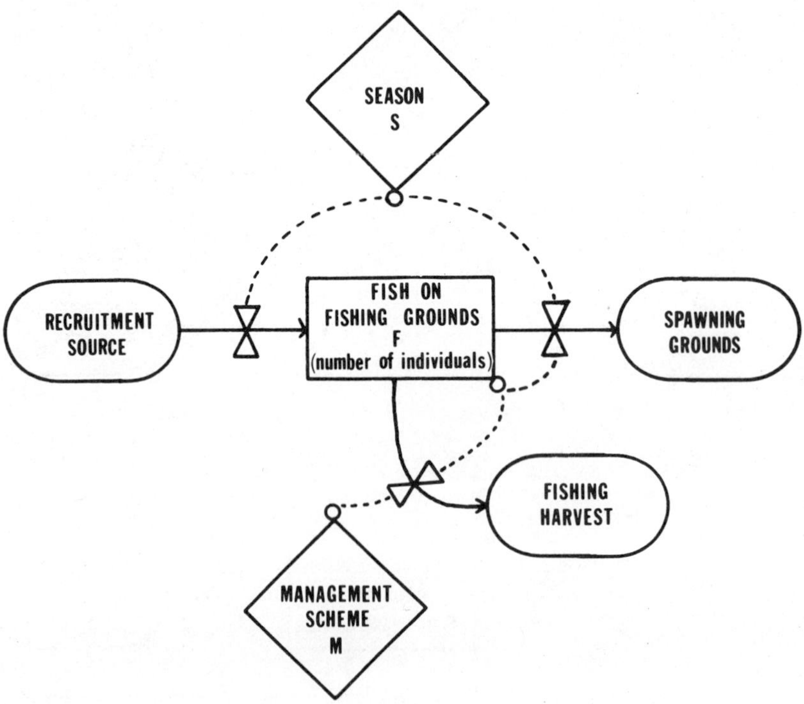

FIGURE 2.3. Diagram of a model predicting harvest of migratory game fish from a specific fishing grounds under different management schemes.

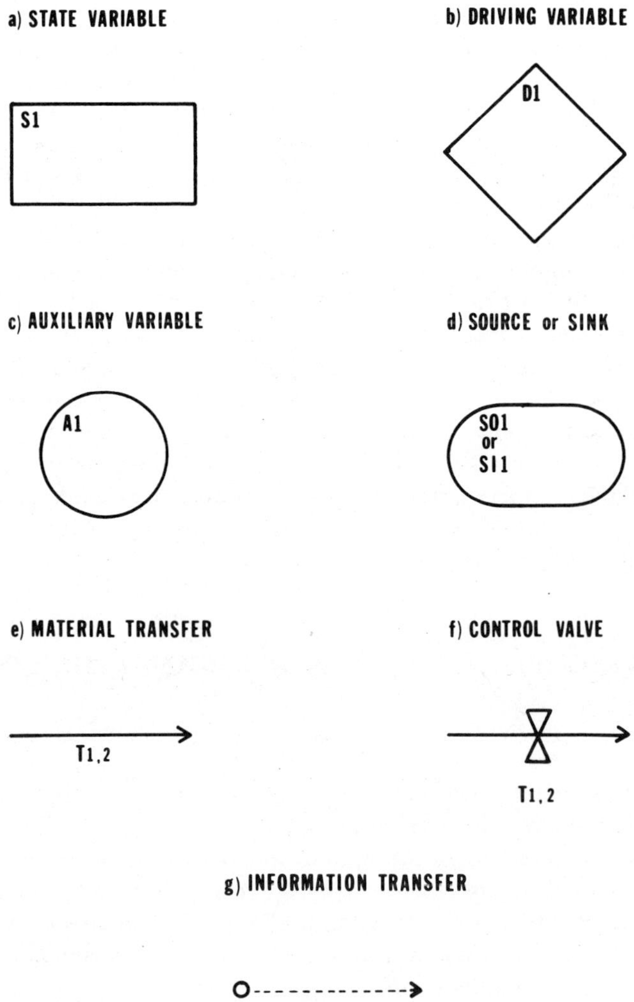

FIGURE 2.4. Symbols used to construct conceptual model diagrams. Modified from Forrester (1961) and Innis (1979).

Consider the problem of predicting harvest of a migratory game fish population from a specific fishing grounds as fish return from the ocean en route to freshwater spawning grounds. Suppose that for our purposes we can assume that the fish are all the same size when they pass through the fishing grounds and that natural mortality is zero during the several weeks of their passage. We might bound the system as indicated in Figure 2.3. We are interested only in the number of fish on the fishing grounds and in the number of fish harvested, perhaps under a variety of management schemes. Rate of recruitment of fish to the fishing grounds is a function of season. Rate of emigration of fish to the spawning grounds is a function of season and number of fish on the grounds. Rate of fishing mortality is a function of the number of fish on the grounds and restrictions on harvest imposed by management. Thus our description of the system is univariate: we are interested only in the number of fish passing through the system.

In some cases we may wish to monitor several attributes of a system simultaneously. For example we may be interested in population dynamics of a particular species. But those dynamics can be represented in terms of several attributes of the population; numbers of individuals, total biomass, or total energy content. It also is possible that different system components may have different attributes in which we are interested. In Figure 2.1 our interest in the animal component of the model was in terms of number of kg, but temperature was monitored in terms of °C. In our later extension of this model to include representation of the number of individuals in the animal population (Figure 2.2) we further increased dimensionality of our description of the system. Thus system descriptions can be multivariate.

2.3. CATEGORIZING COMPONENTS WITHIN THE SYSTEM OF INTEREST

Once the system of interest has been bounded by separating those components that should be included within the system from those that should be excluded and by identifying particular attributes of system components that are of interest, we proceed to step two of conceptual model formulation, categorizing components within the system of interest. System components may not all serve the same purpose in a model. Certainly they all represent important aspects of the system of interest but there may be as many as four fundamentally different categories of system components:

1. state variables,
2. driving variables,
3. auxiliary variables, and
4. sources and sinks.

State Variables

State variables (Figure 2.4a) represent points of accumulation of material within the system. If we are interested in energy flow through an ecosystem, energy contained in plants, energy contained in herbivores, and energy contained in carnivores might be three state variables in the model (Figure 2.5). In the version of our weight fluctuation model described in Figure 2.1, weight of the animal (measured as number of kg) is a state variable. We later expanded this model to include population density of animals (measured as number of individuals/ha) (Figure 2.2), which also is a state variable. In our model predicting harvest of migratory game fish (Figure 2.3) the number of fish on the fishing grounds is the only state variable, representing the single point of accumulation of ''material'' in the system.

Driving Variables

Driving variables (Figure 2.4b) affect but are not affected by the rest of the system. For example we may wish to represent transfer of energy from sun to plants as a function of the amount of rainfall (Figure 2.5). But, of course, amount of rainfall is not affected by plants or by another system component. Season is a driving variable in both our weight fluctuation model (Figures 2.1

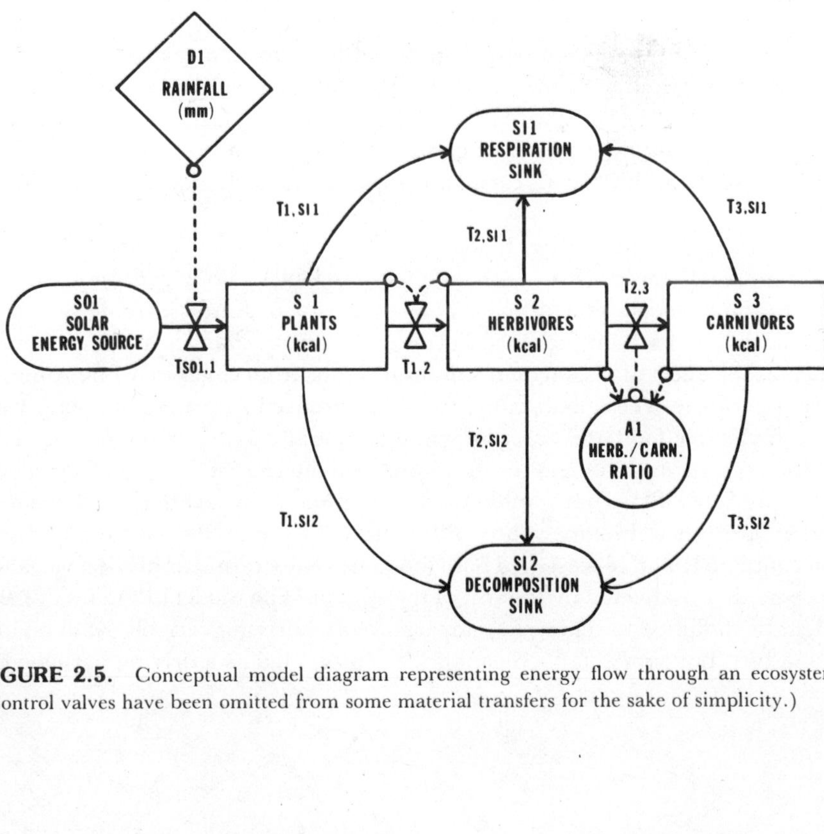

FIGURE 2.5. Conceptual model diagram representing energy flow through an ecosystem. (Control valves have been omitted from some material transfers for the sake of simplicity.)

and 2.2) and our migratory fish model (Figure 2.3). In the fish model, season affects rates of recruitment to and emigration from the fishing grounds. Obviously these rates do not affect season. In the weight model season affects rates of consumption and respiration (indirectly through available food and environmental temperature—both auxiliary variables that we will define shortly). Whereas some driving variables are associated with specific units of measure, such as the rainfall (mm) driving variable in Figure 2.5, others can be unitless. Season, for example, often is described as taking on values of 1, 2, 3, etc., to represent January, February, March, or first quarter, second quarter, third quarter of the year. (The way this works will become clear when we discuss quantitative specification of the model in Chapter 3.) Collectively, levels of accumulation of material within state variables and values of driving variables uniquely describe the state of the system at a point in time.

Auxiliary Variables

Auxiliary variables (Figure 2.4c) arise as part of calculations determining a rate of material transfer or the value of another variable, but represent concepts that we wish to indicate explicitly in the model. Auxiliary variables also may represent an end product of calculations that is of particular interest to us. For example we may wish to represent transfer of energy from herbivores to carnivores as a function of both number of herbivores and number of carnivores. Further, it may be that the ratio of herbivores to carnivores has a special meaning for us over and above the fact that it is an intermediate step in calculations determining the transfer of energy from herbivores to carnivores. Thus we represent the herbivore/carnivore ratio as an auxiliary variable (Figure 2.5).

In our weight fluctuation model (Figures 2.1 and 2.2) available food and environmental temperature are auxiliary variables. Both consumption and respiration by the animal ultimately are a function of season (driving variable). But the effect of season on consumption results from changes in available food whereas the effect of season on respiration results from changes in environmental temperature. We call attention to this difference by creating auxiliary variables (available food and environmental temperature) mediating the effects of season on consumption and respiration. As an alternative version of the model in Figure 2.1, had we not wanted to call attention to the fact that available food and environmental temperature vary seasonally, we might have omitted season completely and represented both food and temperature as driving variables since neither is affected by the rest of the system. The model in Figure 2.2 also might be modified to represent temperature as a driving variable (with no link to season), but available food cannot be represented as a driving variable because it is affected by another system component (the population density state variable).

Sources and Sinks

Sources and sinks (Figure 2.4d) represent origination and termination points, respectively, of material transfers into and out of the system. By definition we are not interested in the level of accumulation of material within sources or sinks. For example we may note that energy enters an ecosystem by being transferred from sun to plants but we may have no interest in the amount of energy in the sun (solar energy source) (Figure 2.5). Likewise we may note that energy is lost from the grazing food chain both through respiration of plants, herbivores, and carnivores, and as plants, herbivores, and carnivores are eaten by decomposer organisms. But we may not be interested in the level of accumulation of respired energy (respiration sink) or of energy in decomposers (decomposition sink) (Figure 2.5).

2.4. IDENTIFYING RELATIONSHIPS BETWEEN COMPONENTS THAT ARE OF INTEREST

Material and Information Transfers

Step three of conceptual model formulation consists of identifying relationships between system components that are of interest. There are basically two ways that system components may be related: by material transfers or by information transfers. A material transfer (Figure 2.4e) represents physical transfer of material over a specific period of time: (1) between two state variables, (2) between a source and a state variable, or (3) between a state variable and a sink.

As energy flows (in kcal/week) through the grazing food chain within an ecosystem it is transferred from plants to herbivores to carnivores (Figure 2.5). As an animal gains weight (in kg/day) biomass is transferred from a food source to the animal to a respiration sink (Figure 2.1). As fish enter and leave fishing grounds (in number of individuals/month) individuals are transferred from a recruitment source to the fishing grounds to spawning grounds (Figure 2.3). Note that units of measure associated with material transfers always include a time dimension such as per hour or per year.

Control valves (Figure 2.4f) and information transfers (Figure 2.4g) represent the control of rates of material transfers by information about the state of the system. We may wish to represent the material transfer of energy from plants to herbivores as a function of both the number of herbivores and the amount of plants (Figure 2.5). Or more strictly speaking, to calculate the rate of energy transfer from the plant state variable to the herbivore state variable we need information about the number of kcal of herbivores and number of kcal of plants in the system. Likewise, to calculate harvest in our fish migration model (the material transfer of individuals from the fish on fishing grounds state variable to the fishing harvest sink) we need information about the number of individuals on the fishing grounds and the management scheme (driving variable) (Figure 2.3).

State variables may be related to other state variables or to sources or sinks via material transfers. State variables, driving variables, and auxiliary variables may be related through their common influences, via information transfers, on the rate of material transfers between state variables or between state variables and sources or sinks. Units of measure of state variables, sources, and sinks connected by material transfers must be the same. Apples cannot become oranges en route.

"Information" transferred within the system refers to information about current values of state variables, driving variables, and derived auxiliary variables. This information is "transferred" for use in determining the rate at which specific material transfers occur. Units of measure of variables affecting a given material transfer need not be the same. Respiration rate in our weight fluctuation model (the material transfer from animal state variable to respiration sink) is determined by information about weight of the animal in kg and environmental temperature in °C.

Submodels

We have seen that the system of interest may be described by more than one attribute but that material transfers can occur only between state variables, sources, and sinks described by the same attribute. This brings us to the concept of submodels within a model.

If we wish to represent transfer of different materials within a system we must represent each set of transfers associated with a given material in a separate submodel. Various submodels, or sets of material transfers, may be connected by information transfers but not by material transfers. Material transfers in Figure 2.2 include both biomass and individuals but they follow different routes within the model. They are in different submodels connected only by an information transfer from the population density state variable to the available food auxiliary variable.

As another example of use of submodels, suppose that we wish to modify our weight fluctuation model (as conceptualized in Figure 2.1) to simulate nitrogen balance as well as weight fluctuations. Further suppose that nitrogen

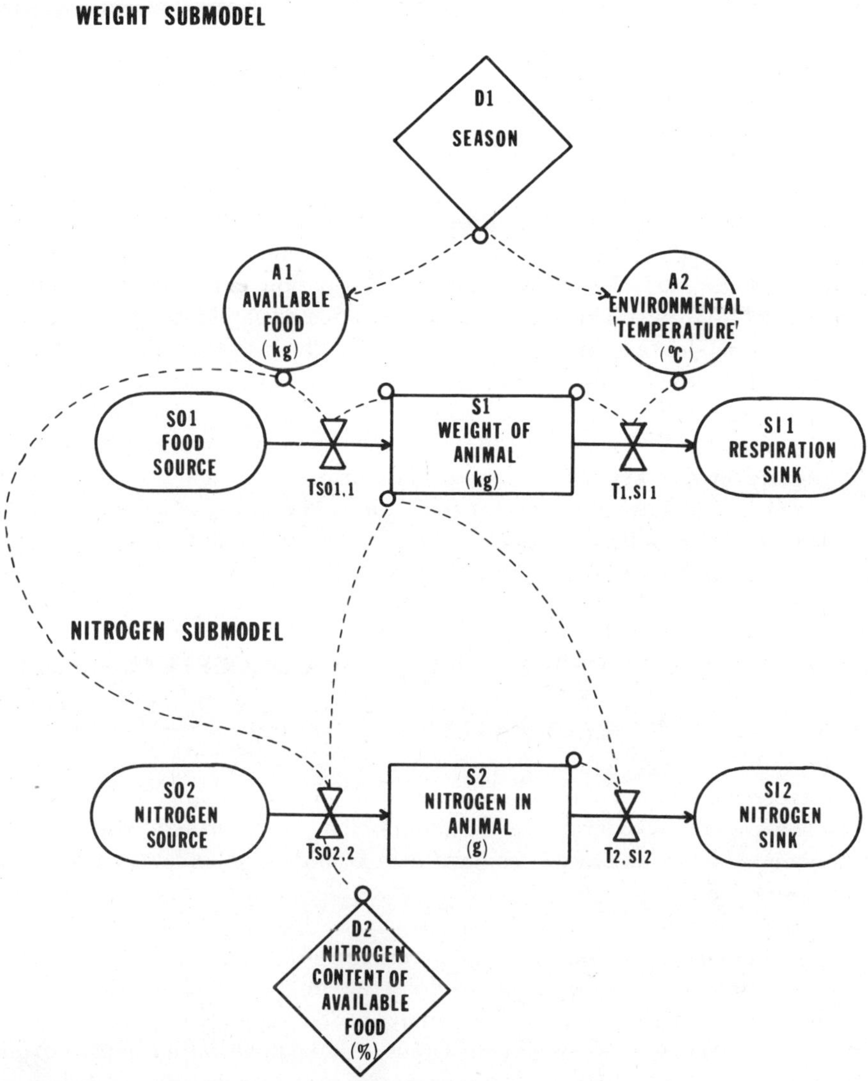

FIGURE 2.6. Conceptual model diagram of the weight fluctuation model (Figure 2.1) modified to include nitrogen balance as well as weight fluctuation submodels.

intake is a function of nitrogen content of available food and food consumption rate, which we already know is a function of animal weight and amount of available food. Nitrogen output is a function of animal weight and nitrogen content of the animal.

We might model this new system by adding a new submodel consisting of a state variable representing g of nitrogen in the animal, a driving variable representing nitrogen content of available food, and a nitrogen source and nitrogen sink (Figure 2.6). We then would use information on the weight of animal state variable $S1$, available food auxiliary variable $A1$, and nitrogen content of available food driving variable $D2$ to determine the material (nitrogen) transfer $T_{SO2,2}$ from the nitrogen source SO2 to nitrogen in animal state variable $S2$. We would use information on the weight of animal state variable $S1$ and nitrogen in animal state variable $S2$ to determine the material (nitrogen) transfer $T_{2,SI2}$ from the nitrogen in animal state variable $S2$ to nitrogen sink SI2 (Figure 2.6). Our model now has a weight submodel and a nitrogen submodel, each representing dynamics of a different attribute of the system of interest, connected only by information transfers or by the "information network."

2.5. FORMALLY REPRESENTING THE CONCEPTUAL MODEL

Conceptual Model Diagrams

Formal representation of the conceptual model most commonly takes the form of a box-and-arrow diagram such as those we have been using to illustrate our example models thus far. As we have seen such diagrams play an important role in modeling by helping us visualize the "big picture" and by facilitating communication between different people who are interested in a particular system. Although we present this as the fourth step in conceptual model formulation, and indeed the conceptual model diagram might be thought of as the end product of the first phase of systems analysis, diagrammatic representation of the conceptual model usually is concurrent with the three earlier steps and aids them greatly. Conceptual model diagrams also provide a framework that facilitates subsequent quantification of the model because equations can be related directly to specific parts of the conceptual model.

A variety of schemes exists for formal representation of the conceptual model. Diagrams based on symbols introduced in Figure 2.4 are particularly useful because they are simple, flexible, and defined in terms identical to those that we have chosen to describe components and relationships within the system of interest. These symbols are similar to those suggested by Forrester (1961) for modeling dynamics of industrial systems although the driving variable symbol was added by Innis (1979) for modeling ecological systems and I have chosen to represent Forrester's "constants" as a special class of driving variables. We will use these symbols to represent models throughout the text.

Matrix of Material Transfers

A matrix of material transfers (Smith 1973) is another schematic representation of model structure that is useful in describing large, highly interconnected models for which box-and-arrow diagrams may be confusing. A matrix of material transfers for our model of energy flow through an ecosystem (Figure 2.5) is presented in Figure 2.7. An "X" in the matrix represents a material transfer from the source or state variable in the indicated row to the state variable or sink in the indicated column. The "Xs" may be replaced by designations such as "T_{12}" that refer to specific material transfers—in this case the transfer from state variable 1 to state variable 2. This facilitates cross-referencing of model equations with the matrix of transfers.

A matrix of material transfers can be a convenient representation of the degree of "connectedness" of a system. By arranging state variables appropriately in the matrix it may also be possible to indicate the existence of submodels within a model more clearly than is possible in a box-and-arrow diagram. If for example we expand our energy flow model in Figure 2.5 to represent nitrogen cycling through the system as well, we might develop a conceptual model diagram similar to the one shown in Figure 2.8. Here in addition to representing four nitrogen state variables we have represented energy fixation by plants as a function of the nitrogen in plants and nitrogen in soil, and energy consumption by both herbivores and carnivores as a function of the nitrogen content of their own bodies. As the model becomes larger and particularly as the degree of

MATERIAL TRANSFER FROM EACH COMPONENT	MATERIAL TRANSFER TO EACH COMPONENT				
	S1	S2	S3	SI1	SI2
SO1	X				
S 1		X		X	X
S 2			X	X	X
S 3				X	X

FIGURE 2.7. Matrix of material transfers representing the conceptual model of energy flow through an ecosystem (Figure 2.5).

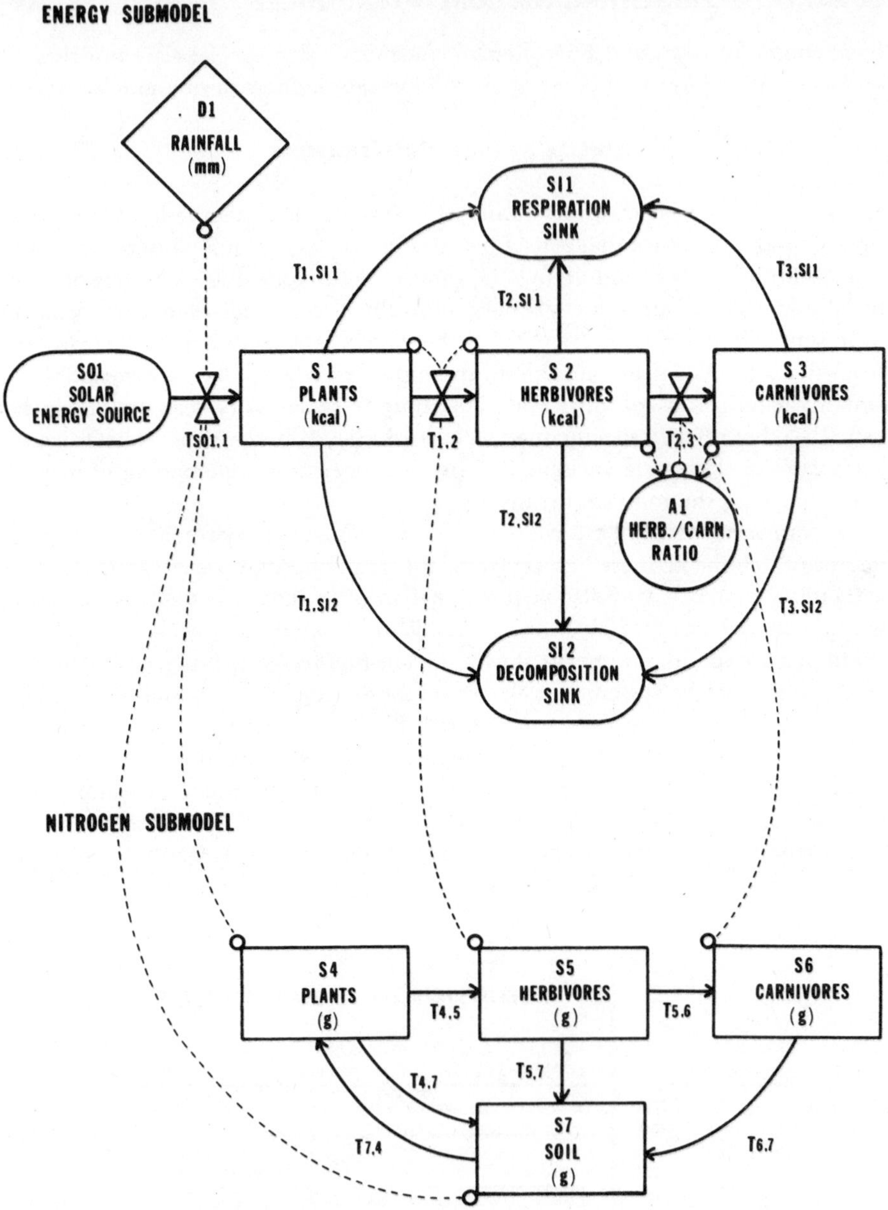

FIGURE 2.8. Conceptual model diagram expanding the representation of energy flow through an ecosystem (Figure 2.5) to include nitrogen cycling through the ecosystem as well. Material transfer units are in kcal/day in energy submodel and in g *N*/day in nitrogen submodel. (Control valves have been omitted from some material transfers for the sake of simplicity.)

MATERIAL TRANSFER TO EACH COMPONENT

MATERIAL TRANSFER FROM EACH COMPONENT	S1	S2	S3	S4	S5	S6	S7	SI1	SI2
S01	X								
S1		X						X	X
S2			X					X	X
S3								X	X
S4					X		X		
S5						X	X		
S6							X		
S7				X					

FIGURE 2.9. Matrix of material transfers representing the conceptual model of energy flow and nitrogen cycling through an ecosystem (Figure 2.8).

connectedness increases the diagram becomes somewhat confusing. Representation of this model as a matrix of material transfers (Figure 2.9) clearly shows the energy and nitrogen submodels of Figure 2.8 through clustering of Xs in the matrix. In this relatively simple example two submodels also are obvious in the box-and-arrow diagram, but such aspects of internal system structure may become lost in large, complex diagrams.

Other Schemes of Model Representation

Energy circuit language provides an alternative diagrammatic scheme for models of systems that can be represented completely in terms of energy flow (Odum 1972). Although this scheme may appear inflexible because of the restriction that all transfers must be represented in terms of energy, it has been used to model a variety of complex ecological systems (Odum 1983).

Analog computer diagrams provide another alternative diagrammatic scheme for models of systems that can be solved on an analog computer (Patten 1971b). Analog computers will be described more fully in Appendix 4 but for now we need only note that they are machines that solve sets of differential equations.

2.6. EXAMPLE OF CONCEPTUAL MODEL FORMULATION: HYPOTHETICAL POND SYSTEM

Description of the Problem and System Structure

We now will demonstrate conceptual model formulation in detail using a hypothetical example which we also will use in Chapters 3, 4, and 5 to demonstrate subsequent steps in systems analysis. The system of interest is a hypothetical 1 ha farm pond in the southern United States that is leased from the landowner on a monthly basis. Each year the pond manager (lessee) stocks and harvests fish. Fish are sold to consumers by bulk weight but individual fish must be heavier than a set weight to be sold legally. The pond manager's objective is to maximize his profit from the operation.

It is not an intensively managed system. No fertilizer or artificial feed is added to the pond. Since stocker fish are obtained free and there is no cost associated with harvest, the only management cost is the lease fee. Historically 75 fish each weighing 0.227 kg have been stocked on April 15 and harvested on November 15. The pond manager wishes to evaluate new management alternatives that involve changing number of fish stocked, when they are stocked, and when they are harvested.

Fish are herbivorous and feed solely on aquatic plants that grow naturally in the pond. No reproduction of fish occurs in the pond; it is strictly a put-and-take operation. Likewise no natural mortality of fish occurs. We know that net production of plants is dependent on biomass of plants and on water temperature. We also know that consumption of plants by fish is a function of biomass of plants, biomass of fish, and size of individual fish, as well as being temperature dependent. Fish excretion and respiration also are dependent on biomass of fish, size of individual fish, and water temperature, with excretion dependent upon assimilation efficiency of fish as well. Finally, we know that natural mortality of plants is a function of plant biomass.

Bounding the System of Interest

The problem to be addressed by the model can be stated quite simply. Which of several stocking and harvesting schemes will result in higher profit for the pond manager? Specific questions of interest include the following:

1. What size harvest might be expected in a "normal" year under the usual scheme of stocking 75 0.227-kg fish on April 15 and harvesting them on November 15, and what profit is associated with this harvest (we might call this the baseline situation)?

2. How variable is the harvest (and therefore profit) from year to year under the baseline stocking and harvesting scheme as the result of inherent year-to-year variability of the system?
3. Would size of the harvest and associated profit be changed by harvesting a month earlier or a month later?
4. Would size of the harvest and associated profit be changed by stocking a month earlier or a month later?
5. Would size of the harvest and associated profit be changed by stocking twice as many fish?

For these last three questions the manager has decided that size of the harvest must be changed by at least 10 percent relative to what he usually harvests (baseline situation) in order for the change to be meaningful to him within a management framework. Changes in size of the harvest of less than 10 percent are of no practical significance to the manager.

Questions such as the first two are common to most wildlife and fisheries modeling projects. Usually we are interested in modeling behavior of the system of interest under normal conditions. How we define normal depends upon our specific objectives, but often normal refers to status quo or to some historical set of conditions under which the system of interest operated. Simulation of system behavior under these normal conditions provides a baseline, or benchmark, set of predictions against which predictions of system behavior under other management schemes or environmental conditions can be compared. It also may provide an opportunity to check model predictions against real world observations of system behavior under similar circumstances and hence serve as one means of validating the model.

Questions such as the last three also are common to most management-oriented modeling projects. The primary purpose of modeling within a management framework usually is to develop a model capable of simulating system behavior under a variety of circumstances in which we are interested. We wish to conduct experiments with the model to answer our questions.

From the general description of the pond system and the list of questions to be addressed by the model we should be able to bound the system of interest more formally. The basic process to be modeled is growth of fish. The quantity of primary interest within the system is weight or biomass since it is the total biomass of fish harvested (and associated profit) that is most interesting to the manager. Fish consume plants, excrete a portion of what they consume depending upon the efficiency with which they assimilate plants, respire another portion of what they consume, and the remainder of what they consume represents growth or secondary production. Thus biomass of plants, biomass of fish, assimilation efficiency, and biomass of fish harvested are components of the system of interest. Since we are interested in fish biomass excreted and respired only to the extent that they represent losses from the system, we will include an excretion sink and a respiration sink as part of the conceptual model.

We also noted that fish consumption is a function of water temperature and weight of individual fish, with smaller fish presumably consuming more per unit of body weight than larger fish. Number of fish stocked (from which we can obtain mean weight of individual fish from total fish biomass) and water temperature therefore are components of the system of interest.

We also must represent the flow of biomass through plants. This means that a biomass source and a natural mortality sink for plants likewise will be included in the conceptual model.

Finally, stocking date and harvest date (which determine length of the lease), initial weight of individual fish (which together with number of fish stocked determines initial fish biomass in the pond), and several economic variables including unit cost of the lease, total cost of the lease, unit price of harvested fish, size limit for selling fish, gross revenue from the harvest, and profit all are system components. Total cost of the lease is a function of length of the lease and unit cost of the lease, and gross revenue from the harvest is a function of biomass of harvested fish, unit price of fish, and size limit for selling fish. Profit is a function of gross revenue from the harvest and total cost of the lease.

Categorizing Components within the System of Interest

Since state variables represent points of accumulation of material—in this case biomass—within the system of interest, plants, fish, and harvested fish are state variables. State of the system relative to the distribution of biomass is defined uniquely if we know the level of biomass in each of these variables at a point in time.

Sources and sinks represent origination and termination points, respectively, of material transfers into and out of the system. The biomass source for plants is a source, and natural mortality of plants, fish excretion, and fish respiration are biomass sinks.

Water temperature, assimilation efficiency, stocking date, and harvest date are driving variables because they affect the rest of the system—in this case the flow of biomass through the system—but are not themselves affected by the rest of the system. Likewise, number of fish stocked, initial size of individual fish, unit cost of the lease, unit price of harvested fish, and size limit for selling fish are driving variables because they affect the rest of the system but are not themselves affected by the rest of the system.

Finally, weight of individual fish is an auxiliary variable since it is calculated directly from number of fish stocked (driving variable) and fish biomass (state variable) as part of calculations determining material transfer rates within the system. Likewise, total cost of the lease and gross revenue from the harvest are auxiliary variables because they are important concepts that arise as part of calculations of another variable, profit. Profit is an auxiliary variable because it is an end product of calculations that is of particular interest to us.

Identifying Relationships between Components That Are of Interest

Relationships between system components that are of interest include both (1) material transfers which represent the physical flow of materials between state variables, sources, and sinks, and (2) information transfers which represent the use of knowledge about values of a state, driving, or auxiliary variable to control a material transfer.

Biomass transfers of interest occur from:

1. biomass source to fish, representing initial stocking of fish,
2. biomass source to plants,
3. plants to fish,
4. fish to harvested fish,
5. plants to natural mortality sink,
6. fish to excretion sink, and
7. fish to respiration sink.

The absence of a material transfer from plants to the respiration sink implies that the material transfer from biomass source to plants is defined as net as opposed to gross primary production. Material transfer units will be biomass per unit of time, such as kg/week or g/day.

Initial introduction of fish into the pond is dependent on the time of stocking and the number and size of fish stocked. The information transfer from stocking date (driving variable), number of fish stocked (driving variable), and initial weight of individual fish (driving variable) controls the material transfer from biomass source to fish (state variable). Net production of plants is dependent on biomass of plants and water temperature. The information transfers from plants (state variable) and water temperature (driving variable) control the material transfer from biomass source to plants. Consumption of plants by fish is a function of biomass of plants, biomass of fish, weight of individual fish, and water temperature. The information transfers from plants, fish, average weight of individual fish (auxiliary variable), and water temperature control the material transfer from plants to fish. Average weight of individual fish is calculated based on information about fish biomass and number of fish stocked. The information transfers run from fish and from number of fish stocked (driving variable) to average weight of individual fish. Fish excretion and respiration are dependent on fish biomass, weight of individual fish, and water temperature. The information transfers from fish, weight of individual fish, and water temperature control the material transfers from fish to excretion sink and from fish to respiration sink. Fish excretion also depends on assimilation efficiency. Information about assimilation efficiency (driving variable) also controls this material transfer. Biomass of fish harvested depends on biomass of fish in the pond and on date of harvest. The information transfers from fish and harvest date (driving variable) control the material transfer from fish to harvested fish. Natural mortality of plants depends on plant biomass. The information transfer from plants controls the material transfer from plants to natural mortality sink.

Total cost of the lease (auxiliary variable) is calculated from information transferred from unit cost of the lease (dollars per day, for example) (driving variable), stocking date (driving variable), and harvest date (driving variable). Gross revenue from the harvest (auxiliary variable) is calculated from information transferred from unit price of harvested fish (dollars per kg, for example) (driving variable), biomass of harvested fish, size limit for selling fish (driving variable), and weight of individual fish, which is needed to assure that fish meet the minimum size limit for legal sale. Finally, profit (auxiliary variable) is calculated from information transferred from total cost of the lease and gross revenue from the harvest.

Formally Representing the Conceptual Model

It is desirable to condense the verbal description of the system of interest presented in the previous section into a box-and-arrow diagram representing our conceptual model of the system (Figure 2.10). Symbols used in Figure 2.10 are defined in Figure 2.4. A matrix of material transfers representing our conceptual model is presented in Figure 2.11. At this point it would be well to reread the previous description of our hypothetical pond system and the manner in which the conceptual model was formulated, relating descriptions in the text to Figures 2.10 and 2.11.

We might characterize our conceptual model of the hypothetical pond system as a relatively simple biomass flow model. We represent the transfer of biomass between three state variables, a source, and three sinks. There are no submodels representing the transfer of materials other than biomass. Our description of the system is multivariate with:

1. water temperature (perhaps in °C),
2. number of fish stocked,

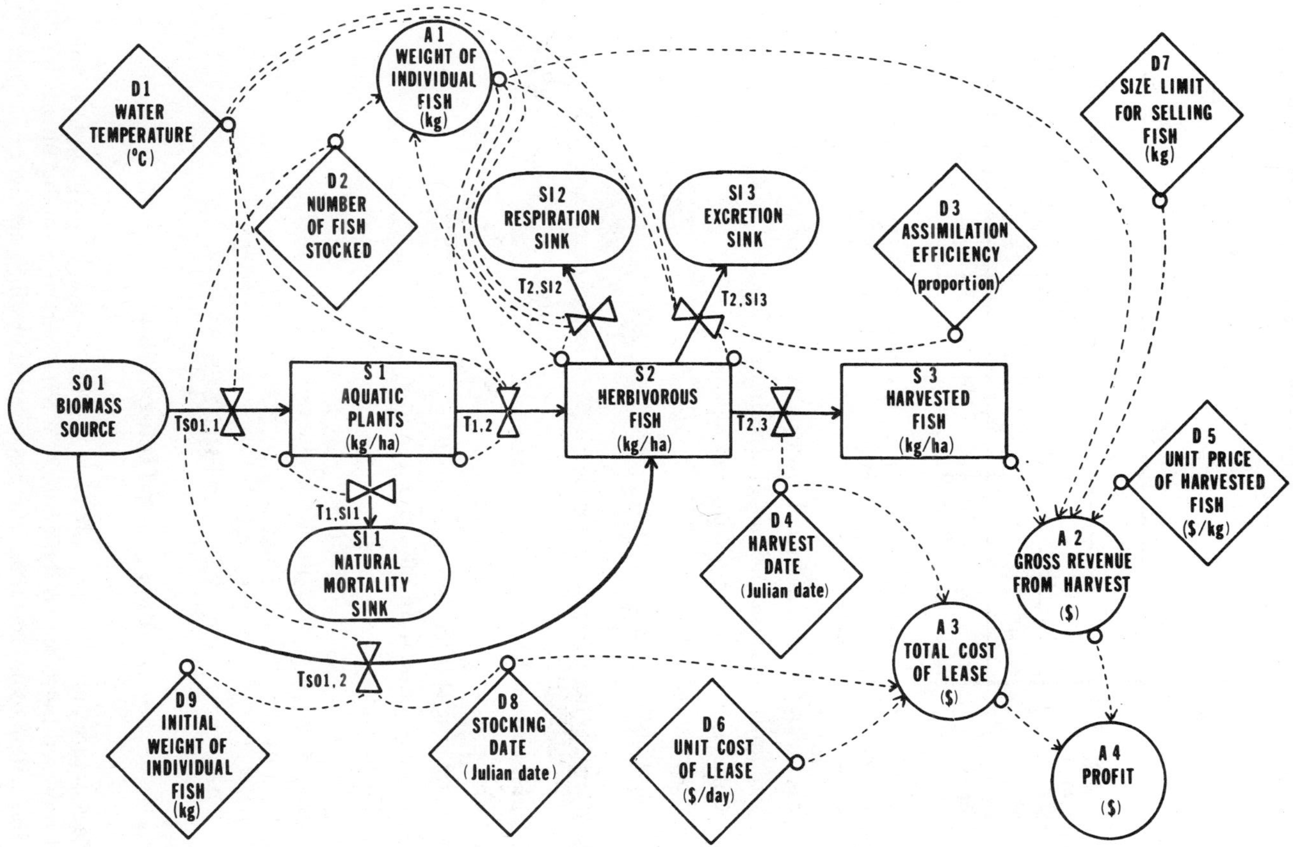

FIGURE 2.10. Conceptual model diagram representing the hypothetical pond system. Symbols are as defined in Figure 2.4.

MATERIAL TRANSFER FROM EACH COMPONENT	MATERIAL TRANSFER TO EACH COMPONENT					
	S1	S2	S3	SI1	SI2	SI3
S01	X	X				
S1		X		X		
S2			X		X	X
S3						

FIGURE 2.11. Matrix of material transfers representing the conceptual model of the pond system (Figure 2.10).

3. assimilation efficiency (perhaps in percent),
4. size limit for selling fish legally (perhaps in kg),
5. initial size of individual fish (perhaps in kg),
6. stocking date, and
7. harvest date

representing driving variables measured in a variety of units, and with

8. unit cost of the lease,
9. unit price of harvested fish,
10. total cost of the lease,
11. gross revenue from the harvest, and
12. profit

representing driving and auxiliary variables measured in dollars.

2.7. EXERCISES

Background Information

Exercises in this and subsequent chapters involve the development and use of a model representing biomass dynamics of a hypothetical plant–herbivore system. The system of interest is an island ecosystem in which herbivores are naturally absent. However, herbivores have been introduced successfully on several islands similar to our system of interest, and island managers are interested in introducing a herbivorous game animal on our island. Before actually introducing herbivores, managers want to develop a model to assess (1) long-term (5-year) impact of herbivores on the island plant community in terms of the effect on biomass of live plants and litter, and (2) biomass of herbivores that could be harvested annually on a sustained yield basis assuming that all harvest occurs on the last day of the year.

It is known that net primary (plant) production is a function of environmental temperature, soil water tension, and biomass of live plants. Plant litter production is solely a function of live plant biomass. Decomposition of plant litter is a function of soil water tension and biomass of litter.

From introductions of herbivores on other islands it is also known that consumption of live plants by herbivores is a function of environmental temperature and the ratio of live plant biomass to herbivore biomass. Both herbivore respiration and excretion are functions of environmental temperature and herbivore biomass. Decomposition of herbivore excreta is a function of soil water tension and biomass of excreta.

Little is known about natality and mortality rates of the herbivores. But it is believed that both rates are relatively constant throughout the year. In any case, observed changes in herbivore population biomass, which are accounted for explicitly by differences between population consumption and population respiration plus excretion, implicitly include the net effects of natality and mortality as well.

Exercise 2.7.1

Based on the description of the plant-herbivore system and the modeling objectives presented above, develop and describe a conceptual model diagram of the system of interest. Justify your choice and categorization of system variables.

Exercise 2.7.2

Construct a matrix of material transfers representing the conceptual model developed in Exercise 2.7.1.

Answers to Chapter 2 exercises are in Appendix 6.

CHAPTER

3

QUANTITATIVE SPECIFICATION OF THE MODEL

3.1. INTRODUCTION

The second phase of systems analysis is quantitative specification of the model. The goal is to quantify all relationships between model components that were identified in a qualitative manner during conceptual model formulation. This involves four steps that we might categorize as:

1. choosing the general quantitative structure for the model (Section 3.2),
2. choosing functional forms of model equations (Section 3.3),
3. choosing the basic time unit for simulations and parameterizing model equations (Section 3.4), and
4. formally presenting and computer coding model equations and executing the baseline simulation (Section 3.5).

A great many quantitative techniques are potentially useful in this phase of systems analysis. When quantitative data from the system of interest are available, the procedures involved in choosing functional forms and parameterizing model equations do not differ fundamentally from the data analysis procedures commonly used outside of a modeling context to establish relationships between variables. Exhaustive coverage of these techniques is neither possible nor desirable within the framework of this book. The specific techniques presented are among the more useful in wildlife and fisheries applications and are used here collectively as a means of describing the general process of quantitatively specifying a model. Once the process itself is understood, additional techniques of data analysis can be mastered as one's needs and interests dictate.

3.2. CHOOSING THE GENERAL QUANTITATIVE STRUCTURE FOR THE MODEL

There are different types of mathematical formats suitable for representing models of dynamic ecological systems: Matrix algebra, differential equations, and difference equations are examples. However, most formats are representatives of a general compartment model structure in which the basic unit, or module, consists of two state variables connected by a material transfer. Material transfer is controlled by information about other parts of the system, including information about state variables, driving variables, and/or auxiliary variables. Of course this is exactly the framework that we used in Chapter 2 to qualitatively represent the conceptual model of a system of interest.

Matrix Algebra

Matrix algebra provides a convenient format for representing the dynamics of populations for which we know age-specific natality and survival rates (Usher 1972), and more generally is useful for representing dynamics of systems in which each material transfer is a constant proportion of the state variable from which it originates, or of the donor state variable (Bledsoe and Van Dyne 1971). An advantage of representing models in this format is that analytical treatment of the resulting matrix of transfer coefficients can provide interesting insights into the dynamics of system behavior without simulation. A description of some relevant properties of matrix algebra and examples of the use of matrix algebra to model wildlife and fisheries systems are presented in Appendix 1.

Differential Equations

Differential equations provide a flexible format for ecological modeling. As with matrix algebra, a potential advantage of differential equations is that simple models composed of certain sets of differential equations can be solved analytically, thus providing insights into the dynamics of system behavior without simulation. But most commonly even relatively simple models formulated as sets of differential equations are solved numerically by simulation, which for practical purposes is quite similar to the method of solution used for difference equation models. A description of some relevant terminology and properties of differential equations, a comparison of differential and difference equations, and examples of the use of differential and difference equations to model wildlife and fisheries systems are presented in Appendix 2.

Difference Equations

Difference equations provide the most flexibility for representing dynamic systems in terms of a general compartment model structure. The price paid for this flexibility is that analytical solutions usually are not possible and therefore models based on difference equations must be solved numerically by simulation. However, it is likely that few systems of direct interest to us can be represented adequately by either matrix models or differential equation models simple enough for analytical solution. Thus it appears that in most cases little is lost, since analytical solutions are at best a remote possibility, and much is gained through use of difference equations. Models presented in remaining chapters will follow the general compartment model structure based on a difference equation format.

General Compartment Model Structure

General compartment model structure is appealing because it is simple and yet can represent large complex models. The strategy is to write an equation or set of equations determining at selected points in time the value of each driving variable, each auxiliary variable, each material transfer (called rate equations), and each state variable. Collectively these equations completely describe the model. They completely describe the behavior of the system of interest.

The computing sequence for model equations each time unit is:

1. driving variable equations,
2. auxiliary variable equations (if present),
3. rate equations, and
4. state variable equations.

Initial conditions of the model, including initial values of all driving variables and state variables, must be specified at time zero. To simulate behavior of the system of interest we ''solve'' the model by calculating equations in sequence each unit of simulated time (Figure 3.1).

Recall our model predicting fluctuations in weight of an animal over time first presented in Chapter 1 (Figure 1.1). We have redrawn the conceptual model diagram to indicate formal designations of the driving and state variables, source and sink, and material transfers as presented in Chapter 2 (Figure 3.2). In Chapter 1 we simulated consumption and respiration of an initially 100 g animal for two consecutive days assuming that environmental temperatures are 20°C and 19°C, respectively. Environmental temperature is a driving variable that is simply assigned a specific value each day: In Chapter 1 we

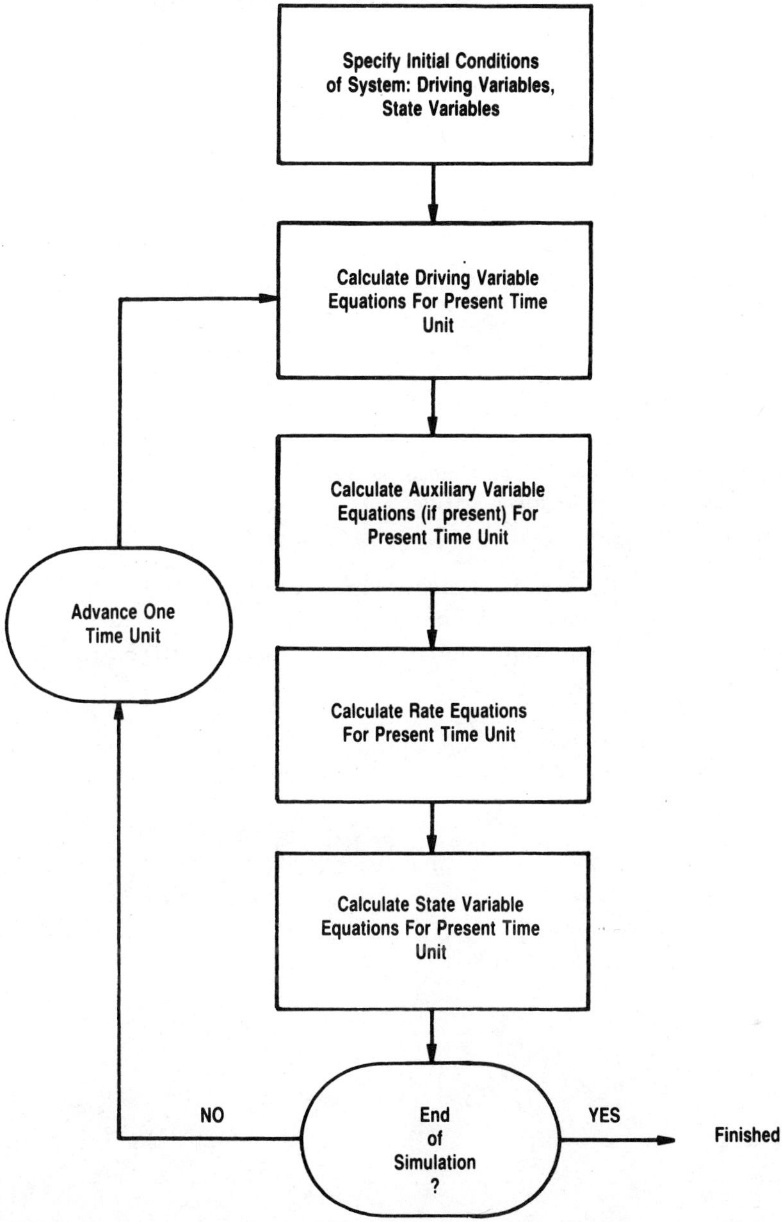

FIGURE 3.1. Computing sequence for model equations within the general compartment model structure.

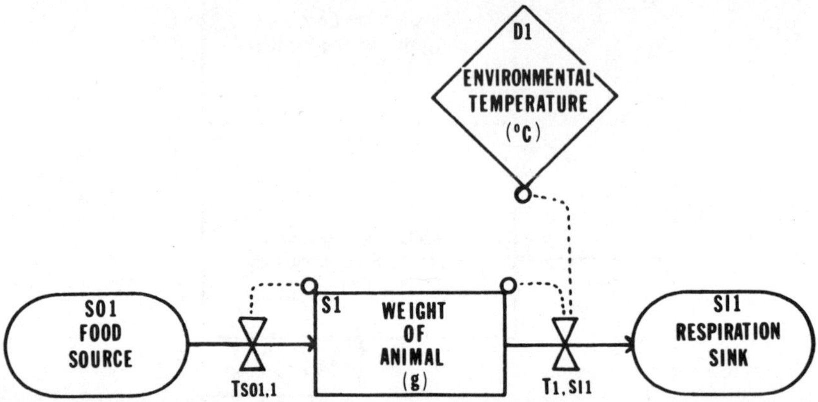

FIGURE 3.2. Conceptual model diagram of the weight fluctuation model (Figure 1.1) indicating formal designations of the driving and state variables, source and sink, and material transfers.

wrote $T_0 = 20$ and $T_1 = 19$, which we now rewrite as $D1 = 20$ if $t = 0$ and $D1 = 19$ if $t = 1$ to conform with our formal designation of variables. These are driving variable equations. There are no auxiliary variable equations. The consumption equation,

$$C_{t,t+1} = 0.05\,W_t$$

and the respiration equation,

$$R_{t,t+1} = 0.05\,W_t\left(\frac{T_t}{20}\right)$$

which we rewrite as

$$T_{\mathrm{SO1},1} = 0.05\,S1$$

and

$$T_{1,\mathrm{SI1}} = 0.5\,S1\left(\frac{D1}{20}\right)$$

are rate equations. Finally

$$W_{t+1} = W_t + \frac{\Delta W}{\Delta t}$$

which we rewrite as

$$S1_{t+1} = S1_t + \frac{\Delta S1}{\Delta t}$$

is the state variable equation, where

$$\frac{\Delta W}{\Delta t} = C_{t,t+1} - R_{t,t+1}$$

or

$$\frac{\Delta S1}{\Delta t} = T_{SO1,1} - T_{1,SI1}$$

(remember that we assumed assimilation efficiency equals 100 percent). For each day of simulated time we first evaluate the driving variable equation, next the two rate equations, and finally the state variable equation.

Note that we usually do not use a time subscript in our scheme of formal designations of variables. It is implicit that we mean the value of a driving, auxiliary, or state variable at time t, and the value of a rate equation (or amount of material transferred) during the time interval t to $t+1$. However, in instances where we need to represent values of variables or rates at other times we do use a time subscript, such as in state variable equations where we need to represent the value of a state variable both at time t ($S1_t$) and at time $t+1$ ($S1_{t+1}$).

3.3. CHOOSING FUNCTIONAL FORMS OF MODEL EQUATIONS

Having decided upon general quantitative structure the next step in quantitative specification of the model is to choose functional forms of model equations. We now specify whether the general forms of equations representing specific relationships within the model are linear, sigmoidal, exponential, etc.

In our model predicting fluctuations in weight of an animal (Figure 3.2) we stated in Chapter 1 that material transfer from the food source to the animal component (consumption, $T_{SO1,1}$) was an increasing linear function of the value of the animal state variable (weight of the animal, $S1$). We chose the functional form of the rate equation representing consumption to be linear. We might indicate the functional form of the consumption equation as

$$T_{SO1,1} = f(S1) = \beta_0 + \beta_1 S1$$

which is the equation for a straight-line, or linear, relationship. Parameters of this equation, β_0 and β_1, must be estimated, but we defer parameter estimation for the moment and concentrate only on specification of the functional forms of equations. The functional form of the other rate equation (representing respiration, $T_{1,SI1}$) was not linear but was of the form

$$T_{1,\mathrm{SI1}} = f(S1, D1) = \beta_0 S1\left(\frac{D1}{\beta 1}\right)$$

where once again β_0 and β_1 are parameters to be estimated. We will look at this equation more closely a bit later. The driving variable equations simply equated environmental temperature ($D1$) with an input datum each time unit. It is common practice to equate driving variables directly with input data although they also can be explicit functions of time. The general functional form of state variable equations is the same for all discrete-time compartment models since these equations are essentially bookkeeping functions to update the state of the system each time unit. The general form is

$$\begin{aligned}(\text{state variable})_{t+1} = {} & (\text{state variable})_t \\ & + (\text{transfers into state variable})_{t,t+1} \\ & - (\text{transfers out of state variable})_{t,t+1}\end{aligned}$$

For our weight fluctuation model the state variable equation is

$$S1_{t+1} = S1_t + T_{\mathrm{SO1},1} - T_{1,\mathrm{SI1}}$$

Information upon Which to Base the Choice of Functional Forms

Information upon which to base the choice of functional forms of model equations is of four general types:

1. quantitative data from direct observation or experimentation with the real system, either collected first hand or from literature,
2. information based on theoretical relationships or generally applicable empirical relationships that are appropriate for the situation being modeled,
3. qualitative information based on expert opinion or on literature, and
4. information gained from experimenting with the model itself.

Quantitative Data

The best information is that based on quantitative experimentation with the real system. If we had data for our weight fluctuation model relating consumption to weight of the animal that appeared as plotted in Figure 3.3, we would feel secure in choosing a linear form for the consumption–weight relationship.

Theoretical or Generally Applicable Empirical Relationships

Sometimes theoretical relationships or generally applicable empirical relationships exist that suggest the functional form of a particular relationship. Suppose with regard to our weight fluctuation model there was a general empirical

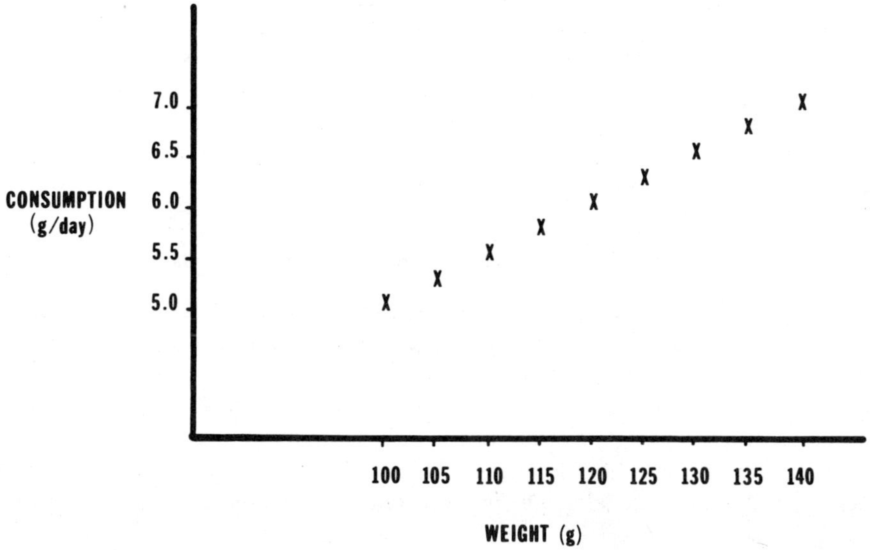

FIGURE 3.3. Hypothetical data for the weight fluctuation model (Figure 3.2) relating consumption ($T_{SO1,1}$) to weight of an animal ($S1$).

relationship relating respiration to environmental temperature that had been established for a variety of animal species closely related to the animal in our model. This relationship suggested that respiration equaled consumption at 20°C but decreased by 5 percent relative to consumption for every 1°C below 20°C and increased by 5 percent relative to consumption for every 1°C above 20°C. Thus at 19°C respiration would be 95 percent of consumption, at 21°C it would be 105 percent of consumption, and so on. Then we might represent the functional form of the respiration relationship, or material transfer from the animal variable ($S1$) to the respiration sink (SI1), as we indicated earlier:

$$T_{1,SI1} = \beta_0 S1\left(\frac{D1}{\beta 1}\right)$$

The $\beta_0 S1$ term is exactly the same as the right-hand side of the consumption equation ($T_{SO1,1} = \beta_0 S1$), and the $D1/\beta_1$ term adjusts respiration upward or downward from this base as environmental temperatures ($D1$) fluctuate above and below 20°C.

Qualitative Information

In some cases there may be no quantitative data or theoretical relationships available upon which to base the choice of functional forms. In such situations we may rely on qualitative information from literature or on opinions of experts in the field to establish assumptions upon which to base the choice of functional forms of some model equations. Without the consumption data in Figure 3.3 we might have based the choice of a linear relationship between consumption and weight of the animal on the opinion of a person with extensive experience in feeding such animals, even though written records concerning rate of consumption never were maintained.

Information from Experimentation with Model

In some cases there may be little known in even a qualitative sense upon which to base our choice of functional forms. In such situations we may be able to gain insight into possible functional forms of a model equation by hypothesizing different functional forms and observing model behavior in response to each. Through such experimentation with the model we may narrow possible choices by excluding those functional forms that produce unreasonable model behavior. Without specific information about the relationship between consumption and weight of the animal in our weight fluctuation model we might have hypothesized an exponential and a sigmoidal as well as a linear relationship. However after parameterizing each equation (parameterization of model equations is discussed below) and running test simulations we may have found that, of these three choices, only the linear relationship produced reasonable predictions of animal weight changes in view of what we know about weight changes of animals in the real system. Obviously the number of equations that can be specified in this manner within any single model necessarily is small. The more equations specified by this trial-and-error method the higher the likelihood that we will obtain apparently reasonable model behavior purely by chance.

Selecting Types of Equations to Represent Chosen Functional Forms

When we choose the functional form for a relationship in our model we often can find a single equation that can be parameterized to represent the appropriate curve. This is not always the case, but knowledge of the types of equations associated with different types of curves is most helpful. Although we probably remember that a straight line can be represented by the equation

$$y = \beta_0 + \beta_1 X$$

and perhaps that an exponential curve can be represented by the equation

$$y = \beta_0 e^{\beta_1 X}$$

there are numerous other useful types of equations. There is no need to commit a large number of equations to memory, but we should be aware that there are

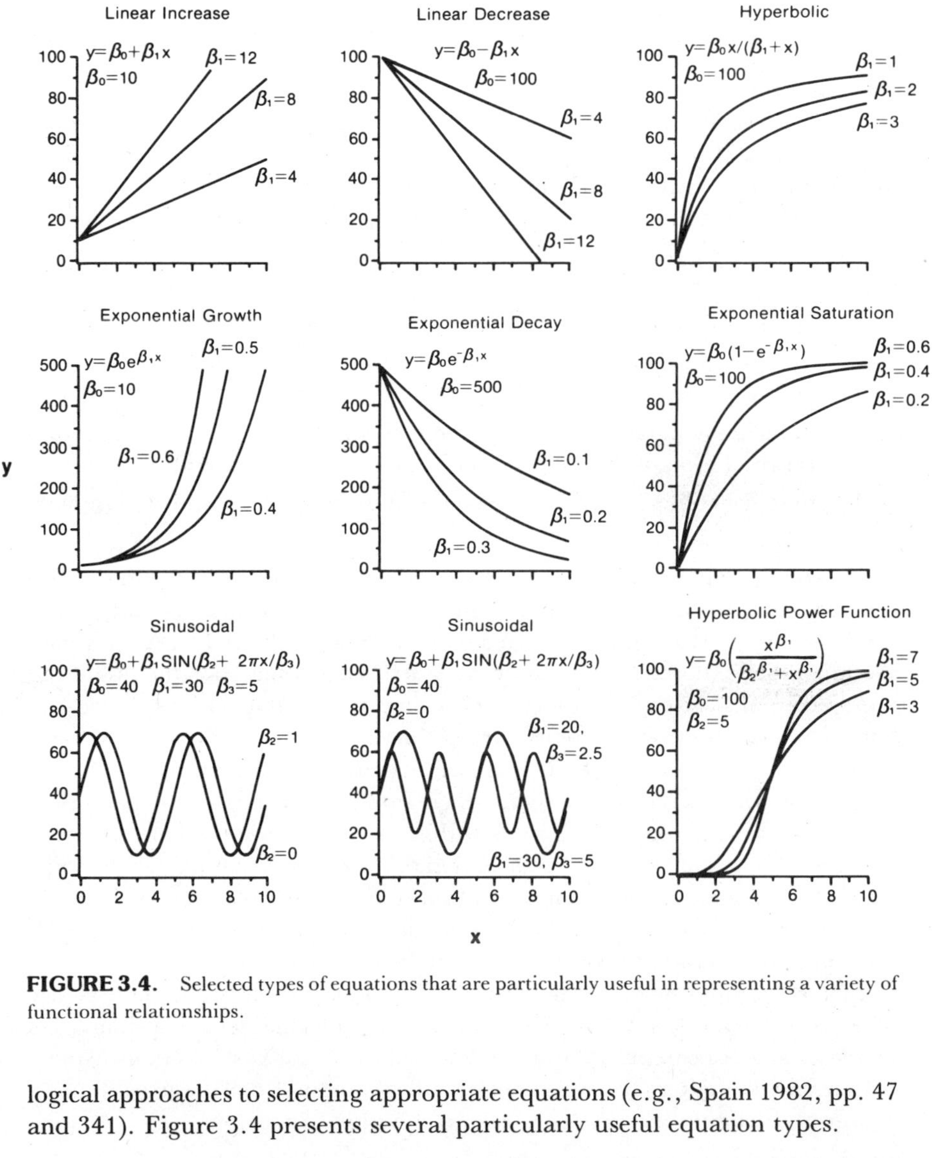

FIGURE 3.4. Selected types of equations that are particularly useful in representing a variety of functional relationships.

logical approaches to selecting appropriate equations (e.g., Spain 1982, pp. 47 and 341). Figure 3.4 presents several particularly useful equation types.

3.4. CHOOSING THE BASIC TIME UNIT FOR SIMULATIONS AND PARAMETERIZING MODEL EQUATIONS

After choosing functional forms of model equations the next step in quantitative specification of the model is to choose the basic time unit for simulations and to estimate parameters of each equation.

Basic Time Unit for Simulations

The basic time unit for simulations is the time interval (Δt) between iterative solutions of model equations. The basic time unit for simulations in our weight fluctuation model as presented in Chapter 1 (p. 13) was 1 day. We solved model equations each day of simulated time.

Choice of the basic time unit depends upon the level of resolution in time needed:

1. for the model to address our questions of interest,
2. to represent appropriately temporal changes in rates at which processes within the system of interest occur,
3. to facilitate parameterization of model equations in view of the temporal resolution of available data or other information, and
4. to make computing costs reasonable in terms of both time and money.

If in our weight fluctuation model we had been interested in questions concerning diurnal fluctuations in consumption and respiration of the animal those questions would have dictated a basic time interval less than $\Delta t = 1$ day. We might have chosen $\Delta t = 1$ hr or $\Delta t = 6$ hr depending upon the resolution required to answer our questions concerning diurnal fluctuations. If we had known or suspected that respiration rates fluctuated through the day in a manner that made us uneasy about representing them as average daily rates we might have chosen Δt smaller than 1 day. Implicit in choice of the basic time unit is the assumption that all rates within the system are constant over any given Δt. If all data and other information that we used to parameterize model equations were based on a temporal resolution of 1 hr we might have chosen $\Delta t = 1$ hr simply to facilitate parameterization of model equations, although this third consideration is not as important as the first two.

Finally, computing costs can become a very real consideration in the choice of Δt for some models. Obviously as the number of time steps in a simulation increases so does computing time and associated dollar costs if one is using a mainframe computer. If one is using a microcomputer dollar costs are not a concern but computing time can become a limiting factor.

Parameterizing Model Equations

Information upon which to base parameterization of model equations is of the same four general types used to choose functional forms of model equations, namely:

1. quantitative data,
2. information based on theoretical or generally applicable empirical relationships,
3. qualitative information, and
4. information gained from experimenting with the model itself.

In fact the choice of functional forms and the parameterization of model equations for any given model most commonly are based on the same information, and the entire process often is an iterative one. Nonetheless, from a theoretical standpoint we should view the choice of functional forms and the parameterization of model equations as two distinct steps because the former usually has more profound implications concerning ecological interpretations of model structure than does the latter.

Quantitative Data

Specific techniques available to estimate parameters of model equations based on quantitative data are diverse, and a substantive discussion covering the spectrum of available techniques is beyond the scope of this text. A variety of techniques for fitting equations to data are presented by Daniel and Wood (1971), and Spain (1982) presents computer programs to fit selected theoretical equations and polynomial equations to data. Problems that arise within the framework of model parameter estimation are identical to those encountered in analysis of data in general, and modelers are well advised to consult with professional statisticians during this step in quantitative specification of the model. Nonetheless the role of the various techniques of data analysis often employed during this step of model development should be clear: They serve as tools allowing us to interpret appropriately quantitative information about important relationships within our system of interest. Among the techniques more widely applicable for our purposes is regression analysis. We present a description of some relevant aspects of regression analysis and examples of its application to modeling wildlife and fisheries systems in Appendix 3.

As an example of parameterization of model equations based on quantitative data we return again to our weight fluctuation model (Figure 3.2). Parameterization of the consumption equation, the functional form of which is

$$T_{SO1,1} = \beta_0 + \beta_1 S1$$

involves estimation of β_0 and β_1. If the data plotted in Figure 3.3 relating consumption to animal weight were available to us we might run a simple linear regression analysis with consumption as the dependent variable and weight as the independent variable to estimate β_0 and β_1. In this particular case we have specified hypothetical data such that all data points lie exactly on the line

$$T_{SO1,1} = 0.05S1.$$

Thus $b_0 = 0$, $b_1 = 0.05$ (where b_0 and b_1 represent our estimates of parameters β_0 and β_1, respectively), and the proportion of the variability in consumption accounted for by variability in weight equals one ($r^2 = 1.0$). Of course the r^2 of regressions based on real data will always be less than 1.0, even if the true relationship is perfectly linear, due to experimental error in data measurements.

Theoretical or Generally Applicable Empirical Relationships

In cases where data describing a relationship within our system of interest are not available we may be able to rely on theoretical or generally applicable empirical relationships established based on information about similar relationships in other systems. Earlier we assumed the respiration relationship in our weight fluctuation model was based on a general empirical relationship that suggested respiration equaled consumption at 20°C but decreased by 5 percent relative to consumption for every 1°C below 20°C and increased by 5 percent for every 1°C above 20°C. The functional form of the equation was specified as

$$T_{1,SI1} = \beta_0 S1\left(\frac{D1}{\beta_1}\right)$$

where β_0 must be estimated such that b_0S1 numerically equals consumption, and β_1 must be estimated such that $D1/b_1$ causes respiration to equal consumption at 20°C and makes respiration 5 percent larger or smaller than consumption as environmental temperatures either rise or fall 1°C above or below 20°C. Thus b_0 must equal 0.05 since consumption is represented as

$$T_{SO1,1} = 0.05S1$$

and b_1 must equal 20 since this causes appropriate shifts in respiration as temperature varies.

Qualitative Information

In cases where no quantitative data are available we may be able to rely upon qualitative information from literature or on opinions of experts to establish assumptions upon which to base our estimates of model parameters. Without the consumption data in Figure 3.3 we might have based our estimates of parameters of the consumption equation on the opinion of an expert on animal consumption. He or she may have given us approximate consumption estimates for both relatively light and relatively heavy animals, thus providing two points through which we could fit a straight line and estimate β_0 and β_1.

Information from Experimentation with the Model

In cases where little is known even in a qualitative sense upon which to base our parameter estimates we may be able to bound the range of reasonable parameter estimates by hypothesizing different estimates and observing model behavior in response to each. Those parameter estimates resulting in unreasonable model behavior can be excluded from further consideration. Suppose that we lacked both consumption data and expert opinion concerning consumption of light and heavy animals for our weight fluctuation model. However, we did know that consumption was some linear function of weight and we knew the relationship of respiration to consumption at different environmental temperatures. Given this information we could bound our parameter estimates for the consumption equation simply through trial-and-error experimentation with the model itself. As with the choice of functional forms, the number of parameters in any single model estimated by this trial-and-error method must be relatively small or we run the risk of obtaining apparently reasonable model predictions purely by chance.

Deterministic versus Stochastic Model Parameterization

One final consideration regarding model parameterization is whether the model is deterministic or stochastic. So far we have talked in terms of point estimates of variables and transfer rates within models, thus implying that we are dealing with deterministic models. However, the same parameterization procedures just presented are equally applicable to stochastic models except that some additional information is needed to represent inherent randomness of variables or transfer rates occurring within the model.

Representation of a random variable or transfer rate in a stochastic model requires specification of either a statistical distribution or an empirical frequency distribution from which values of the variable or transfer rate are selected randomly, rather than specification of a single point estimate such as is required for deterministic models. As before, information upon which to base specification of the statistical or empirical distributions can be:

1. quantitative data,
2. information based on theoretical relationships or generally applicable empirical relationships,
3. qualitative information, or
4. information gained from experimenting with the model itself.

To illustrate the differences in information required to parameterize deterministic versus stochastic models, consider possible representations of the environmental temperature driving variable in our weight fluctuation model. Recall that initially we equated temperature ($D1$) directly with an input datum for each of two days that were simulated: $D1 = 20°C$ if $t = 0$ and $D1 = 19°C$ if $t = 1$. This representation is deterministic. If we wish to represent environmental temperature as a stochastic (random) variable rather than deterministically specifying its value each day there are at least three alternatives. We can select random temperature variables from:

1. a frequency distribution of historical data,
2. a statistical distribution fitted to historical data, or
3. a set of statistical distributions generated from regression of temperature on day of the year.

Frequency Distribution of Historical Data

If temperature data from the real system are available from several years, we might use all historical temperatures available for a given date to construct a frequency distribution of temperatures for that date (Figure 3.5a). We then randomly select temperature for the first day of simulated time ($t = 0$) from the day 0 frequency distribution rather than simply specifying it to be 20°C as in the deterministic model, and we select temperature for the second day of simulated time ($t = 1$) from the day 1 frequency distribution rather than specifying it to be 19°C as before.

To randomly select temperature from the frequency distributions in Figure 3.5a we first construct the associated cumulative frequency distributions (Figure 3.5b). Height of each bar in the cumulative frequency distributions represents the proportion of observed temperatures less than the value indicated beneath the bar. Thus for day 0 one tenth of the observed temperatures were less than 18°C, three tenths were less than 19°C, and so on (Figure 3.5b). Next we select a uniform random variate on the interval zero to one (a random number between zero and one) from a table of random numbers (Rohlf and Sokal 1969, p. 152). Or, we might obtain a uniform random variate from a random number generator on a computer (Law and Kelton 1982, p. 219). We then find the temperature in the appropriate cumulative frequency distribution corresponding to the frequency whose value equaled the selected random number. If we were using the day 0 distribution and if the random number selected was 0.1, the corresponding temperature would be 18°C (Figure 3.5b). If the random number was 0.3 the corresponding temperature would be 19°C, and so on.

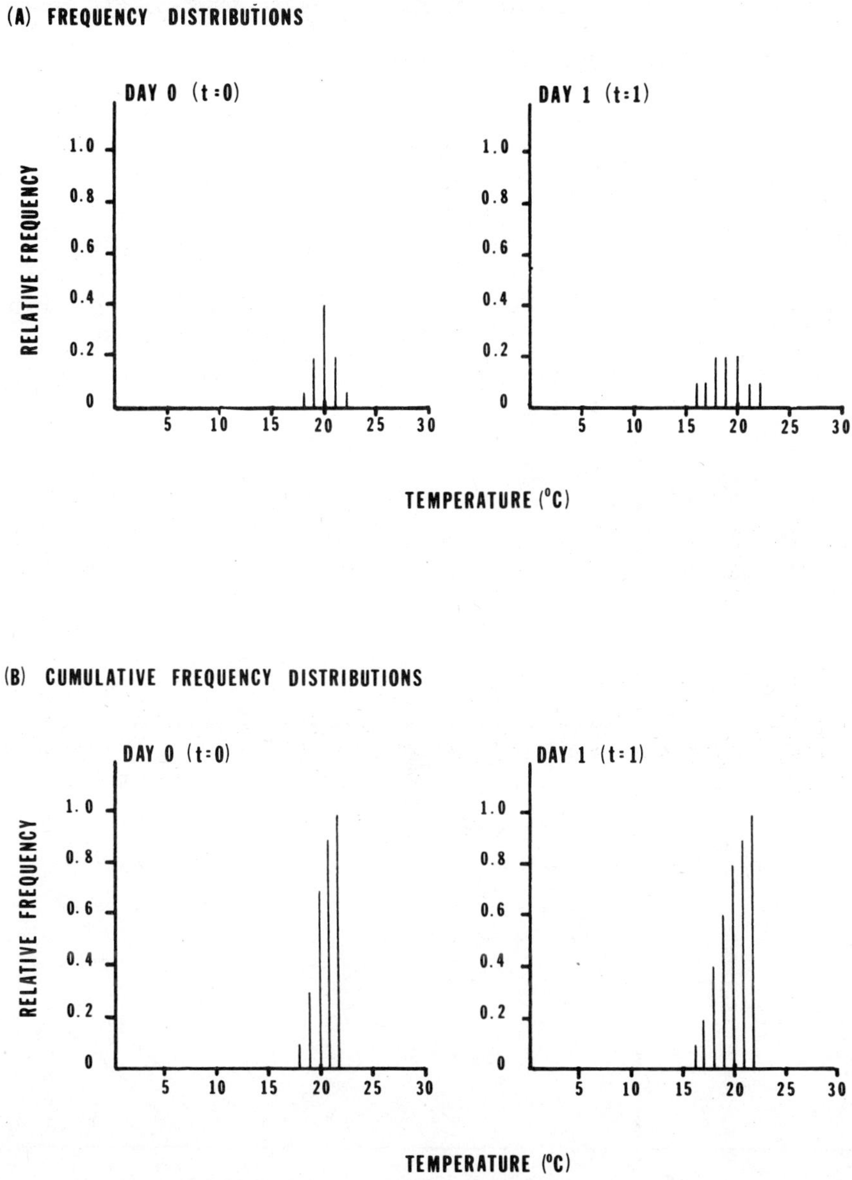

FIGURE 3.5. (A) Frequency distributions of hypothetical environmental temperature data for two days and (B) the associated cumulative frequency distributions.

Statistical Distribution Fitted to Historical Data

An alternative way to represent environmental temperature as a random variable is to fit a statistical distribution to historical real system temperature data. We might find that temperatures for each date are approximately normally distributed. We then might fit a normal distribution to temperatures for each date by estimating the parameters of a normal distribution (mean μ and standard deviation σ) based on temperatures for each date (Figure 3.6). Next we randomly select temperature for the first day of simulated time ($t = 0$) from the

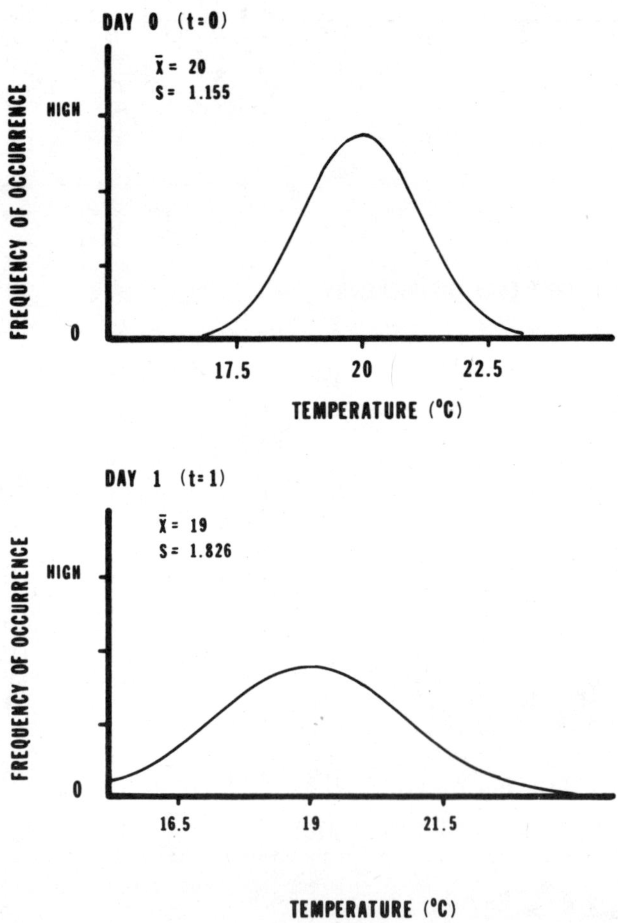

FIGURE 3.6. Normal distributions fitted to the hypothetical environmental temperature data in Figure 3.5a.

day 0 normal distribution and temperature for the second day of simulated time ($t = 1$) from the day 1 normal distribution.

To randomly select temperature from normal distributions in Figure 3.6 we first select 12 uniform random variates on the interval zero to one from a table of random numbers. We then use these 12 uniform random variates to calculate a standard normal random variate (a random number from a standard normal distribution) using the formula:

$$Z \cong \sum_{i=1}^{i=12} R_i - 6$$

(Hastings and Peacock 1975, p. 100), where Z is the calculated standard normal random variate and R_i is the ith uniform random variate selected from a random numbers table. Twelve uniform random variates are considered sufficient to calculate a standard normal random variate for most practical purposes (Hastings and Peacock 1975). Accuracy of the approximation increases as number of uniform random variates used increases but the formula becomes more complex. Other methods for calculating normal random variates and tables of normal random variates are also available (Law and Kelton 1982). Next we convert the calculated standard normal random variate to a random variate from the specific normal distribution in which we are interested, using the formula

$$X = \mu + \sigma Z$$

where X is the normal random variate from the specific normal distribution of interest, μ is the mean and σ is the standard deviation of the specific normal distribution of interest (which we estimate as $\bar{X}$ and s, respectively), and Z is the standard normal random variate. If we want a random temperature from the day 0 normal distribution (Figure 3.6) and if the 12 uniform random variates are 0.48, 0.15, 0.73, 0.74, 0.52, 0.37, 0.60, 0.86, 0.67, 0.65, 0.77, and 0.38, we calculate

$$Z \cong \sum_{i=1}^{i=12} R_i - 6 = 6.16 - 6 = 0.16$$

assuming that $\mu = 20$ and $\sigma = 1.155$.

We then calculate

$$X = \mu + \sigma Z = 20 + (1.155)(0.16) = 20.185$$

Thus 20.185°C is the random temperature for day 0.

Of course there are numerous statistical distributions other than the normal that might be fitted to historical data. The uniform and Poisson, for example, are two distributions that also find wide application in ecology. The degree to which a data set resembles a particular statistical distribution, or the goodness of fit to that distribution, can be evaluated using standard statistical procedures (Johnson and Kotz 1969). Algorithms for generating random variates from a variety of statistical distributions that we might wish to use in stochastic simulation models are available (Hastings and Peacock 1975).

Statistical Distributions Generated from Regression

Yet another way to represent environmental temperature as a random variable is to generate the required statistical distributions from a regression of temperature on day of the year. Suppose that we are interested in simulating animal weight fluctuations for a particular 30-day period and that real system data suggest that environmental temperatures decrease linearly during this period from 20°C to 15°C (Figure 3.7a). We could calculate a regression of temperature on day and use the mean squared error (MSE) of regression to generate a normal distribution about the temperature predicted for each day. Since the MSE of regression is an unbiased predictor of variance (σ^2), the normal distribution associated with each predicted daily temperature has the temperature predicted by regression as its mean and a variance equal to the MSE. For this example the regression equation is temperature $= 20 - 0.167t$ and the MSE $= 2.0006$ (Figure 3.7a). As before, we randomly select temperature each day from the normal distribution for that day. The normal distribution of temperature generated for day 15 ($t = 15$) based on the MSE of regression is shown in Figure 3.7b. Actually if we do not know the slope of the regression line exactly, the variance about predicted temperatures will increase with increasing distance from the overall mean temperature. It is possible to obtain the variance associated with each predicted temperature (Snedecor and Cochran 1967, p. 154) and these separate variances could be used instead of the MSE of regression to specify the temperature distributions. However, assuming that our intent is to represent in the model the general variability of temperature, rather than the variability associated with temperatures predicted from a particular data set, the MSE of regression probably is the better choice.

Note in the preceding discussion the relationship between statistical analyses, or statistical models, and simulation models. We commonly employ statistical techniques to help us determine appropriate quantitative representations for particular relationships in simulation models when data pertaining to those relationships are available. Often the statistical models used to analyze the data, or, more strictly speaking, the results of the statistical analyses, actually become a part of the simulation model. If, for example, we use a linear regression model to relate consumption to animal weight based on the data in Figure 3.3, the right-hand side of the regression equation (of the form $\beta_0 + \beta_1 S1$) becomes the right-hand side of the rate equation describing consumption in our weight fluctuation model ($T_{SO1,1} = \beta_0 + \beta_1 S1$).

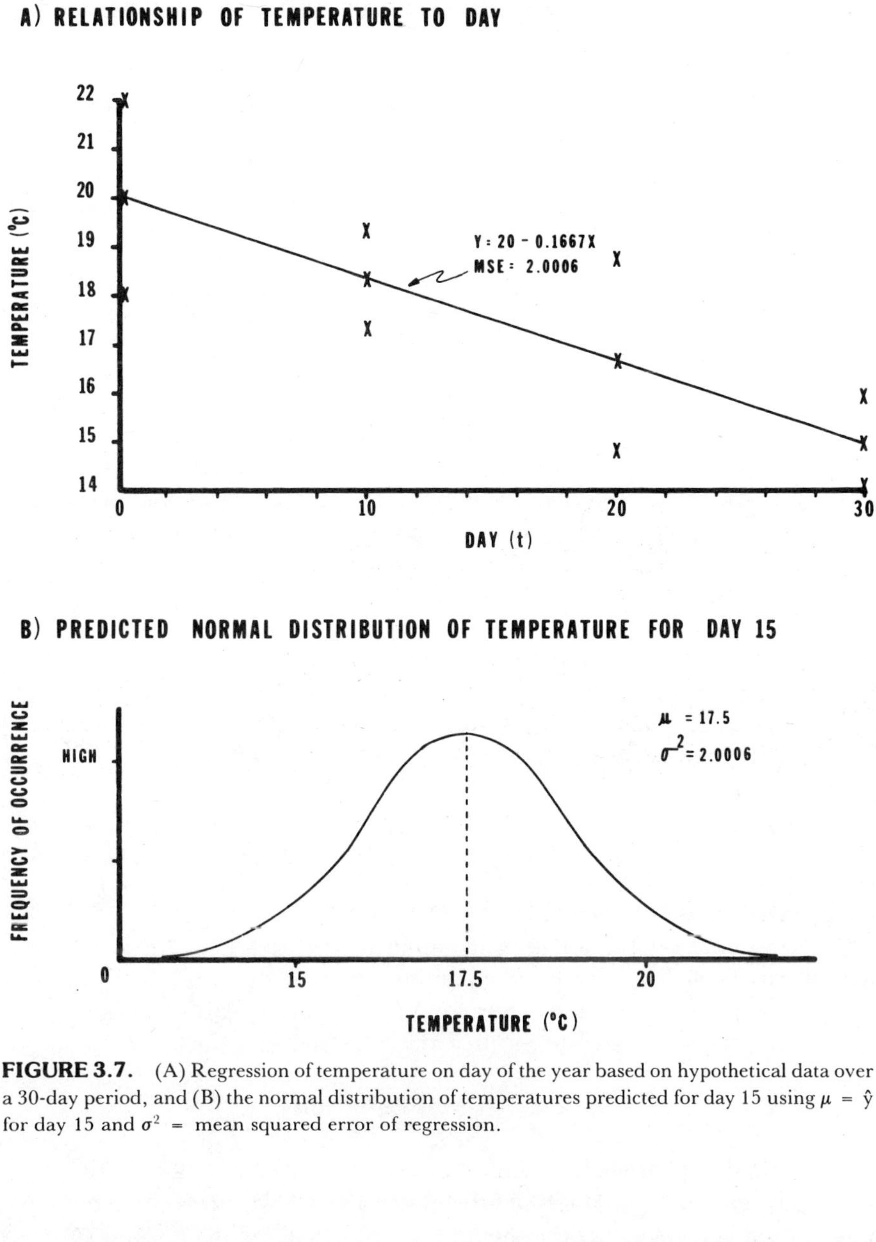

FIGURE 3.7. (A) Regression of temperature on day of the year based on hypothetical data over a 30-day period, and (B) the normal distribution of temperatures predicted for day 15 using $\mu = \hat{y}$ for day 15 and σ^2 = mean squared error of regression.

3.5. FORMALLY PRESENTING AND COMPUTER CODING MODEL EQUATIONS AND EXECUTING THE BASELINE SIMULATION

After model equations have been parameterized the next step in quantitative specification of the model involves:

1. formally presenting model equations,
2. coding equations for solution on a digital computer, and
3. executing the baseline simulation.

Formal presentation of model equations aids in ultimately solving the model and also facilitates the unambiguous description of model structure to others. Coding of equations on a digital computer likewise can serve the dual role of facilitating both the solution and description of the model, but most commonly the primary role is in solving the model. Baseline simulation represents the behavior of the system of interest, or the solution of the model, under the particular set of conditions that we wish to use as a benchmark or standard against which to compare changes in system behavior resulting from other sets of conditions of interest to us.

Formally Presenting Model Equations

An effective manner in which to present model equations formally is in a table organized by the computing sequence of equation types. Driving variable equations are presented first, followed by auxiliary variable equations, rate equations, and state variable equations. Initial conditions of the model also must be identified. Our weight fluctuation model equations are represented formally in Table 3.1. This set of equations uniquely describes the model and relates model equations directly to a conceptual model diagram through corresponding variable designations as in Figure 3.2 and Table 3.1.

Coding Model Equations

Coding model equations for solution on a digital computer can be done in a variety of computer languages. FORTRAN and BASIC are among the most widely used languages. Regardless of the language used the general strategy for programming a simulation model on a digital computer involves:

1. constructing a flow diagram indicating the steps involved in solving model equations and simulating model behavior,

2. writing computer code in the chosen language to accomplish those steps,
3. debugging the computer program, or correcting the syntax or logical errors that prevent the program from compiling, and
4. verifying that the program is doing what the modeler intended.

A brief description of digital (and analog) computers, with references to manuals on several general programming languages and special-purposes programming packages, is presented in Appendix 4.

Constructing a Flow Diagram

A flow diagram indicating the general steps involved in solving a simulation model on a digital computer is presented in Figure 3.8. We first read the input data defining model parameters or constants and initial conditions of the model into the computer. Then the input data required each time unit of simulation are read in for the first Δt of simulated time. The next steps involve calculating in turn driving variable equations, auxiliary variable equations, rate equations, and state variable equations. Then we write, or output, results in which we are interested for the first Δt. If the simulation is not finished we advance simulated time one Δt and return to read the input data required for the next time unit. If the simulation is finished we write any summary results of the

TABLE 3.1

Formal presentation of the weight fluctuation model equations written to correspond to variable designations in Figure 3.2 and organized by computing sequence of equation types

Initial Conditions of the Model

$S1 = 100$

Driving Variable Equations

$D1 = 20$ if $t = 0$
$D1 = 19$ if $t = 1$

Rate Equations

$T_{SO1,1} = 0.05S1$
$T_{1,SI1} = 0.05S1(D1/20)$

State Variable Equations

$S1_{t+1} = S1_t + T_{SO1,1} - T_{1,SI1}$

entire simulation in which we are interested. If it is a stochastic model we ask if enough replicate simulations have been completed. If more replication is required we cycle back to begin a new simulation.

Writing Computer Code

The strategy for writing computer code is simply to follow the outline or logical flow specified in the computer flow diagram and fill in appropriate statements in the chosen language to accomplish the task. The procedure is not unlike converting an essay outline into prose. FORTRAN and BASIC programs for our weight fluctuation model are presented in Table 3.2.

Debugging the Computer Program

Debugging the computer program consists of examining the program for errors in logic or syntax that prevent it from compiling or that prevent the computer from interpreting instructions provided in the program. Most compilers provide diagnostic messages that help identify the type of error made and the location in the program where the error occurred.

Verifying the Computer Program

Verification consists of examining the debugged program to be sure that it is accomplishing what the modeler intended. Unfortunately the best way to verify that the program is doing what it should is to perform all calculations by hand for at least two time units and compare these results with those of the computer.

Executing the Baseline Simulation

Baseline simulation is the end product of the second phase of systems analysis just as the conceptual model diagram is the end product of the first phase of systems analysis. Baseline simulation represents the solution of the model under a particular set of conditions. Most commonly, baseline conditions are chosen such that the model simulates the normal situation found in the real system. As we will see in Chapter 4 model validation involves close examination of the baseline simulation and comparisons of baseline simulation predictions of various system attributes with corresponding attributes observed in the real system. As we will see in Chapter 5 model use involves comparison of baseline simulation results with results of simulations representing management policies or environmental situations that we wish to evaluate.

Suppose that the normal, or baseline, situation that we wish to simulate with our weight fluctuation model consists of the 30-day period for which we have data from the real system and during which temperatures decreased linearly

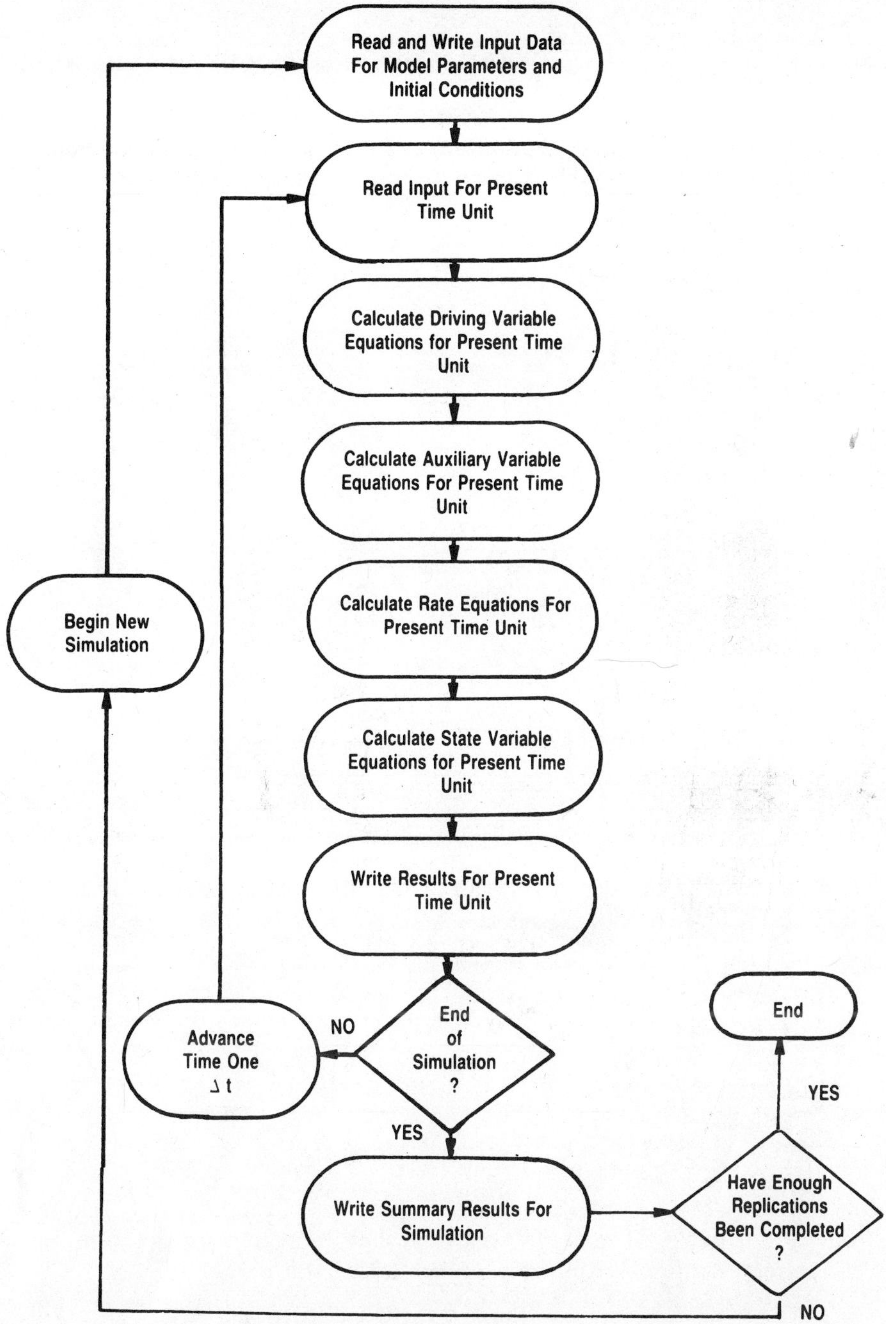

FIGURE 3.8. Computer programming flow diagram indicating the general steps involved in solving a simulation model on a digital computer.

TABLE 3.2

(A) FORTRAN and (B) BASIC programs to solve equations of the weight fluctuation model presented in Table 3.1 and to simulate model behavior

A. FORTRAN Program

```
      INTEGER T
      S1=100
      PRINT 1000
      DO 10 T=0, 1
      IF(T.EQ.0) THEN
       D1=20
      ELSE
       D1=19
      END IF
      T1=S1*0.05
      T2=(S1*0.05)*(D1/20)
      S1=S1+T1-T2
      PRINT 2000,T,S1,D1
10    CONTINUE
      STOP
1000  FORMAT (T6,'DAY',T16,'WEIGHT',T35,'TEMP')
2000  FORMAT (T6,I3,T16,F7.3,T35,F7.3)
      END
```

B. BASIC Program

```
      S1=100
      LPRINTTAB(6);'DAY';TAB(16);'WEIGHT';TAB(35);'TEMP'
      FOR T=0 TO 1
      IF T=0 THEN D1=20 ELSE D1=19
      T1=S1*0.05
      T2=(S1*0.05)*(D1/20)
      S1=S1+T1-T2
      LPRINTTAB(5);T;TAB(15);S1;TAB(35);D1
      NEXT T
      END
```

from 20°C to 15°C (Figure 3.7a). Recall that the regression equation representing temperature as a function of day during this period was:

$$\text{temperature } (^\circ\text{C}) = 20 - 0.167t$$

where $0 \leq t \leq 30$ represented day. We modify our weight fluctuation model by using this new equation to calculate the environmental temperature driving variable:

$$D1 = 20 - 0.167t$$

Other model equations remain as in Table 3.1. Results of the baseline simulation indicate that the animal gains weight throughout the 30-day period, but that rate of weight gain both in absolute terms (g/day) and as a proportion of animal weight (g/g-day) increases as environmental temperature decreases (Figure 3.9).

Baseline Simulations for Stochastic Models

Whereas there is only one baseline simulation for deterministic models, baseline simulation for stochastic models actually consists of a set of replicate simulations, each representing baseline conditions. Thus for stochastic models an additional consideration arises at this point, namely how many replicate simulations should be run. This question is directly analogous to the question of how many samples should be obtained in an experiment in the real world.

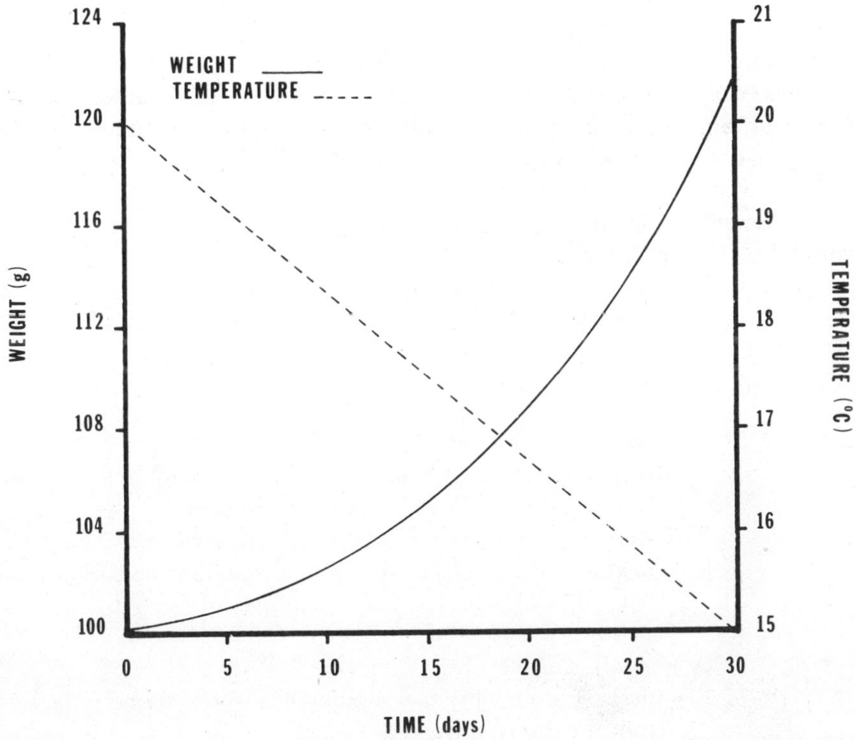

FIGURE 3.9. Results of the baseline simulation of the weight fluctuation model (Figure 3.2).

There is no simple answer to this sample size question, but perhaps one of the most useful answers can be found if we consider the framework within which the model ultimately is to be used and rephrase the question as follows: How large a sample must we obtain in order to show that a true difference of magnitude δ between alternative management policies or environmental situations is statistically significant at a significance level α with a probability P that the difference will be detected if it exists (Sokal and Rohlf 1969, p. 246)?

We can answer this question if we know two things:

1. inherent variability of items that we are sampling and
2. magnitude of difference between different samples of items (representing different management policies or different environmental situations) that we consider practically significant relative to the problem at hand.

We can obtain an estimate of the variability of items that we are sampling by running relatively many stochastic simulations of baseline conditions and calculating the variance of the model output in which we are interested. Although it is difficult to define "relatively many," as we increase the number of these initial simulations the estimate of variance should decrease to a stable level. We must determine the magnitude of differences between samples that we consider practically important independent of the model based on our knowledge of what size differences are relevant in the real system.

The formula for calculating the number of samples needed to detect a given true difference between sample means, assuming that we have an estimate of variability within samples, is

$$n \geq 2\left(\frac{\sigma}{\delta}\right)^2 (t_{\alpha,\gamma} + t_{2(1-P),\gamma})^2$$

(Sokal and Rohlf 1969, p. 247)

where
- n = number of samples
- σ = true standard deviation, which we estimate as the square root of estimated variance within samples
- δ = smallest true difference that we desire to detect
- γ = degrees of freedom of the sample standard deviation with b groups of samples and n samples per group, or $\gamma = b(n-1)$
- α = significance level
- P = desired probability that a difference will be found to be significant if it is as small as δ

$t_{\alpha,\gamma}$ and
$t_{2(1-P),\gamma}$ = values from a two-tailed t-table with γ degrees of freedom and corresponding to probabilities of α and $2(1 - P)$, respectively

We cannot solve for n directly because we do not know γ, which is a function of n. Therefore we:

1. guess a value for n,
2. calculate γ, and
3. solve the equation for n.

We then compare the calculated n with our guess of n. If the calculated n rounded to the nearest whole number is not equal to the guessed n, we adjust our guess for n accordingly and repeat the procedure.

Suppose that we wish to represent temperature as a random variable in our weight fluctuation model and use the stochastic version to simulate the same baseline situation that we simulated previously using the deterministic version (Figure 3.9). We modify the calculation of the temperature driving variable such that we now randomly choose temperature for each day from a normal distribution with a mean equal to the temperature predicted by the regression equation representing temperature as a function of day and a variance equal to the MSE of that regression (Figure 3.7). Other model equations remain as in Table 3.1. Further suppose that our primary use of the model is to compare the weight of an initially 100 g animal after 30 days exposure to baseline conditions with the weight of an initially 100 g animal after 30 days exposure to a warmer temperature regime. We want to detect a true difference in weight of 3 g (δ) at a statistical significance level of $\alpha = 0.05$ with probability $P = 0.80$ that the difference will be detected if it exists. Fifty preliminary stochastic baseline simulations suggest that 7.14 g is a good estimate of the inherent variability (σ^2) of weight after 30 days exposure to baseline conditions. Assuming that this also is a good estimate of the inherent variability of weight after 30 days exposure to the warmer temperature regime we can estimate the appropriate number of stochastic simulations to run as:

$$n \geq 2\left(\frac{\sigma}{\delta}\right)^2 (t_{\alpha,\gamma} + t_{2(1-P),\gamma})^2$$

If we estimate $n = 14$, then

$$\sigma = \sqrt{7.14} = 2.67$$

$$\delta = 3$$

$$\gamma = 2(14 - 1) = 26$$

$$\alpha = 0.05$$

$$P = 0.80$$

$$t_{\alpha,\gamma} = t_{0.05,26} = 2.056$$

and

$$t_{2(1-P),\gamma} = t_{0.40,26} = 0.856$$

Therefore the formula becomes

$$14 \geq 2\left(\frac{2.67}{3}\right)^2 (2.056 + 0.856)^2$$

or

$$14 \geq 13.43$$

and we should run 14 replicate stochastic simulations.

For practical purposes the improvement in our estimate of n gained by iteration is often unimportant. Usually we arbitrarily set $\gamma = \infty$ and use the initial calculated value of n as our estimate of required sample size, realizing that it is a slight underestimate. Had we done this for the example above, we would have calculated $n \geq 12.44$, and probably would have chosen a sample size of 13 or 14.

3.6. EXAMPLE OF QUANTITATIVE SPECIFICATION OF THE MODEL: HYPOTHETICAL POND SYSTEM

Choosing the General Quantitative Structure for the Model

We now return to our hypothetical pond system described initially at the end of Chapter 2. General quantitative structure of the pond model will consist of a discrete-time compartment model format including a set of driving variable equations, a set of auxiliary variable equations, a set of rate equations, and a set of state variable equations. Referring back to the conceptual model in Figure 2.10 we see that the pond model will have equations for nine driving variables, four auxiliary variables, seven material transfers, and three state variables. Each time step these equations will be solved in the indicated order.

Information Available for the Hypothetical Pond System

Three types of information on the "real" pond system are available:

1. information on historical states of the system,
2. information on processes occurring within the system, and
3. information on economic conditions and legal constraints pertaining to pond management.

Information on Historical States of the System

This includes data on standing crop biomass of plants and fish, and average weight of individual fish at various times of the year for several years (Table 3.3a). Data are also available on water temperature recorded at several different locations within the pond on selected dates during a normal year (Table 3.3b). It has been observed that seasonality of changes in water temperature is relatively invariable from year to year but that some years are uniformly warmer or colder than normal. Frequency of occurrence of relatively hot and cold years follows the normal distribution in Figure 3.10, in which years are characterized by the number of °C that their mean annual water temperature is displaced from normal, or long-term, mean annual water temperature. Finally, we know that historically 75 fish each weighing 0.227 kg are stocked on April 15 and harvested on November 15.

Information on Processes Occurring within the System

This includes data on rate of net production of plants as a function of water temperature (Table 3.4a) and rate of consumption of plants by fish as a function of water temperature and weight of individual fish (Table 3.4b). It is also known that consumption by fish becomes limited by plant biomass when plant biomass falls below 20,000 kg/ha and that "desired" consumption rate is reduced by roughly one-half for every 5000 kg/ha decrease in plant biomass below that point. Finally, it is known that the natural mortality rate of plants is density-dependent. Although no data are available experts have suggested that the curve presented in Figure 3.11 is a good approximation of this relationship.

There is no direct information available on excretion or respiration of fish in this system. However, fish excretion can be calculated from consumption since assimilation efficiency is known to be 2 percent. It is known that fish respiration can be represented as a function of the weight of fish and water temperature by the generally applicable empirical relationship

$$VO_2 \text{ (mg } O_2 \text{ respired/g fish-hr)} = A\text{wt}^{-0.2} e^{0.07\,T}$$

TABLE 3.3

Data on historical states of the pond system including (A) standing crop biomass of plants and fish, and average weight of individual fish on various dates during several years, and (B) water temperature at selected locations within the pond on various dates during a "normal" year

A. Standing Crop Biomass of Plants and Fish, and Average Weight of Individual Fish

		Standing Crop Biomass (kg/ha)		Weight of Individual Fish			Standing Crop Biomass (kg/ha)		Weight of Individual Fish
Year	Date	Plants	Fish	(kg)	Year	Date	Plants	Fish	(kg)
1975	Apr. 15	39,973	17	0.23	1979	Apr. 15	39,973	17	0.23
	May 15	39,273	18	0.24		May 15	39,295	20	0.26
	Jun. 15	38,534	30	0.40		Jun. 15	38,399	37	0.49
	Jul. 15	36,947	60	0.80		Jul. 15	36,207	78	1.03
	Aug. 15	34,006	109	1.46		Aug. 15	32,336	141	1.88
	Sep. 15	29,925	172	2.29		Sep. 15	27,604	208	2.78
	Oct. 15	26,512	216	2.88		Oct. 15	23,919	250	3.33
	Nov. 15	24,976	222	2.97		Nov. 15	22,154	258	3.44
1976	Apr. 15	39,981	17	0.23	1980	Apr. 15	39,978	17	0.23
	May 15	39,259	29	0.39		May 15	39,318	23	0.31
	Jun. 15	37,347	67	0.89		Jun. 15	38,084	48	0.64
	Jul. 15	32,453	147	1.96		Jul. 15	34,779	107	1.43
	Aug. 15	26,227	227	3.03		Aug. 15	29,655	185	2.47
	Sep. 15	21,191	267	3.56		Sep. 15	24,608	245	3.27
	Oct. 15	17,878	276	3.68		Oct. 15	21,106	273	3.64
	Nov. 15	16,212	270	3.60		Nov. 15	19,104	281	3.74
1977	Apr. 15	39,976	17	0.23	1981	Apr. 15	39,976	17	0.23
	May 15	39,296	21	0.27		May 15	39,306	22	0.30
	Jun. 15	38,341	39	0.52		Jun. 15	38,176	45	0.60

Year	Date				Year	Date			
	Jul. 15	36,006	82	1.09		Jul. 15	35,240	98	1.30
	Aug. 15	31,923	148	1.97		Aug. 15	30,504	172	2.29
	Sep. 15	27,190	214	2.86		Sep. 15	25,452	236	3.15
	Oct. 15	23,630	253	3.38		Oct. 15	21,980	268	3.57
	Nov. 15	21,774	262	3.49		Nov. 15	20,058	276	3.67
1978	Apr. 15	39,968	17	0.23	1982	Apr. 15	39,968	17	0.23
	May 15	39,291	21	0.28		May 15	39,291	21	0.28
	Jun. 15	38,348	39	0.52		Jun. 15	38,360	38	0.51
	Jul. 15	36,041	81	1.08		Jul. 15	36,090	80	1.06
	Aug. 15	31,894	148	2.00		Aug. 15	32,090	145	1.93
	Sep. 15	27,078	216	2.88		Sep. 15	27,242	213	2.84
	Oct. 15	23,499	254	3.39		Oct. 15	23,659	252	3.37
	Nov. 15	21,631	263	3.51		Nov. 15	21,797	262	3.49

B. Water Temperature

		Temperature (°C) at Location							Temperature (°C) at Location				
Year	Date	1	2	3	4	5	Year	Date	1	2	3	4	5
1983	Jan. 1	13.9	15.2	16.0	15.2	14.7	1983	Jul. 1	29.7	30.7	29.1	30.8	29.4
	Jan. 15	15.6	14.4	15.1	15.0	14.8		Jul. 15	30.5	30.2	31.1	29.4	28.8
	Feb. 1	15.0	14.5	18.3	13.1	14.3		Aug. 1	31.2	29.5	31.8	28.2	29.3
	Feb. 15	15.4	12.9	13.8	15.1	17.8		Aug. 15	31.8	28.4	28.4	32.4	29.1
	Mar. 1	14.4	14.4	16.7	14.8	14.7		Sep. 1	29.9	29.4	29.9	30.8	30.1
	Mar. 15	16.1	14.5	15.5	13.0	15.7		Sep. 15	28.8	30.0	32.0	29.6	29.5
	Apr. 1	14.9	14.5	14.2	15.4	15.7		Oct. 1	25.1	27.3	26.5	23.0	27.0
	Apr. 15	13.9	13.7	14.1	15.3	17.8		Oct. 15	24.0	21.0	23.6	21.4	22.3
	May 1	21.5	17.6	19.9	17.9	17.6		Nov. 1	17.7	19.1	19.5	17.5	18.5
	May 15	21.0	22.8	19.3	23.3	25.5		Nov. 15	13.4	15.1	14.5	15.5	16.3
	Jun. 1	26.4	26.3	28.1	27.3	24.5		Dec. 1	13.1	13.8	14.9	18.0	14.6
	Jun. 15	29.7	30.9	28.3	30.3	30.6		Dec. 15	15.6	13.9	13.8	17.9	13.6

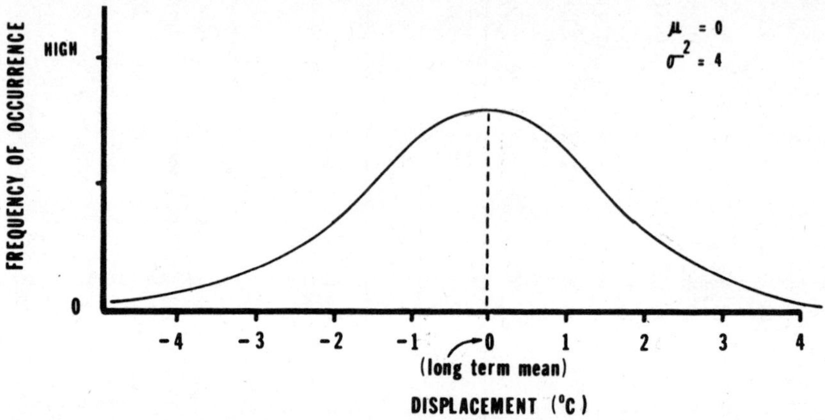

FIGURE 3.10. Distribution of displacements of mean annual water temperature from normal, or long-term, mean annual water temperature.

where VO_2 is respiration rate, wt is live weight of fish in g, T is water temperature in °C, and A is an empirically derived constant that equals 0.0142.

Information on Economic Conditions and Legal Constraints

The pond can be leased from the landowner on a monthly basis for 0.50 dollar/day, harvested fish can be sold for 1 dollar/kg, but only fish larger than 3 kg can be sold legally.

Choosing Functional Forms, Parameterizing Model Equations, and Choosing the Basic Time Unit for Simulations

Functional forms of model equations can be hypothesized and hypothesized equations can be parameterized based on data and other information just presented on the real system. This information includes examples of three of the four general types of information mentioned in Section 3.3. Experimental data on growth rate of plants (Table 3.4a) and on rate of consumption of plants by fish (Table 3.4b), and observational data on standing crop biomass (Table 3.3a) and water temperature in the pond (Table 3.3b) represent quantitative data from direct observation or experimentation with the real system. The equation predicting fish respiration as a function of the weight of fish and water temperature is based on a generally applicable empirical relationship. Density-dependent mortality of plants is postulated by experts (Figure 3.11). In a moment we will use this information first to choose the functional form of and then

TABLE 3.4

Data on processes occurring within the pond system including (A) net production rate of plants as a function of water temperature and (B) rate of consumption of plants by fish as a function of water temperature and weight of individual fish

A. Net Production Rate of Plants (g produced/kg plant biomass-day)

Water Temperature (°C)	Net Production Rep 1	Rep 2	Rep 3	Rep 4	Rep 5
10	0.225	0.422	0.072	0.358	0.002
15	0.639	0.601	0.578	0.143	0.231
20	0.595	0.353	1.180	0.377	1.555
25	0.814	1.028	1.176	1.223	1.748
30	1.362	2.223	1.612	1.459	1.374
35	2.540	1.436	1.779	2.255	2.009

B. Rate of Consumption of Plants by Fish of Different Sizes (kg consumed/kg fish biomass-day)

Water Temperature (°C)	Weight of Fish (kg): 0.227			1.500			3.636		
	Rep 1	Rep 2	Rep 3	Rep 1	Rep 2	Rep 3	Rep 1	Rep 2	Rep 3
15	0.001	0.340	0.387	0.043	0.056	0.243	0.376	0.083	0.086
20	0.730	0.534	1.026	0.360	0.166	0.605	0.132	0.278	0.254
25	2.071	1.270	1.124	0.810	0.710	0.674	0.135	0.816	0.265
30	2.631	1.925	2.284	1.000	1.804	0.718	0.782	0.464	0.553
35	2.631	3.099	2.889	1.402	1.684	1.420	1.053	0.366	1.001

to parameterize each model equation in turn, beginning with rate equations and proceeding to driving variable equations, auxiliary variable equations, and finally state variable equations. But first we turn attention to choice of the basic time unit for simulations.

Basic Time Unit for Simulations

We will use 1 day as the basic time unit for simulations. This choice is arbitrary but certainly is reasonable in view of the four criteria upon which to base such decisions suggested in Section 3.4. $\Delta t = 1$ day provides adequate temporal resolution to address our questions of interest, allows us to represent appropri-

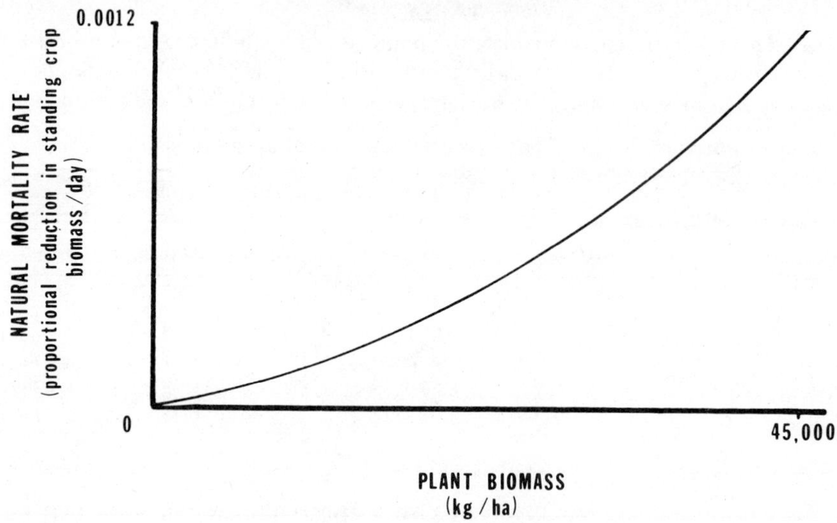

FIGURE 3.11. Curve based on expert opinion relating natural mortality rate of plants to plant biomass.

ately temporal changes in rates at which processes within the system of interest occur, and also facilitates parameterization of model equations in view of temporal resolution of available data. It also is unlikely that computing costs will become a serious concern with $\Delta t = 1$ day in a model of this size since one simulation will consist of roughly 210 days (recall that fish usually are stocked on April 15 and harvested on November 15 (p. 52). One might legitimately argue that $\Delta t = 2$ days or even $\Delta t = 1$ week could have been chosen as the basic time unit. But appropriate representation of temporal changes in process rates within the system might become difficult with Δt much longer than 1 week, and the resolution attained with Δt shorter than 1 day is not needed. Further, energy or biomass flow through individual animals is commonly presented in terms of daily energy budgets.

Rate Equations

We now turn attention to the choice of functional forms and parameterization of rate equations representing material transfers:

1. from biomass source to plants ($T_{SO1,1}$),
2. from biomass source to fish ($T_{SO1,2}$),
3. from plants to fish ($T_{1,2}$),
4. from fish to respiration sink ($T_{2,SI2}$),

5. from fish to excretion sink ($T_{2,SI3}$),
6. from plants to natural mortality sink ($T_{1,SI1}$), and
7. from fish to harvested fish ($T_{2,3}$) (Figure 2.10)

Transfer from Biomass Source to Plants ($T_{SO1,1}$)

Material transfer from biomass source (SO1) to plants ($S1$) is a function of water temperature ($D1$) and the biomass of plants ($S1$) (Figure 2.10), or

$$T_{SO1,1} = f(D1, S1)$$

A plot of available data relating plant growth in g net primary production per kg of live plant biomass per day to water temperature in °C (Table 3.4a) suggests a linear functional form (Figure 3.12). A simple linear regression model of the form

$$Y = \beta_0 + \beta_1 X + \varepsilon$$

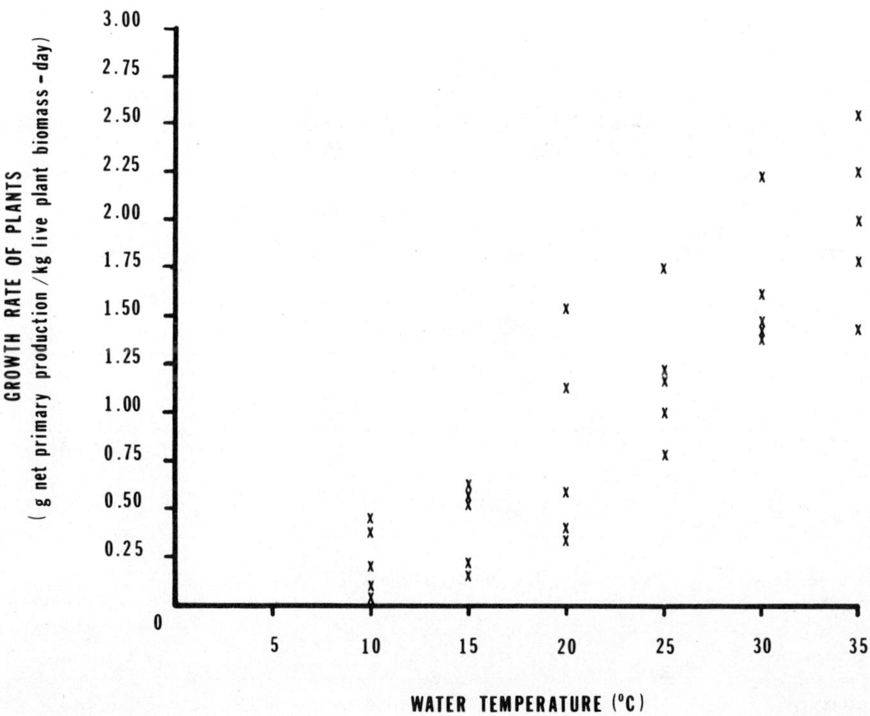

FIGURE 3.12. Plot of data from Table 3.4a relating growth of plants to water temperature.

with Y representing plant growth and X representing temperature can be fitted to data in Table 3.4a to parameterize the relationship. (See Appendix 3, p. 256 for a numerical example of linear regression calculations.) Parameterization through simple linear regression results in the equation

$$Y = -0.604 + 0.073X$$

The regression equation is significant ($P < 0.0001$) and water temperature accounts for 78 percent of variability in plant growth ($r^2 = 0.78$) (Table 3.5). Since regression predicts plant growth in terms of g net primary production per kg of live plant biomass per day, we must multiply the right-hand side of the equation by number of kg of live plant biomass in the system ($S1$) at time t and divide this by 1000 to calculate material transfer from the biomass source to plants from time t to $t + 1$ in kg/day, or

$$T_{SO1,1} = \frac{(-0.604 + 0.073D1)S1}{1000}$$

We now have a deterministic representation of plant production. However, data (Table 3.4a) suggest that plant production per kg of live plant biomass is

TABLE 3.5

Results of simple linear regression of growth rate of plants (g net primary production/kg live plant biomass-day) on water temperature (°C), based on data in Table 3.4a. Meanings of table entries are explained in Appendix 3

Regression Equation

growth rate = $-0.604 + 0.073$ temperature
($r^2 = 0.78$)

Analysis of Variance

Source of Variation	Degrees of Freedom	Sum of Squares	Mean Square	F Value	$P > F$
Regression	1	11.755	11.755	100.4	0.0001
Error	28	3.280	0.117		
Total	29	15.035			

Parameter Estimates

Parameter	Estimate	T for $H_0: B = 0$	$P > T$	Standard Error of Estimate
Intercept	−0.604	−3.428	0.0019	0.176
Temperature	0.073	10.017	0.0001	0.0073

not always the same at a given temperature. At 25°C plant production varies between 0.814 and 1.748 g/kg-day. Thus it is appropriate to represent this inherent randomness of plant growth explicitly in the model. As described in Section 3.4 one way to do this is to generate a normal distribution about each mean value (in this case of plant growth) predicted by regression using the mean squared error (MSE) as an unbiased estimate of variance. The MSE for regression of plant growth on water temperature is 0.117 (Table 3.5). Therefore we calculate

$$T_{\mathrm{SO}1,1} = \frac{(\mathrm{RV}1)S1}{1000}$$

where RV1 is a random variable drawn from a normal distribution with mean

$$\mu = -0.604 + 0.073D1$$

and variance

$$\sigma^2 = 0.117$$

Transfer from Biomass Source to Fish ($T_{SO1,2}$)

Material transfer from the biomass source (SO1) to fish ($S2$) (Figure 2.10) is zero except for stocking day at which time fish biomass transferred into the pond is a function of the number ($D2$) and initial weight ($D9$) of fish stocked. This transfer is represented by the equation

$$T_{\mathrm{SO}1,2} = D8 \times D2 \times D9$$

where $D8$ is the value of the stocking date driving variable, which equals zero except on stocking day when it equals 1.

Transfer from Plants to Fish ($T_{1,2}$)

Material transfer from plants ($S1$) to fish ($S2$) is a function of water temperature ($D1$), average weight of individual fish ($A1$), biomass of fish ($S2$), and biomass of plants ($S1$) (Figure 2.10), or

$$T_{1,2} = f(D1, A2, S2, S1)$$

A plot of available data relating fish consumption in kg of plants consumed per kg of fish biomass per day to water temperature in °C and weight of individual fish in kg live weight (Table 3.4b) suggests that, for a given size fish, consumption is related linearly to water temperature (Figure 3.13). But the slope of the relationship is steeper for small fish and flatter for large fish. Likewise a plot of data relating fish consumption at 30°C to fish size suggests a relationship not far from linear (Figure 3.14). Although the functional form representing si-

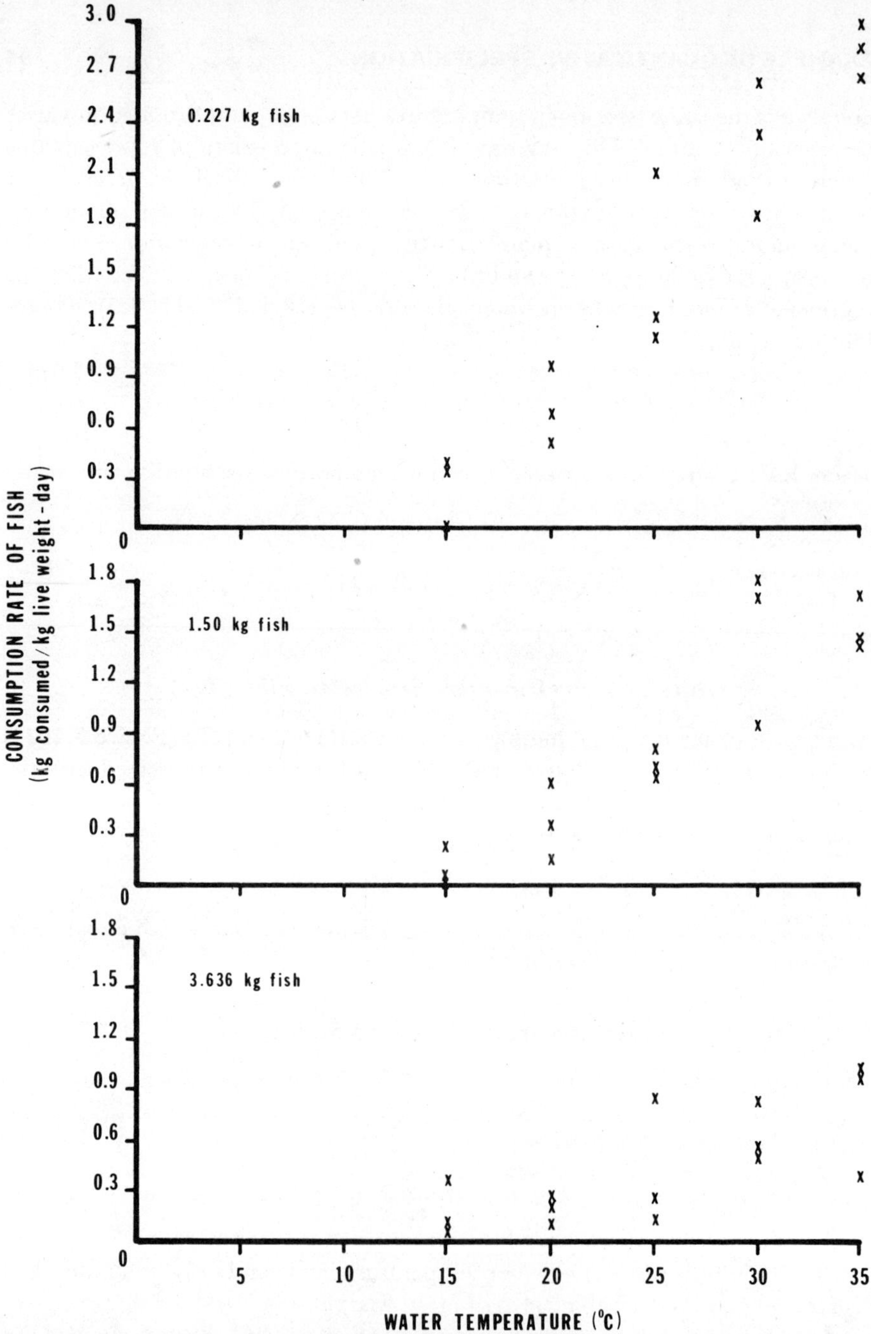

FIGURE 3.13. Plot of data from Table 3.4b relating fish consumption to water temperature and size of fish.

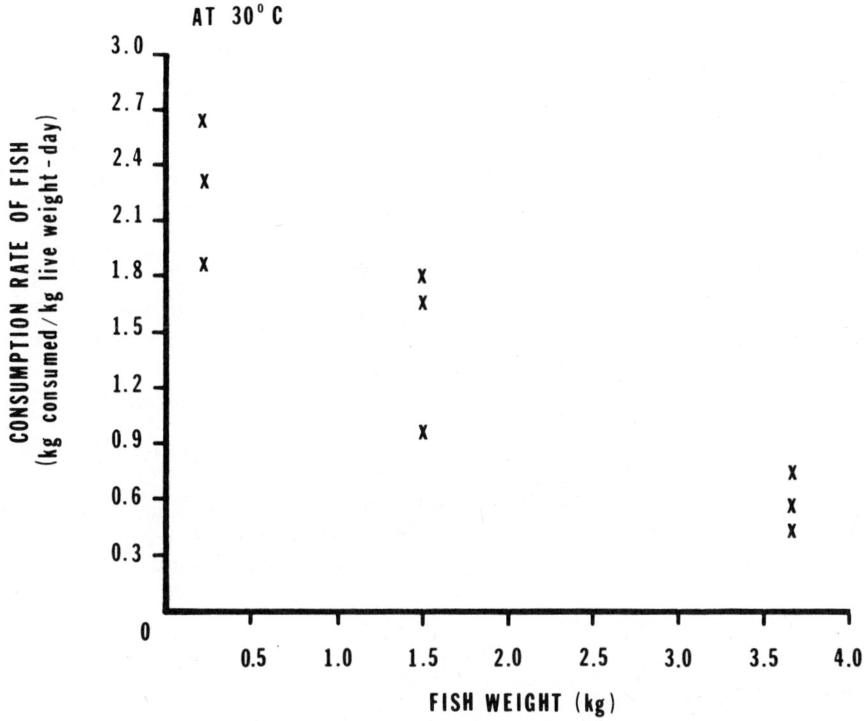

FIGURE 3.14. Plot of data from Table 3.4b relating fish consumption at a water temperature of 30°C to size of fish.

multaneous effects of size of fish and water temperature on consumption is difficult to visualize, the fact that both isolated effects appear linear, and the fact that there appears to be an interaction between water temperature and size, suggest that a linear model including both effects and their interaction is an appropriate functional form.

A multiple linear regression model of the form

$$Y = \beta_0 + \beta_1 X_1 + \beta_2 X_2 + \beta_3 X_1 X_2 + \varepsilon$$

with Y representing individual fish consumption, X_1 representing water temperature, X_2 representing fish weight, and X_1X_2 representing the interaction of water temperature and fish weight, can be fitted to data in Table 3.4b to parameterize the relationship. The interaction term is included because it appears that water temperature affects consumption of different size fish differently. Parameterization through multiple linear regression results in the equation

$$Y = -1.834 + 0.132X_1 + 0.422X_2 - 0.029X_1X_2$$

The regression equation is significant ($P < 0.0001$) and water temperature, individual fish weight, and their interaction together account for 86 percent of the variability in fish consumption ($r^2 = 0.86$) (Table 3.6). Since regression predicts consumption of plants by fish in terms of kg of plants consumed per kg of fish per day, we must multiply the right-hand side of the equation by number of kg of fish in the system ($S2$) at time t to calculate material transfer from plants to fish from time t to $t + 1$ in kg/day, or

$$T_{1,2} = (-1.834 + 0.132D1 + 0.422A1 - 0.029D1 \times A1)S2$$

We have represented consumption by fish as a function of water temperature, size of individual fish, and total fish biomass. This represents consumption when available plant biomass is not limiting. However, recall that when plant biomass falls below 20,000 kg/ha consumption by fish becomes limited by

TABLE 3.6

Results of multiple linear regression of consumption rate of fish (kg consumed/kg fish biomass-day) on fish weight (kg), water temperature (°C), and on the interaction of fish weight and water temperature, based on data in Table 3.4b. Meanings of table entries are explained in Appendix 3

Regression Equation

fish consumption = − 1.834 + 0.132 temperature + 0.422 weight − 0.029 (temperature × weight)

($r^2 = 0.86$)

Analysis of Variance

Source of Variation	Degrees of Freedom	Sum of Squares	Mean Square	F Value	$P > F$
Regression	3	26.205	8.735	85.7	0.0001
Error	41	4.180	0.102		
Total	44	30.385			

Parameter Estimates

Parameter	Estimate	T for $H_0: B = 0$	$P > T$	Standard Error of Estimate
Intercept	− 1.834	− 6.483	0.0001	0.283
Temperature	0.132	12.092	0.0001	0.011
Weight	0.422	3.395	0.0015	0.124
Temperature × weight	− 0.029	− 6.052	0.0001	0.005

availability of plant biomass and desired consumption rate is reduced by roughly one-half for every 5000 kg/ha decrease in plant biomass below that point (p. 89). If we let p represent the proportion of desired consumption actually realized when plant biomass ($S1$) is limiting, p takes the form of an exponentially decreasing function of plant biomass:

$$p = \beta_0 e^{-\beta_1 S1}$$

or in this particular case

$$p = e^{-0.00013863(20{,}000 - S1)} \text{ when } S1 \leq 20{,}000 \text{ kg/ha}$$

(Figure 3.15). Thus transfer of biomass from plants to fish ($T_{1,2}$) must be represented by two rate equations with the one used at any particular time dependent upon the amount of plant biomass in the system:

$$T_{1,2} = (-1.834 + 0.132D1 + 0.422A1 - 0.029D1 \times A1)S2 \text{ if } S1 > 20{,}000$$

$$T_{1,2} = (-1.834 + 0.132D1 + 0.422A1 - 0.029D1 \times A1)S2e^{-0.00013863(20{,}000 - S1)} \text{ if } S1 \leq 20{,}000$$

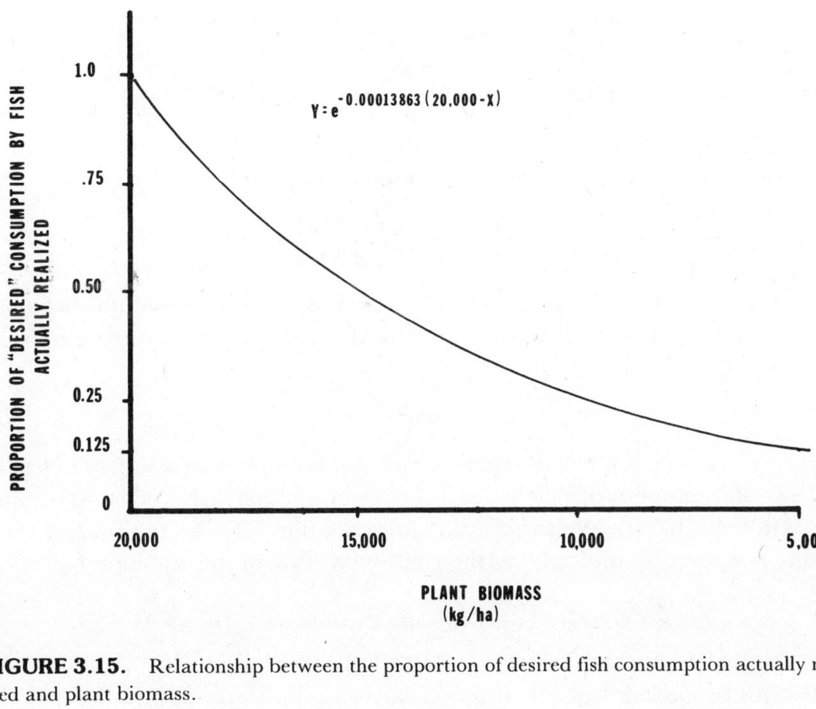

FIGURE 3.15. Relationship between the proportion of desired fish consumption actually realized and plant biomass.

We now have a deterministic representation of fish consumption. However, data (Table 3.4b) suggest that fish consumption per g of fish biomass is not always the same at a given temperature for a given size fish. At 25°C consumption by a 1.5-kg fish varies between 0.674 and 0.810 kg/kg-day. Thus as with plant production it is appropriate to represent the inherent randomness of fish consumption explicitly within the model. We do this in the same manner described for plant production. We generate a normal distribution about each predicted mean value of fish consumption using the mean squared error (MSE) of the regression of fish consumption on water temperature and fish weight as an unbiased estimate of the variance (σ^2). The MSE is 0.102 (Table 3.6). Therefore we calculate

$$T_{1,2} = \mathrm{RV2} \times S2 \text{ if } S1 > 20{,}000$$

or

$$T_{1,2} = \mathrm{RV2} \times S2\, e^{-0.00013863(20{,}000 - S1)} \text{ if } S1 \leq 20{,}000$$

where RV2 is a random variable drawn from a normal distribution with mean

$$\mu = -1.834 + 0.132D1 + 0.422A1 - 0.029D1 \times A1$$

and variance

$$\sigma^2 = 0.102$$

Transfer from Fish to Respiration Sink ($T_{2,SI2}$)

Material transfer from fish ($S2$) to respiration sink (SI2) is a function of water temperature ($D1$), average weight of individual fish ($A1$), and biomass of fish ($S2$) (Figure 2.10), or

$$T_{2,SI2} = f(D1, A1, S2)$$

We know that fish respiration can be represented as a function of water temperature and weight of individual fish by the generally applicable empirical equation

$$VO_2 = A\mathrm{wt}^{-0.2} e^{0.07\,T}$$

where VO_2 is mg of O_2 respired per g of fish per hour, wt is live weight of fish in g, T is water temperature in °C, and A is an empirically derived constant that equals 0.0142 (p. 89). We can convert this equation into the appropriate form for our purposes by multiplying the right-hand side of the equation by:

1. 3.4 cal/mg O_2 to convert from mg O_2/g-hr to cal/g-hr,
2. 1 kcal/1000 cal to convert from cal/g-hr to kcal/g-hr,
3. 1 g live wt. of fish/1 kcal to convert from kcal/g-hr to g/g-hr,

4. 24 hr/day to convert from g/g-hr to g/g-day, which is the same as kg/kg-day, and
5. number of kg of fish in the system at time t, which yields kg respired per day.

Thus

$$T_{2,\mathrm{SI2}} = (0.0142A1^{-0.2}e^{0.07D1})(3.4)\left(\frac{1}{1\ 000}\right)(24)S2$$

or

$$T_{2,\mathrm{SI2}} = (0.0142A1^{-0.2}e^{0.07D1})(0.0816)S2$$

Transfer from Fish to Excretion Sink ($T_{2,SI3}$)

Material transfer from fish ($S2$) to the excretion sink (SI3) is a function of water temperature ($D1$), average weight of individual fish ($A1$), assimilation efficiency of fish ($D3$), and biomass of fish ($S2$) (Figure 2.10), or

$$T_{2,\mathrm{SI3}} = f(D1, A1, D3, S2)$$

We know that assimilation efficiency of fish is 2 percent (p. 89). Therefore we know that 98 percent of the food consumed ($T_{1,2}$) must be excreted, or

$$T_{2,\mathrm{SI3}} = 0.98\,T_{1,2} = 0.98\mathrm{RV2} \times S2 \text{ if } S1 > 20{,}000$$

or

$$T_{2,\mathrm{SI3}} = 0.98\mathrm{RV2} \times S2e^{-0.00013863(20{,}000 - S1)} \text{ if } S1 \leq 20{,}000$$

Technically a transfer rate cannot be a function of another transfer rate because both occur simultaneously. But in all cases the same information about the state of the system that is used to calculate one transfer rate is available for calculation of other transfer rates and thus there is no problem.

Transfer from Plants to Natural Mortality Sink ($T_{1,SI1}$)

Material transfer from plants ($S1$) to the natural mortality sink (SI1) is a function of biomass of plants ($S1$) (Figure 2.10), or

$$T_{1,\mathrm{SI1}} = f(S1)$$

The curve based on expert opinion relating natural mortality to plant biomass is an exponentially increasing function (Figure 3.11). One equation that represents an exponentially increasing curve is

$$Y = \beta_0 e^{\beta_1 X} - 1$$

where β_0 represents the y-intercept plus 1, and β_1 controls the degree of curvature, or the rate of increase, of X relative to Y. Since we know the y-intercept ($\beta_0 = 0 + 1$) and another point that the curve passes through ($X = 45{,}000$, $Y = 0.0012$) (Figure 3.11), we can solve the equation for b_1.

$$Y = b_0 e^{b_1 X} - 1$$

$$0.0012 = (1)e^{b_1 \times 45{,}000} - 1$$

$$1.0012 = e^{45{,}000\, b_1}$$

$$\ln 1.0012 = 45{,}000\, b_1$$

$$b_1 = \frac{\ln 1.0012}{45{,}000}$$

$$b_1 = 0.00000002665$$

Thus the parameterized equation passing through the points (0,0) and (45,000,0.0012) is,

$$Y = e^{0.00000002665\, X} - 1$$

Since this equation represents the proportion of plants dying per day (Y) as a function of plant biomass (X, or $S1$), we must multiply the right-hand side of the equation by the number of kg of plants in the system ($S1$) at time t to calculate material transfer from plants to the natural mortality sink from time t to $t + 1$ in kg/day,

$$T_{1,SI1} = (e^{0.00000002665\, S1} - 1)S1$$

Transfer from Fish to Harvested Fish ($T_{2,3}$)

Material transfer from fish ($S2$) to harvested fish ($S3$) (Figure 2.10) is zero except on harvest day at which time all fish biomass is transferred to harvested fish. This transfer is represented by the equation

$$T_{2,3} = D4 \times S2$$

where $D4$ is the value of the harvest date driving variable, which equals zero except on harvest day when it equals 1.

Driving Variable Equations

We have hypothesized functional forms and parameterized rate equations for seven material transfers within the system. We now turn attention to equations describing driving variables which include:

1. water temperature ($D1$),
2. number of fish ($D2$),

3. assimilation efficiency (*D*3),
4. harvest date (*D*4),
5. unit price of harvested fish (*D*5),
6. unit cost of lease (*D*6),
7. size limit for selling fish (*D*7),
8. stocking date (*D*8), and
9. initial weight of individual fish (*D*9)

(Figure 2.10).

Water Temperature (D1)

A plot of available data on water temperature during a normal year (Table 3.3b) suggests that water temperature increases linearly from April 15 to June 15, remains relatively constant from June 15 to September 15, decreases linearly from September 15 to November 15, and remains relatively constant from November 15 to April 15 (Figure 3.16). One simple way to represent water temperature as a function of time of year is to use a piecewise linear

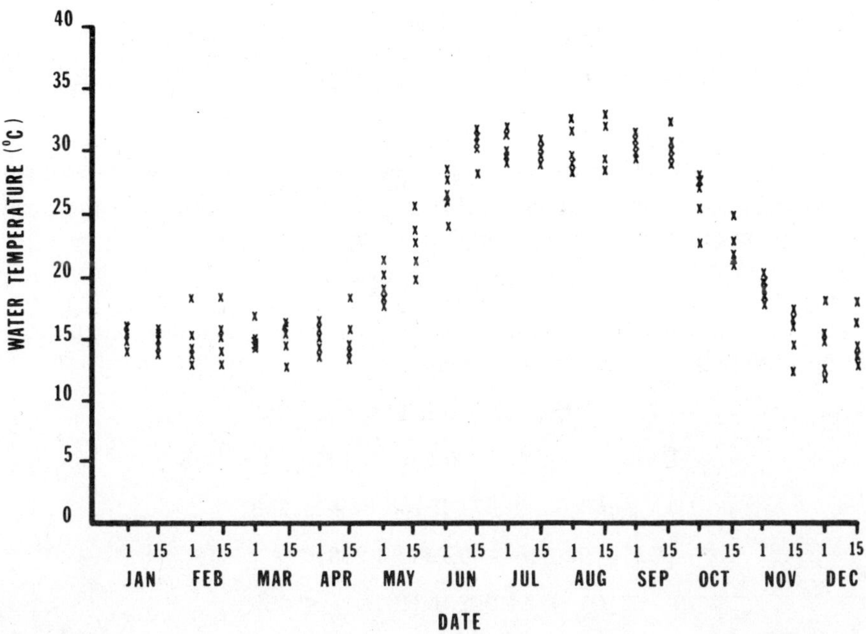

FIGURE 3.16. Plot of data from Table 3.3b indicating water temperature on various dates during a normal year (1983).

approximation. We estimate five simple linear regressions of water temperature ($D1$) on day of year or Julian date (t):

1. using data from January 1 ($t = 1$) to April 15 ($t = 105$),
2. using data from April 15 to June 15 ($t = 166$),
3. using data from June 15 to September 15 ($t = 258$),
4. using data from September 15 to November 15 ($t = 319$), and
5. using data from November 15 to December 31 ($t = 365$).

The first regression is not significant ($P > 0.918$) (Table 3.7a), therefore our best predictor of water temperature is simply the mean water temperature for this time period, or

$$D1 = 15.015$$

The second regression is significant ($P < 0.0001$) and yields the equation

$$D1 = -10.845 + 0.246t$$

with $r^2 = 0.92$ (Table 3.7b). The third regression is not significant ($P >$ 0.928) (Table 3.7c), therefore our best predictor of water temperature is the mean water temperature for this time period, or

$$D1 = 29.871$$

The fourth regression is significant ($P < 0.0001$) and yields the equation

$$D1 = 92.846 - 0.244t$$

with $r^2 = 0.95$ (Table 3.7d). The fifth regression is not significant ($P > 0.999$) (Table 3.7e), therefore our best predictor of water temperature is the mean water temperature for this time period, or

$$D1 = 14.952$$

Thus water temperature is represented by a set of five equations, one for each of five different time periods during the year:

$$D1 = 15.015 \text{ if } 1 \leq t \leq 105$$

$$D1 = -10.845 + 0.246t \text{ if } 105 < t < 166$$

$$D1 = 29.871 \text{ if } 166 \leq t \leq 258$$

$$D1 = 92.846 - 0.244t \text{ if } 258 < t < 319$$

$$D1 = 14.952 \text{ if } 319 \leq t \leq 365$$

Piecewise linear approximations often are useful in quantitatively representing an environmental driving variable such as temperature based on an

TABLE 3.7

Results of simple linear regression of 1983 water temperature in the pond (°C) on Julian date (t) for the period (A) January 1 (t = 1) to April 15 (t = 105), (B) April 15 to June 15 (t = 166), (C) June 15 to September 15 (t = 258), (D) September 15 to November 15 (t = 319), and (E) November 15 to December 31 (t = 365) based on data in Table 3.3b. Meanings of table entries are explained in Appendix 3

A. January 1 to April 15

Regression Equation

temperature = 15.015 − 0.0006t
(r^2 = 0.0003)

Analysis of Variance

Source of Variation	Degrees of Freedom	Sum of Squares	Mean Square	F Value	$P > F$
Regression	1	0.016	0.016	0.01	0.918
Error	38	55.354	1.457		
Total	39	55.370			

Parameter Estimates

Parameter	Estimate	T for $H_0: B = 0$	$P > T$	Standard Error of Estimate
Intercept	15.015	42.531	0.0001	0.353
t	−0.0006	−0.102	0.920	0.006

B. April 15 to June 15

Regression Equation

temperature = −10.845 + 0.246t
(r^2 = 0.92)

Analysis of Variance

Source of Variation	Degrees of Freedom	Sum of Squares	Mean Square	F Value	$P > F$
Regression	1	708.135	708.135	283.7	0.0001
Error	23	57.407	2.496		
Total	24	765.542			

TABLE 3.7 (*Continued*)

Parameter Estimates

Parameter	Estimate	T for $H_0: B = 0$	$P > T$	Standard Error of Estimate
Intercept	− 10.845	− 5.40	0.0001	2.007
t	0.246	16.84	0.0001	0.014

C. June 15 to September 15

Regression Equation

temperature = 29.871 + 0.0006t
(r^2 = 0.0002)

Analysis of Variance

Source of Variation	Degrees of Freedom	Sum of Squares	Mean Square	F Value	$P > F$
Regression	1	0.010	0.010	0.01	0.928
Error	33	40.385	1.224		
Total	34	40.395			

Parameter Estimates

Parameter	Estimate	T for $H_0: B = 0$	$P > T$	Standard Error of Estimate
Intercept	29.871	23.03	0.0001	1.297
t	0.0006	0.091	0.928	0.006

D. September 15 to November 15

Regression Equation

temperature = 92.846 − 0.244t
(r^2 = 0.954)

Analysis of Variance

Source of Variation	Degrees of Freedom	Sum of Squares	Mean Square	F Value	$P > F$
Regression	1	698.40	698.40	481.98	0.0001
Error	23	33.33	1.449		
Total	24	731.73			

TABLE 3.7 *(Continued)*

Parameter Estimates

Parameter	Estimate	T for $H_0: B = 0$	$P > T$	Standard Error of Estimate
Intercept	92.846	28.82	0.0001	3.221
t	– 0.244	– 21.95	0.0001	0.011

E. November 15 to December 31

Regression Equation

temperature = 14.952 – 0.0000t
(r^2 = 0.000)

Analysis of Variance

Source of Variation	Degrees of Freedom	Sum of Squares	Mean Square	F Value	$P > F$
Regression	1	0.000	0.000	0.000	0.999
Error	13	32.293	2.484		
Total	14	32.293			

Parameter Estimates

Parameter	Estimate	T for $H_0: B = 0$	$P > T$	Standard Error of Estimate
Intercept	14.952	1.39	0.188	10.767
t	– 0.0000	– 0.000	0.999	0.032

erratically fluctuating time series of observed values. An alternative approach if the data are not fluctuating too erratically is to fit a single equation to the time series. Sometimes a higher-order polynomial of the form

$$Y = \beta_0 + \beta_1 X + \beta_2 X^2 + \ldots + \beta_n X^n$$

can be used, or general trends in the time series may suggest another particular type of equation. (Actually, general trends in our water temperature data (Figure 3.16) suggest that we might have fitted a sine function to this time series. In fact,

$$D1 = 21.3 + 8.9 \sin(4.2 + (2\pi(t - 1)/364))$$

is a viable alternative to our set of piecewise linear approximations.)

We now have a deterministic representation of water temperature. However, data (Table 3.3b) suggest that water temperature varies from location to location within the pond for any given date. April 15 temperatures ranged from 13.7 to 17.8°C from place to place within the pond. As with plant production and fish consumption, we wish to represent this inherent randomness of water temperature explicitly within the model. As before we generate a normal distribution about each predicted mean value of water temperature using the mean squared error (MSE) of the appropriate regression of water temperature on Julian date (t) as an unbiased estimate of variance (σ^2). The MSEs of the five regressions are 1.457, 2.496, 1.224, 1.449, and 2.484, respectively (Table 3.7). Therefore we calculate $D1$ as a stochastic variable as:

$$D1 = \text{RV3 if } 1 \leq t \leq 105$$

$$D1 = \text{RV4 if } 105 < t < 166$$

$$D1 = \text{RV5 if } 166 \leq t \leq 258$$

$$D1 = \text{RV6 if } 258 < t < 319$$

$$D1 = \text{RV7 if } 319 \leq t \leq 365$$

where RV3 through RV7 are random variables drawn from the following normal distributions:

$$\text{RV3}, \mu = 15.015, \sigma^2 = 1.457$$

$$\text{RV4}, \mu = -10.845 + 0.246t, \sigma^2 = 2.496$$

$$\text{RV5}, \mu = 29.871, \sigma^2 = 1.224$$

$$\text{RV6}, \mu = 92.846 - 0.244t, \sigma^2 = 1.449$$

$$\text{RV7}, \mu = 14.952, \sigma^2 = 2.484$$

We have represented seasonal changes in water temperature including inherent randomness of temperatures from place to place within the pond during a normal year. We now wish to represent year-to-year variability in water temperature since some years are uniformly warmer or colder than normal. Recall that the frequency of occurrence of relatively hot and cold years as indicated by the displacement of their mean annual temperature from the normal mean annual temperature follows the normal distribution in Figure 3.10. Thus one way to represent the occurrence of hot and cold years is to select a temperature displacement randomly from a normal distribution with $\mu = 0$ and $\sigma^2 = 4$ (Figure 3.10) at the beginning of each replicate stochastic simulation, and then adjust each daily temperature predicted during a given simulation by the amount of the randomly selected displacement. We might randomly select a temperature displacement of $-1.7°C$, simulating a relatively cool year. Each day during that simulation we would calculate water temperature using the appropriate equation above (representing water temperature on that day during a normal year) and then subtract 1.7°C to obtain the water temperature for that day during the cooler year. In terms of model equations this means that we add one term (RV8) to each water temperature equation:

$$D1 = \text{RV3} + \text{RV8 if } 1 \leq t \leq 105$$

$$D1 = \text{RV4} + \text{RV8 if } 105 < t < 166$$

$$D1 = \text{RV5} + \text{RV8 if } 166 \leq t < 258$$

$$D1 = \text{RV6} + \text{RV8 if } 258 < t < 319$$

$$D1 = \text{RV7} + \text{RV8 if } 319 \leq t \leq 365$$

where RV8 is a random variable drawn from a normal distribution with $\mu = 0$ and $\sigma^2 = 4$.

Stocking Date (D8)

The stocking date driving variable controls date of stocking and therefore serves only to turn the material transfer from the biomass source to fish ($T_{SO1,2}$) (Figure 2.10) on or off in an all-or-none fashion. Thus

$$D8 = 1 \text{ if } t = t_{\text{begin}}$$

and

$$D8 = 0 \text{ if } t > t_{\text{begin}}$$

where t = day of simulated time and t_{begin} represents stocking day or the beginning of the simulation.

Harvest Date (D4)

Similarly, the harvest date driving variable controls date of harvest and therefore serves only to turn the material transfer from fish to harvested fish ($T_{2,3}$) (Figure 2.10) on or off in an all-or-none fashion. Thus

$$D4 = 0 \text{ if } t < t_{\text{end}}$$

and

$$D4 = 1 \text{ if } t = t_{\text{end}}$$

where t = day of simulated time and t_{end} represents harvest day or the end of the simulation.

Other Driving Variables: D2, D3, D5, D6, D7, and D9

The number of fish stocked ($D2$), assimilation efficiency ($D3$), unit price of harvested fish ($D5$), unit cost of the lease ($D6$), size limit for selling fish ($D7$), and initial weight of individual fish ($D9$) represent a special class of driving variables in that their values do not change over the course of one simulation. They actually are constants. For the typical baseline situation described earlier equations for these driving variables are:

$$D2 = 75 \text{ (fish stocked)}$$

$$D3 = 0.02 \text{ (assimilation efficiency)}$$

$$D5 = 1.0 \text{ (dollar/kg)}$$

$$D6 = 0.50 \text{ (}\tfrac{1}{2}\text{ dollar/day)}$$

$$D7 = 3.0 \text{ (kg)}$$

$$D9 = 0.227 \text{ (kg)}$$

Auxiliary Variable Equations

We now turn attention to equations describing auxiliary variables, which include:

1. weight of individual fish ($A1$),
2. gross revenue from the harvest ($A2$),
3. total cost of the lease ($A3$), and
4. profit ($A4$)

(Figure 2.10). Average weight of individual fish ($A1$) is calculated based on the total biomass of fish ($S2$) and the number of fish ($D2$):

$$A1 = \frac{S2}{D2}$$

Gross revenue from the harvest ($A2$) is calculated based on biomass of harvested fish ($S3$), unit price of harvested fish ($D5$), weight of individual fish ($A1$), and size limit for selling fish ($D7$):

$$A2 = S3 \times D5 \text{ if } A1 \geq D7$$

$$A2 = 0 \text{ if } A1 < D7$$

Total cost of the lease ($A3$) is calculated based on stocking date (t_{begin}), harvest date (t_{end}), and unit cost of the lease ($D6$):

$$A3 = D6(t_{\text{begin}} - t_{\text{end}})$$

Profit ($A4$) is calculated based on gross revenue from the harvest ($A2$) and total cost of the lease ($A3$):

$$A4 = A2 - A3$$

State Variable Equations

Finally we write three state variable equations that update the level of each state variable each time step. These equations are essentially bookkeeping equations and simply add and subtract appropriate material transfers (rate equations) from each state variable each Δt. For the pond model

$$S1_{t+1} = S1_t + T_{\text{SO1},1} - T_{1,\text{SI1}} - T_{1,2}$$

$$S2_{t+1} = S2_t + T_{\text{SO1},2} + T_{1,2} - T_{2,\text{SI2}} - T_{2,\text{SI3}} - T_{2,3}$$

$$S3_{t+1} = S3_t + T_{2,3}$$

Formally Presenting and Computer Coding Model Equations and Executing Baseline Simulations

Formally Presenting Model Equations

Equations developed above collectively describe the pond model. Collectively they are the model. In Table 3.8 these equations are presented formally by computing sequence of equation types:

1. driving variable equations,
2. auxiliary variable equations,
3. rate equations for material transfers, and
4. state variable equations.

Initial conditions of the model are also presented. (We know that initially $S1 = 40{,}000$ from Table 3.3 and that $S2 = 17$ since 75 0.227 kg fish are stocked. Stocking date is April 15 and harvest date is November 15 thus $t_{\text{begin}} = 105$ and $t_{\text{end}} = 319$.) This form of presentation is useful in relating equations to the flow diagram of the system (Figure 2.10) and to the matrix of material transfers (Figure 2.11). It is also useful in detailed examinations of model structure. Finally, it lends itself well to programming equations on a digital computer.

Computer Coding Model Equations

Having formally presented model equations the next step is to code, or program, equations on a digital computer. First we develop a flow chart similar to

TABLE 3.8

Formal presentation of pond model equations organized by computing sequence of equation types. Variables are identified in Figure 2.10

Initial Conditions of Model

$S1 = 40{,}000$
$S2 = 17.0$
$S3 = 0.0$
$t_{\text{begin}} = 105$ (April 15)
$t_{\text{end}} = 319$ (November 15)

Driving Variable Equations

$D1 = \text{RV3} + \text{RV8}$ if $1 \leq t \leq 105$
where RV3 = normal random variate (n.r.v.) from distribution with $\mu = 15.015$, $\sigma^2 = 1.457$ and RV8 = n.r.v. from distribution with $\mu = 0$, $\sigma^2 = 4$

$D1 = \text{RV4} + \text{RV8}$ if $105 < t < 166$
where RV4 = n.r.v. from distribution with $\mu = -10.845 + 0.246t$, $\sigma^2 = 2.496$

$D1 = \text{RV5} + \text{RV8}$ if $166 \leq t \leq 258$
where RV5 = n.r.v. from distribution with $\mu = 29.871$, $\sigma^2 = 1.224$

$D1 = \text{RV6} + \text{RV8}$ if $258 < t < 319$
where RV6 = n.r.v. from distribution with $\mu = 92.846 - 0.244t$, $\sigma^2 = 1.449$

$D1 = \text{RV7} + \text{RV8}$ if $319 \leq t \leq 365$
where RV7 = n.r.v. from distribution with $\mu = 14.952$, $\sigma^2 = 2.484$

$D2 = 75$
$D3 = 0.02$
$D4 = 0$ if $t < 319$
$D4 = 1$ if $t = 319$
$D5 = 1$
$D6 = 0.5$
$D7 = 3$
$D8 = 0$ if $t > 105$
$D8 = 1$ if $t = 105$
$D9 = 0.227$

Auxiliary Variable Equations

$A1 = S2/D2$
$A2 = S3 \times D5$ if $A1 \geq D7$
$A2 = 0$ if $A1 < D7$
$A3 = D6\,(t_{\text{begin}} - t_{\text{end}})$
$A4 = A2 - A3$

TABLE 3.8 (*Continued*)

Rate Equations

$T_{SO1,1} = RV1 \times S1/1000$
where $RV1$ = n.r.v. from distribution with $\mu = -0.604 + 0.073D1$, $\sigma^2 = 0.117$
$T_{SO1,2} = D8 \times D2 \times D9$
$T_{1,2} = RV2 \times S2$ if $S1 > 20{,}000$
where $RV2$ = n.r.v. from distribution with $\mu = -1.834 + 0.132D1 + 0.422A1 - 0.029D1 \times A1$, $\sigma^2 = 0.102$
$T_{1,2} = RV2 \times S2e^{-0.00013863(20{,}000-S1)}$ if $S1 \leq 20{,}000$
$T_{1,SI1} = (e^{0.00000002665S1} - 1)S1$
$T_{2,SI2} = (0.0142A1^{-0.2}e^{0.07D1})(0.0816)S2$
$T_{2,SI3} = 0.98RV2 \times S2$ if $S1 > 20{,}000$
$T_{2,SI3} = 0.98RV2 \times S2e^{-0.00013863(20{,}000-S1)}$ if $S1 \leq 20{,}000$
$T_{2,3} = D4 \times S2$

State Variable Equations

$S1_{t+1} = S1_t + T_{SO1,1} - T_{1,SI1} - T_{1,2}$
$S2_{t+1} = S2_t + T_{SO1,2} + T_{1,2} - T_{2,SI2} - T_{2,SI3} - T_{2,3}$
$S3_{t+1} = S3_t + T_{2,3}$

the one in Figure 3.8, but identifying at each step what variables are involved for our particular model (Figure 3.17).

Input data for model parameters and initial conditions include the initial conditions of state variables ($S1$, $S2$, $S3$), beginning and ending times for simulation (t_{begin}, t_{end}), and also values for six driving variables ($D2$, $D3$, $D5$, $D6$, $D7$, $D9$). Although number of fish ($D2$), assimilation efficiency ($D3$), unit price of harvested fish ($D5$), unit cost of the lease ($D6$), size limit for selling fish ($D7$), and initial weight of individual fish ($D9$) are driving variables, we noted earlier that they represent a special class of driving variable in that they do not vary throughout simulation. Thus $D2$, $D3$, $D5$, $D6$, $D7$, and $D9$ need be defined only once and can be defined most easily when initial conditions are defined.

We have no input data needed each time unit. An example of this type of input might be daily water temperature data if we had chosen to represent water temperature as a time series of input data rather than as a set of functions within the model.

Driving variable equations include equations for $D1$, $D4$, and $D8$. Auxiliary variable equations include equations for $A1$, $A2$, $A3$, and $A4$. Rate equations include equations for $T_{SO1,1}$, $T_{SO1,2}$, $T_{1,2}$, $T_{2,3}$, $T_{1,SI1}$, $T_{2,SI2}$, and $T_{2,SI3}$. State variable equations include equations for $S1$, $S2$, and $S3$.

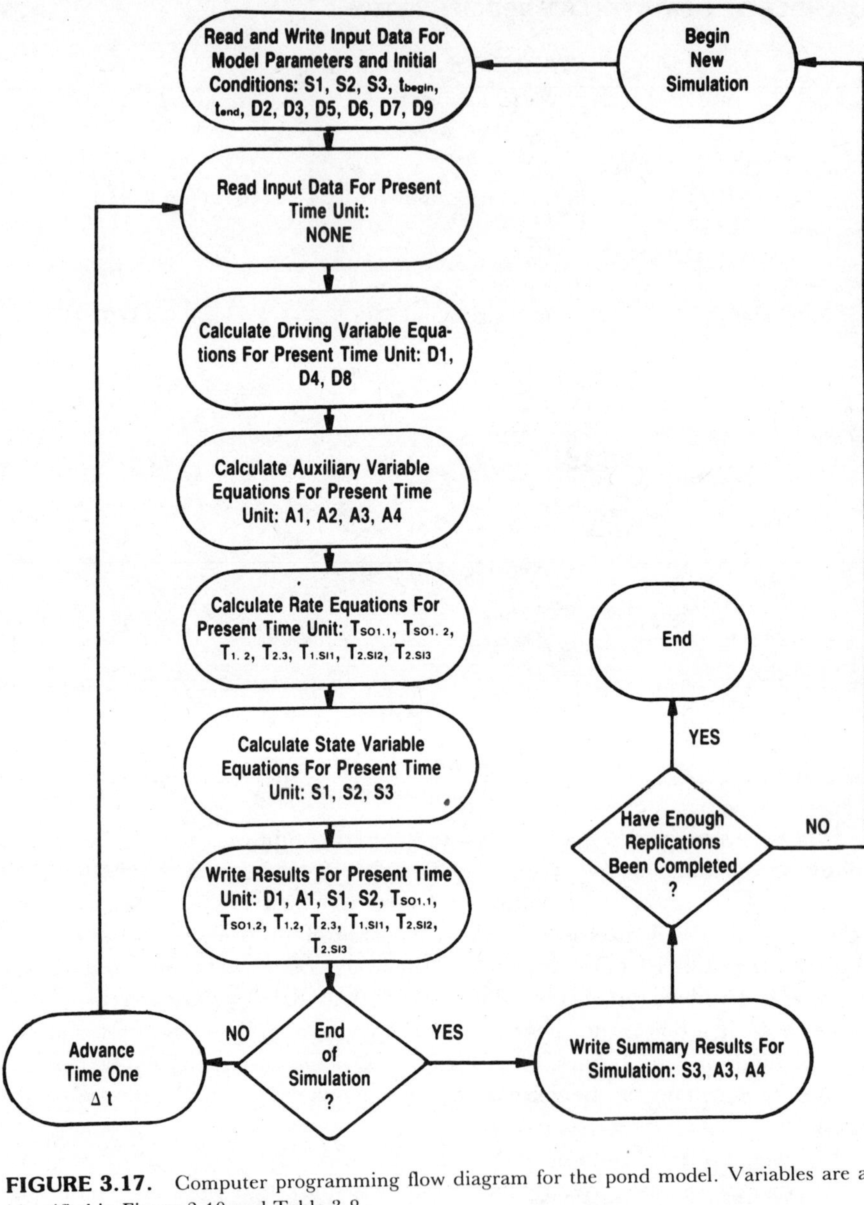

FIGURE 3.17. Computer programming flow diagram for the pond model. Variables are as identified in Figure 2.10 and Table 3.8.

Results written each time step potentially include values of all driving variables, auxiliary variables, state variables, and material transfer rates. But what we actually look at depends upon whether we are debugging, validating, or using the model. When debugging the computer program we may look at each of these values each time unit. Later, when validating the model we may be interested only in those variables or rates that we can compare to data or other information from the real system. Finally, when using the model for policy evaluation we may be interested only in a few key variables or rates that are applicable directly to the problem being addressed. In Figure 3.17 we have listed the values of two state variables (plant biomass, *S*1, and fish biomass, *S*2), all material transfer rates, one driving variable (water temperature, *D*1), and one auxiliary variable (weight of individual fish, *A*1) as desired output each time unit.

We are interested in three additional summary results for the entire simulation:

1. biomass of harvested fish (*S*3),
2. profit (*A*4), and
3. total cost of the lease (*A*3)

on the last day of simulated time.

Another example of this type of summary information might be total net primary production during the simulation, which is represented by the sum of all material transfers from the biomass source to plants. Finally, we ask if enough replicate simulations have been completed.

The strategy for writing computer code for the model is simply to tell the computer in the chosen language to do each operation outlined in Figure 3.17. An example of BASIC code that calculates rate equations for the pond model is presented in Table 3.9.

The BASIC code in Table 3.9 has been debugged and verified as described in Section 3.5. A type of error commonly found when debugging a program is simply misspelling or misplaced commas or periods in a program statement. In line number 16970 of the BASIC program in Table 3.9 we might originally have typed Z2 instead of S2. This would have created a new variable (*Z*2) for which the computer would have had no value. When the computer tried to compile the statement on line number 16970 it would have detected this problem and the program would not have compiled successfully. A type of error commonly found when verifying a program is a logical error causing the computer to perform calculations in a manner that we did not intend. In the BASIC program in Table 3.9 we might originally have omitted the equation on line 16980. This is the equation that reduces fish consumption when plant biomass is less than 20,000 kg/ha. The program would have compiled and run

TABLE 3.9

Example of BASIC code which calculates rate equations for the pond model

```
16860 ' Compute transfers
16870 '
16880 ' Source to plants
16890      LET F1=(Y(9)+Y(10)*D1)*S1
16900      LET F1=F1/1000
16910      IF STCH(6) THEN
           F1=FN RN(F1, SD(6))
16940 '
16950 ' Plants to fish
16960      LET F2=Y(11)+Y(12)*D1+Y(13)*A1+Y(14)*D1*A1
16970      LET F2=F2*S2
16980      IF S1<Y(35) THEN
           F2=F2*EXP(Y(37)*(Y(35)-S1))
16990      IF STCH(7) THEN
           F2=FN RN(F2, SD(7))
17020 '
17030 ' Plants to mortality
17040      IF PM=2 THEN 17070
17050      LET F3=(Y(15)+Y(16)*S1)*S1
17060      GOTO 17080
17070      LET F3=(EXP(Y(15)*S1)-1)*S1
17100 '
17110 ' Fish to respiration
17120      LET Y(0)=0
17130      IF A1<>0 THEN Y(0)=A1[Y(20)
17140      LET F5=S2*Y(19)*Y(0)*EXP(Y(21)*D1)*Y(22)/Y(23)
17170 '
17180 ' Fish to excreta
17190      LET F6=(1-D3)*F2
```

without line 16980, but would not have performed calculations in the manner that we had intended if plant biomass became less than 20,000 kg/ha.

Executing Baseline Simulations

As our baseline conditions for the pond model we have chosen normal conditions of the real system as described in Chapter 2 (p. 52). Seventy-five fish ($D2 = 75$) each weighing 0.227 kg ($A1 = 0.227$) are stocked on April 15 ($t_{\text{begin}} = 105$, $D8 = 0$ if $t > 105$, $D8 = 1$ if $t = 105$) and harvested on November 15 ($t_{\text{end}} = 319$, $D4 = 0$ if $t < 319$, $D4 = 1$ if $t = 319$), and water temperature is "normal" throughout the year.

To determine how many replicate baseline simulations to run we first run relatively many preliminary baseline simulations to estimate the variability of model results in which we are most interested. We examine the variability in biomass of harvested fish ($S3$) since both profit ($A4$) and weight of individual fish ($A1$) are related directly to $S3$. Figure 3.18 shows the estimated variance of harvested fish biomass under baseline conditions as we increase the number of preliminary simulations used to estimate variance. Each point in Figure 3.18 represents the variance estimated from an independent sample of n replicate stochastic simulations,

$$s^2 = \frac{\Sigma X_i^2 - (\Sigma X_i)^2/n}{n - 1}$$

where s^2 = estimated variance

X_i = biomass of harvested fish from the ith replicate stochastic simulation

n = number of replicate stochastic simulations used to estimate the variance

Estimates of the variance (σ^2) stabilize at about 380 kg. This is our best estimate of the inherent variability in biomass of harvested fish.

Next we specify the magnitude of difference in harvest that we consider practically significant relative to our interests. Recall from Chapter 2 (p. 53) that the pond manager decided profit must change at least 10 percent relative to baseline conditions for it to be meaningful within a management framework. Mean annual biomass of harvested fish predicted by preliminary baseline simulations is 258 kg. Thus harvest must change by 25.8 kg, or let's say 25 kg, to be practically significant. Suppose the manager would like to show that a true difference of 25 kg in harvest between different management policies is statistically significant at the 95 percent significance level ($\alpha = 0.05$) with an 80 percent probability that the difference will be detected if it exists ($P = 0.80$, or type II error $\beta = 0.20$). Then the sample size needed, or number of replicate simulations to run, can be calculated based on the relationship presented in Section 3.5:

$$n \geq 2\left(\frac{\sigma}{\delta}\right)^2 (t_{\alpha,\gamma} + t_{2(1-P),\gamma})^2$$

In our case $\sigma^2 = 380$ or $\sigma = 19.49$, $\delta = 25$, $\alpha = 0.05$, and $P = 0.80$. Recall that there are six management schemes including baseline that the manager wishes to compare (p. 52). If we assume the variability in predicted biomass of fish harvested is the same for all six versions of the model, which is the case, and if our initial guess for the correct sample size is 10, then

$$\gamma = 6(10 - 1) = 54$$

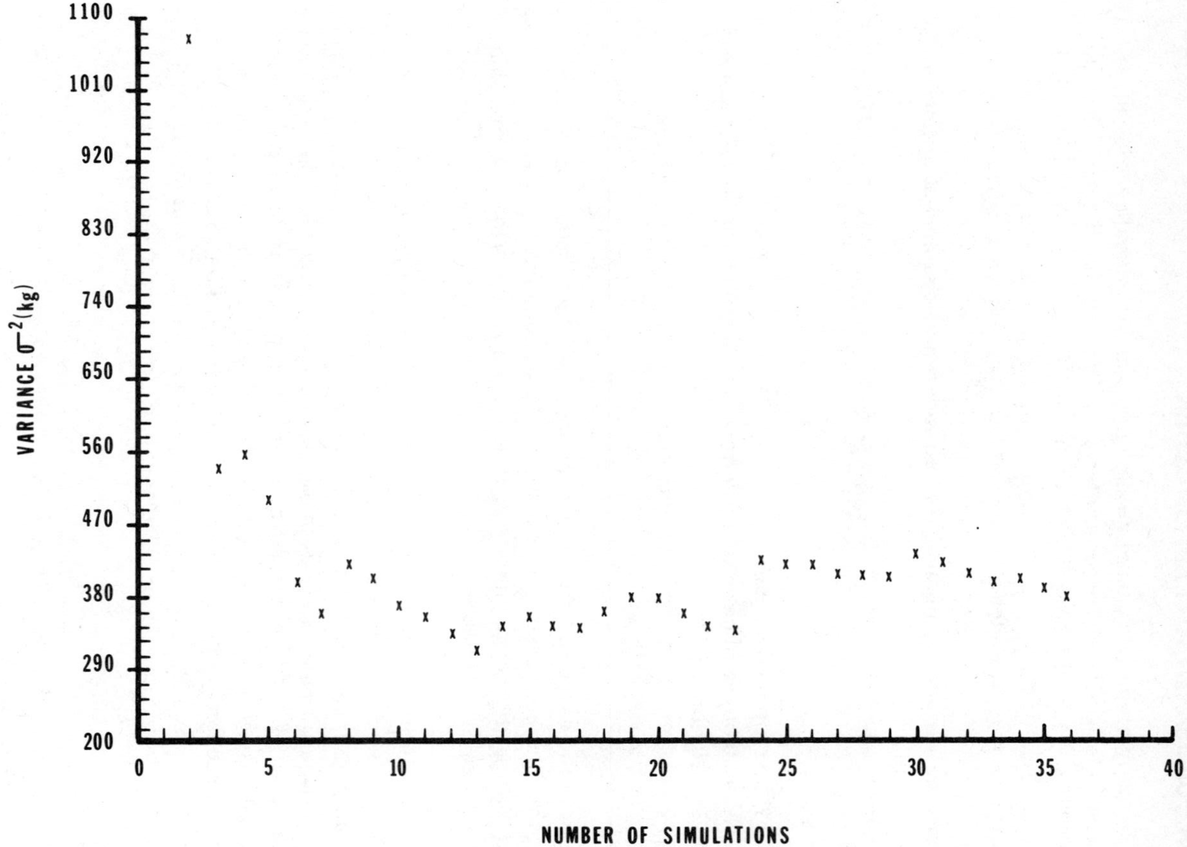

FIGURE 3.18. Relationship of the estimated variance of harvested fish biomass (*S*3) under baseline conditions to the number of simulations used to estimate variance.

Thus

$$10 \geq 2\left(\frac{19.49}{25}\right)^2 (t_{0.05,54} + t_{0.40,54})^2$$

or

$$10 \geq 2\left(\frac{19.49}{25}\right)^2 (2.000 + 0.848)^2$$

and

$$10 \geq 9.86$$

and the correct sample size rounded to the nearest whole number is 10.

Results of ten stochastic replications of the baseline version of the model are summarized in Table 3.10 and Figure 3.19. Mean (± 1SE) biomass of plants and harvested fish, mean (± 1SE) weight of individual fish, and mean (± 1SE) profit on November 15 (harvest day) are 21,254 (± 637) kg/ha, 263 (± 6) kg/ha, 3.51 (± 0.08) kg, and 156 (± 6) dollars, respectively. We will defer further discussion of these results until we return to the pond model example at the end of the next chapter on model validation.

A computer program (POND) for the pond simulation model example is presented in Appendix 5. POND is an interactive, user-friendly program writ-

TABLE 3.10

Summary of results of ten replicate stochastic baseline simulations of the pond model, including biomass of plants and harvested fish, weight of individual fish, and profit on November 15 (harvest date)

Replication	Plant Biomass (kg/ha)	Fish Biomass (kg/ha)	Weight of Individual Fish (kg)	Profit (dollars)
1	20,147	275	3.67	168
2	24,482	229	3.05	122
3	22,590	253	3.37	146
4	19,081	281	3.74	173
5	19,511	279	3.72	171
6	21,115	268	3.58	161
7	19,795	278	3.70	170
8	24,032	236	3.14	128
9	19,275	280	3.73	172
10	22,510	255	3.40	147
$\bar{X}$(± 1SE)	21,254(± 637)	263(± 6)	3.51(± 0.08)	156(± 6)

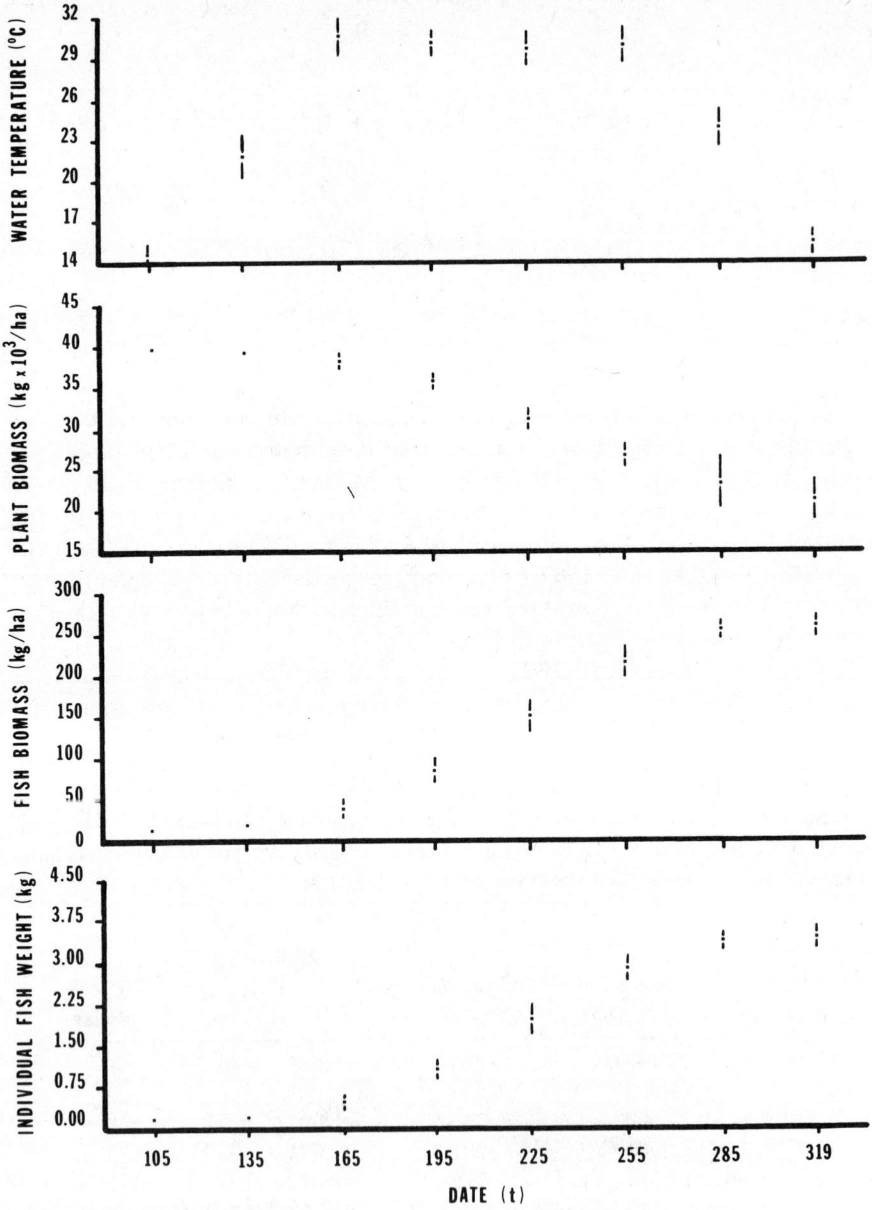

FIGURE 3.19. Results of ten replicate stochastic baseline simulations of the pond model. Vertical bars represent ± 1 standard error of the mean.

ten in BASIC for use on a TRS80 microcomputer. POND is intended to provide hands-on experience in quantifying, validating, and using the pond simulation model. It would be well at this point to use POND as it relates to material covered in this chapter.

3.7. EXERCISES

Additional Information

Available information with which to quantitatively specify our plant–herbivore model consists of:

1. data on several processes occurring within the system (data gathered during the first year post-introduction of herbivores on those islands where herbivores have been introduced (p. 59)),
2. information from literature on one process within the system, and
3. expert opinion concerning the seasonal dynamics of driving variables and the general form of negative feedback mechanisms influencing two processes.

Data on processes occurring within the system pertain to net primary (plant) production (Table 3.11), consumption of plants by herbivores, herbivore res-

TABLE 3.11
Data relating net primary (plant) production to environmental temperature and soil water tension in the plant-herbivore system

Net Primary Production (g of production per g of live plant biomass per day)	Environmental Temperature (°C)	Soil Water Tension (– bars)
0.02	11	0.8
0.02	20	1.5
0.01	29	2.1
0.007	35	2.6
0.005	39	2.9
0.005	40	3.0
0.006	38	2.9
0.01	34	2.5
0.01	27	2.0
0.02	18	1.4

piration, and herbivore excretion (Table 3.12), and decomposition of plant litter and herbivore excreta (Table 3.13). Information from literature indicates that 0.01 g of plant litter is produced each day for every 1 g of live plant biomass in the system, regardless of environmental conditions.

Expert opinion suggests that the seasonal dynamics of both environmental temperature and soil water tension can be represented as half sine waves with peak values occurring during the middle of the year (Figure 3.20). Note that one functional form that can be used to generate the curves in Figure 3.20 is

$$Y = \beta_0 \sin\left(\frac{t}{\beta_1}\right)$$

where Y = environmental temperature or soil water tension
t = Julian date
β_0 = parameter controlling height of curve
β_1 = parameter controlling position of peak of curve relative to x-axis

Also note that the larger the value of soil water tension in − bars the drier the soil. When soil water tension is zero the soil is at field capacity.

Expert opinion also suggests that net primary production and consumption of plants by herbivores are both constrained by negative feedback mechanisms which are not detectable from data in Tables 3.11 and 3.12. Net primary production (g produced/g live plant biomass-day) is reduced due to competition

TABLE 3.12

Data relating consumption of plants by herbivores, herbivore respiration, and herbivore excretion to environmental temperature in the plant-herbivore system

Consumption (g consumed per g of herbivore per day)	Respiration (g respired per g of herbivore per day)	Excretion (g excreted per g of herbivore per day)	Environmental Temperature (°C)
0.00	0.01	0.00	10
0.07	0.02	0.05	12
0.21	0.04	0.14	16
0.25	0.05	0.20	18
0.32	0.07	0.26	20
0.47	0.09	0.35	24
0.56	0.11	0.43	27
0.68	0.14	0.49	30
0.80	0.17	0.63	35
0.98	0.20	0.76	40

TABLE 3.13

Data relating decomposition of (A) plant litter and (B) herbivore excreta to soil water tension in the plant-herbivore system

A. Plant Litter

Decomposition (g decomposed per g of plant litter per day)	Soil Water Tension (– bars)
0.10	0
0.08	0.3
0.07	0.5
0.07	1
0.03	2
0.02	3
0.01	4
0.008	5
0.006	6
0.0001	15

B. Herbivore Excreta

Decomposition (g decomposed per g of herbivore excreta per day)	Soil Water Tension (– bars)
0.20	0
0.17	0.3
0.15	0.5
0.14	1
0.07	1.5
0.07	2
0.05	2.5
0.04	3
0.02	5
0.0001	15

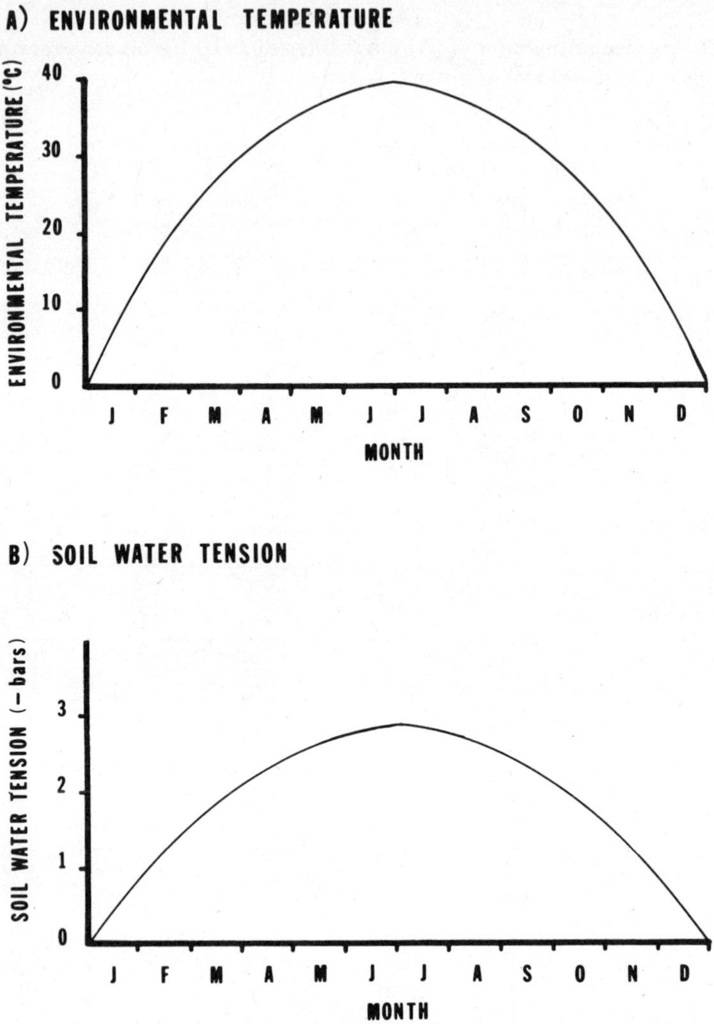

FIGURE 3.20. Seasonal dynamics of (A) environmental temperature and (B) soil water tension in the plant–herbivore system based on expert opinion.

when live plant biomass exceeds 50,000 kg/ha. The functional form of this constraint can be represented as

$$Y' = Y\left(\frac{10{,}000}{S1}\right) \text{ if } S1 > 50{,}000$$

where Y' = net primary production constrained by competition (g/g-day)
Y = net primary production unconstrained by competition (g/g-day)
$S1$ = live plant biomass (kg/ha)

Consumption of plants by herbivores (g consumed/g herbivore biomass-day) is reduced due to competition when the ratio of live plant biomass to herbivore biomass falls below 1500. The functional form of this constraint can be represented as

$$Y' = Y\left(\frac{A1}{1500}\right) \text{ if } A1 < 1500$$

where Y' = consumption of plants by herbivores constrained by competition (g/g-day)
Y = consumption of plants by herbivores unconstrained by competition (g/g-day)
$A1$ = ratio of live plant biomass to herbivore biomass (unitless)

Exercise 3.7.1

Based on the information presented above, quantitatively specify the plant-herbivore model by choosing functional forms, parameterizing model equations, and choosing the basic time unit for simulation. Use a discrete-time compartment model format, including a set of driving variable equations, a set of auxiliary variable equations, a set of rate equations, and a set of state variable equations. Justify your choice of functional forms for model equations and the basic time unit for simulation, and document each step in parameterization of equations.

Exercise 3.7.2

Present model equations formally in a table organized by computing sequence of equation types and write a BASIC program (see Appendix 4) to solve the model on a digital computer. Assume that initially live plant biomass is 50,000 kg/ha, herbivore biomass is 1 kg/ha, plant litter biomass is 6600 kg/ha, herbivore excreta biomass is 0.1 kg/ha, and harvested herbivore biomass is zero.

Also assume that $t_{begin} = 1$ and $t_{end} = 365$ and that no herbivores are harvested ($D3 = 0$).

Exercise 3.7.3

Run a one-year baseline simulation on a digital computer. Debug and verify the program. (We will postpone examination of results of the baseline simulation, which may appear unreasonable at this point, until the Chapter 4 exercises.)

Answers to Chapter 3 exercises are in Appendix 6.

CHAPTER 4

MODEL VALIDATION

4.1. INTRODUCTION

Model validation is a set of procedures undertaken to determine if a model is indeed a valid representation of the system of interest. A variety of interesting discussions regarding both the general philosophy and specific techniques used in model validation is available (Caswell 1976, Mankin et al. 1977, Holling 1978, Steinhorst 1979). As pointed out in Chapter 1, model validation is directly analogous to scientific hypothesis testing in that the general strategy involves attempting to refute, or disprove, or invalidate, a model. Failure to invalidate a model increases our confidence that it may indeed be valid. The scientific procedures used in model validation may vary depending upon the type of model being validated. Validation of an empirical, or correlative, model emphasizes the correspondence between model behavior and behavior of the real system, whereas validation of a mechanistic, or explanatory, model emphasizes interpretability and irrefutability, that is, validity of model structure and individual model mechanisms. In any case a model can be validated or invalidated only in view of the objectives of the overall project of which it is a part.

We might identify five steps in model validation:

1. examining the capability of the model to address the problem of interest (Section 4.2),
2. examining the reasonableness of model structure and individual model mechanisms (Section 4.3),
3. examining the qualitative reasonableness of overall model behavior (Section 4.4),

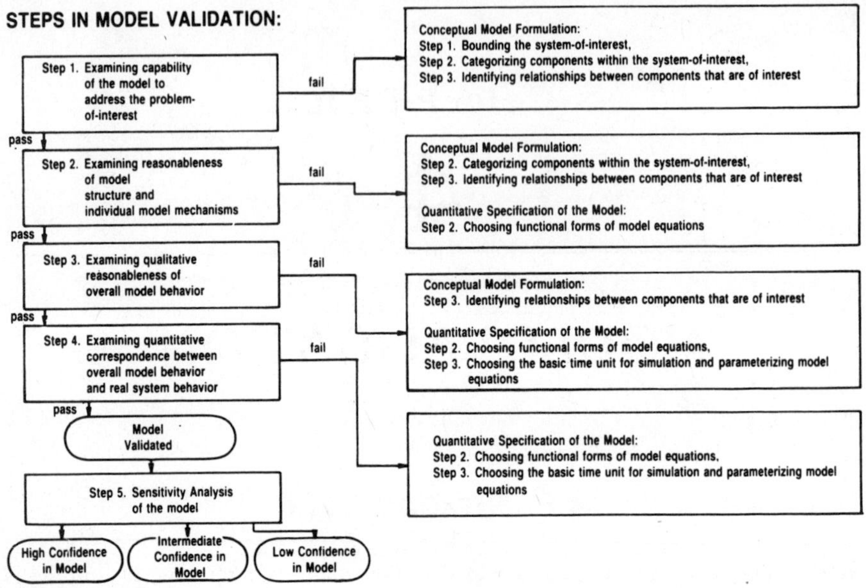

FIGURE 4.1. Steps in model validation and the steps of systems analysis to which we should return if our model fails to pass a given validation step.

4. examining the quantitative correspondence between overall model behavior and real system behavior (Section 4.5), and
5. sensitivity analysis of the model (Section 4.6).

Step 2 may be omitted for purely empirical, or correlative, models and step 4 may be omitted for purely mechanistic, or explanatory, models. But since most models are somewhere on the continuum between these two extremes we frequently wish to consider each of these five steps as part of model validation. If we refute the model during any of these steps we must return to an earlier step in systems analysis to make appropriate modifications to the model. The step to which we return depends upon the reasons for which the model was invalidated (Figure 4.1).

4.2. EXAMINING CAPABILITY OF THE MODEL TO ADDRESS THE PROBLEM OF INTEREST

The first step in validation of any model is examination of the capability of the model to address the problem of interest. A valid model must first and foremost be a useful aid in addressing the questions that are of interest to us. As an abstract concept validity has no meaning. A model that is valid for one purpose may be invalid for another purpose. To be useful to us a model must be able to represent that range of system conditions and respond to that range of input conditions that encompass our interests. A model also must provide the type of output or predictions that allow us to address our questions.

This first step in model validation begins long before the baseline version of the model finally is formalized. Indeed, the model should be designed from the beginning and checked at each stage of development to ensure its capability to aid us in addressing the questions in which we are interested. We should expect our model to pass this validation step with ease. However, it is well to consider this first validation step formally before model use because subtle shifts in our interests or priorities may have occurred during model development as we further refined our questions about the system of interest. If the model fails to pass this validation step we should reconsider the early steps in conceptual model formulation including bounding the system of interest, categorizing components within the system of interest, and identifying relationships between components that are of interest (Figure 4.1).

In the original version of our model predicting fluctuations in weight of an animal over time (Figure 1.1), we represented consumption as a function of weight of the animal and respiration as a function of environmental temperature and weight of the animal. Recall we assumed that assimilation efficiency was 100 percent. Perhaps our original objective was to simulate weight fluctuations in the laboratory under a variety of temperatures with food provided *ad libitum*. The model as presented in Figure 1.1 appears capable of addressing this objective and would pass the first step in model validation. However, if our interests changed such that we wished to simulate weight fluctuations under field conditions in which food availability limited consumption during some seasons, we would reject, or invalidate, the model as incapable of addressing our new interests. We would return to the initial step in conceptual model formulation and modify system bounds such that food availability and season were included within the system of interest. We might recategorize components within the system of interest (step 2 in conceptual model formulation) such that available food and environmental temperature are auxiliary variables and season is a driving variable, and identify relationships between components that are of interest (step 3 in conceptual model formulation) such that available food and environmental temperature are a function of season, consumption is a function of weight of the animal and available food, and respiration is a function of environmental temperature and weight of the animal. This revised version of the model (Figure 2.1) appears capable of addressing our new interests and would pass the first step in model validation.

4.3. EXAMINING REASONABLENESS OF MODEL STRUCTURE AND INDIVIDUAL MODEL MECHANISMS

The second step in model validation is to examine the reasonableness of model structure and individual model mechanisms. This step receives particular emphasis with mechanistic, or explanatory, models and might be omitted for strictly empirical, or correlative, models in which there is no concern about the correspondence between model and real system mechanisms. The procedure involves attempting to refute aspects of model structure and individual model mechanisms based on their lack of correspondence with the real system and is exactly the same as scientific hypothesis testing. The hypotheses tested are our hypotheses about how individual model mechanisms work and about the structural relationships among individual mechanisms within the model. Hypothesis tests are based on the best information available about the structure and mechanisms in the real system. If any aspect of model structure or any individual mechanism within the model can be shown to be an inadequate representation of the corresponding aspect or mechanism in the real system (at the appropriate level of resolution), then that aspect or mechanism is refuted. Criteria for failing hypothesis tests may be qualitative or quantitative depending upon the type of information available from the real system and upon objectives of the modeling project.

As with the first step in model validation, this second step begins long before the baseline version of the model finally is formalized. Each aspect of overall model structure and each mechanistic relationship within the model should be examined rigorously based on the best information available at the time of its proposed inclusion in the model. Thus we also should expect our model to pass this validation step with ease. But formal consideration of this second step before model use is particularly important to ensure that the model cannot be refuted by any new information about the system of interest that recently has become available. If the model fails to pass this validation step we should reconsider steps 2 and 3 in conceptual model formulation, categorizing components within the system of interest and identifying relationships between components that are of interest, and then proceed to reconsider step 2 in quantitative specification of the model, choosing functional forms of model equations (Figure 4.1).

Recall that we initially represented consumption in our weight fluctuation model (Figure 1.1) as a linear function of weight of the animal. Presumably this representation was consistent with available information when we first included it in the model. But suppose that subsequently during model validation we discover new information, based on data collected over a wider range of animal weights than had previously been examined, suggesting that consumption per unit body weight is actually a decreasing function of animal weight.

Thus consumption is some curvilinear function of weight and we reject the hypothesis of a linear relationship. In this case our conceptual model of the system still appears adequate, but we must return to step 3 of quantitative specification of the model and choose a new functional form for the consumption equation.

4.4. EXAMINING QUALITATIVE REASONABLENESS OF OVERALL MODEL BEHAVIOR

The third step in model validation is to examine the reasonableness of overall model behavior in a qualitative sense. We look for obvious impossibilities or implausibilities in the baseline simulation(s) such as negative values for state variables or transfer rates that by definition must be positive, or impossibly high or low values for state variables or transfer rates. In addition to examining the baseline simulation for unreasonable behavior, we might also examine model behavior over a wide range of input conditions (wide range of values for driving variables) to attempt to expose additional inadequacies in the model. It may seem strange to mention such obvious shortcomings in model behavior, but models that initially show gross inconsistencies with our knowledge about the real system are common. At this point these gross inconsistencies may result from a fundamental misconception about the nature of relationships within the system of interest, in which case changes in the conceptual model are required. Presumably errors in coding the model or errors in model logic have been corrected already as the computer code for the model was debugged and verified.

When a model no longer exhibits obviously implausible behavior we turn our attention to a closer examination of system variables (state, driving, and auxiliary variables) and material transfer rates. General dynamics of each system variable should be reasonable in terms of the timing of maximum and minimum values, the relative amplitude and periodicity of fluctuations, and its relationship to dynamics of other system variables. Likewise each material transfer should be reasonable in terms of timing of maximum and minimum values, relative amplitude and periodicity of fluctuations, and its relationship to other transfer rates. Both state variables and transfer rates should vary in the right direction and by the right order of magnitude in response to changes in values of driving variables. Model inadequacies detected as a result of this closer examination of state variables and material transfer rates still may be caused by a fundamental misconception about the nature of relationships within the system of interest, but at this stage it is likely that inadequacies are caused by inclusion of erroneous parameter estimates or perhaps by inclusion of incorrect functional forms of equations.

In Section 4.2 we suggested that our weight fluctuation model, which was developed initially to simulate animal growth under laboratory conditions with food provided *ad libitum*, could not be used to simulate weight fluctuations under food-limited field conditions. We claimed that since we knew food availability limited consumption during some seasons, the model was incapable of addressing our questions and should be refuted on those grounds. But suppose we had no reason to suspect that consumption would ever be food-limited over the range of field conditions in which we were interested, and thus, failed to reject the model during step 1 of validation. We might first discover this mistake in the model when we examine the qualitative reasonableness of overall model behavior and find that simulated animal weights are unreasonably high. Perhaps a closer examination reveals that our simulated animals gain weight throughout the year, whereas animals in the field seldom gain weight during a particular season (a phenomenon that we had erroneously attributed, say, to changes in environmental temperature). In this case inadequacies in our model cannot be remedied by modification of parameter estimates or functional forms of equations. We must correct a fundamental error in our conceptual model of the system of interest.

Tuning the Model

At this point, assuming that no gross inadequacies in model behavior remain, it may be possible to further improve the qualitative reasonableness of overall model behavior by adjusting some model parameters or by altering the functional forms of some equations. This general procedure is referred to as tuning the model. It is a legitimate activity during this step of model validation, but is an activity that must be confined strictly in terms of:

1. choosing parameters or functional forms to be altered,
2. determining when to end tuning activities, and
3. limiting the number of parameters or functional forms tuned.

Choosing Parameters or Functional Forms to Be Altered

Discretion must be used in choosing the parameters or functional forms to be altered and in choosing the types of alterations to be made. Alterations will be arbitrary since no new information external to the model is being drawn upon to make them, although the nature of the unreasonableness in baseline model behavior may suggest specific alterations. But we must alter only those parameters or functional forms that we have specified in a tentative fashion in the baseline version of the model and the type of alteration made must not result in a parameter value or functional form that could have been refuted in the previous model validation step.

Suppose that we have estimated two parameters in a given model. For one parameter we have strong evidence that the value must be 8. For the other parameter we have information suggesting only that the value probably lies between 10 and 20; perhaps we have chosen a value of 15 for inclusion in the model. Since we have strong evidence supporting our estimate of the first parameter we should not "tune" or alter that parameter. But since our current estimate of the second parameter (15) was chosen over other possible estimates between 10 and 20 based on little or no information, we might legitimately tune or alter this parameter within the range of 10 to 20 in order to improve the qualitative reasonableness of overall model behavior.

Likewise, suppose that we have specified functional forms of two material transfer equations within a given model. For one transfer we have strong evidence that the functional form of the relationship governing the transfer is linear. For the other transfer we have information suggesting only that the curve relating the rate of transfer (Y) to the variable governing the transfer (X) passes through the point (X_1, Y_1) and the point (X_2, Y_2). Perhaps we have chosen a linear relationship passing through these two points for inclusion in the model. Since we have strong evidence supporting our choice of a linear functional form for the first transfer it should not be altered. But since our current functional form of the equation representing the second transfer was chosen over other possible functional forms that pass through (X_1, Y_1) and (X_2, Y_2) based on little or no information, we might legitimately tune this transfer by specifying another functional form for the equation. We might see if an exponential functional form of the equation improves the qualitative reasonableness of overall model behavior.

Determining When to End Tuning Activities

All tuning of the model must be completed before proceeding to the next model validation step, which is examination of the quantitative correspondence between overall model behavior and real system behavior. If statistical comparisons are made between model behavior and real system behavior, then clearly tuning must be completed before statistical tests are performed. If quantitative comparisons between model behavior and real system behavior are non-statistical, tuning still must be completed before formal comparisons are performed. Admittedly for non-statistical comparisons it is more difficult to identify the point at which tuning is no longer legitimate. Most often the information upon which we base our decision that overall model behavior is qualitatively unreasonable, and therefore that we would like to tune the model, is the same information upon which we ultimately base our quantitative comparisons in the next model validation step. But in good faith we must terminate tuning before the point at which our comparisons of model behavior and real system behavior become formal.

Limiting Number of Parameters or Functional Forms Tuned

We must limit the number of parameter values or functional forms of equations that we tune in any given model. There is no set limit on the number of parameters or functional forms that can be tuned. But obviously the more that are tuned the more possible combinations of alterations there are and the greater the possibility that any improvement in the qualitative reasonableness of model behavior will result solely by chance from a fortuitous combination of alterations.

In practice it is more common to tune parameter values than to tune functional forms of equations. This is understandable because the choice of functional forms usually has more profound implications concerning ecological interpretations of model structure than does the choice of parameter values. We are likely to have based our original choice of general functional forms on a relatively sound understanding of the nature of relationships in the real system, whereas we are less likely to have based our original choice of specific parameter values on an equally sound basis.

If the model fails to pass this third validation step we should reconsider steps 2 and 3 in quantitative specification of the model, choosing functional forms of model equations and choosing the basic time unit for simulations and parameterizing model equations, and perhaps even reconsider step 3 in conceptual model formulation, identifying relationships between components that are of interest (Figure 4.1).

There is an interesting philosophical point here with regard to whether we are tuning a model or developing a new model after having invalidated the old model. Perhaps the most practical distinction is based on the degree of our dissatisfaction with the qualitative behavior of the model. If our dissatisfaction is relatively minor we may tune the model as first described without use of new information external to the model. If our dissatisfaction is major we may decide that the model is invalid and begin developing a new model by drawing upon new information about the system of interest or by changing basic assumptions about the structure or function of the system.

Suppose that initial predictions of our weight fluctuation model indicated that respiration exceeds consumption at all environmental temperatures and thus weight changes are always negative. Since animals in the real system do gain weight, we conclude that model behavior is unreasonable. This unreasonable behavior might be remedied by arbitrarily adjusting, or tuning, parameters of the consumption or respiration equations, or perhaps by arbitrarily altering their functional forms, assuming that this is appropriate. However, it may be that such tuning, within the constraints imposed upon us by our current knowledge about the real system, does not improve model behavior. In the latter case we must reject, or invalidate, the model and reconsider quantitative specification of the model, or perhaps even conceptual model formulation, based on new information about the system of interest or on new assumptions about the system.

4.5. EXAMINING QUANTITATIVE CORRESPONDENCE BETWEEN OVERALL MODEL BEHAVIOR AND REAL SYSTEM BEHAVIOR

The fourth step in model validation is to examine the quantitative correspondence between overall model behavior and behavior of the real system. This step receives particular emphasis with empirical, or correlative, models and theoretically might be omitted for strictly mechanistic, or explanatory, models, although in practice we almost always are interested to some extent in a model's predictive abilities. Strictly speaking, data from the real system that are used in model validation must be independent of data used to develop the model. If the same data that are used in quantitative specification of the model are used in model validation we can hardly expect to reject the model, because we already would have examined and failed to reject any aspects of the model that were quantified based on those data. The situation is directly analogous to quantifying a regression model using a given data set, concluding that we will tentatively accept the regression model because it has an r^2 of 0.90, and then "validating" the model simply by observing that it does indeed predict values in that given data set well. The appropriate validation for the regression model is to use the model, as parameterized based on the first data set, to predict values of the dependent variable in a new data set based on values of the independent, or regressor, variables in that new data set. Likewise, for this step in simulation model validation to be meaningful model predictions must be compared to real system data that were not used to quantify model equations.

The manner in which we quantitatively compare model predictions to real system data depends upon:

1. the type of model predictions in which we are interested, single-value or time-series,
2. the type of model, deterministic or stochastic, and
3. the type of data available from the real system, replicated or non-replicated (Table 4.1).

TABLE 4.1

Summary of appropriate quantitative comparisons of model predictions to real-system data and associated procedures, depending upon (1) type of model predictions, (2) type of model, and (3) type of data available from real system

				Appropriate Procedure		
				Statistical		
Type of Model Predictions	Type of Model	Type of Real-System Data	Appropriate Comparison	Hypothesis Tested	Test Statistic	Non-statistical
Single-value	Stochastic	Replicated	Compare mean of model predictions (μ_{MP}) to mean of real-system data (μ_{RSD})	$\mu_{MP} = \mu_{RSD}$	$t = (\bar{X}_{MP} - \bar{X}_{RSD})/\sqrt{(s^2_{MP} + s^2_{RSD})/n}$ See Table 4.2 for example calculations and symbol definitions	
		Non-replicated	Compare mean of model predictions to single real-system datum (which represents μ_{RSD})	$\mu_{MP} = \mu_{RSD}$	$t = (\bar{X}_{MP} - X_{RSD})/(s_{MP}/\sqrt{n})$ See Table 4.3 for example calculations and symbol definitions	
	Deterministic	Replicated	Compare single model prediction (which represents μ_{MP}) to mean of real-system data	$\mu_{MP} = \mu_{RSD}$	$t = (X_{MP} - \bar{X}_{RSD})/(s_{RSD}/\sqrt{n})$ See Table 4.3 for example calculations and symbol definitions	
		Non-replicated	Compare single model			

			prediction to single real-system datum		X
Time-series	Stochastic	Replicated	Compare mean curve predicted by model to mean curve based on real-system data with regard to timing of maximum and minimum values and amplitude and periodicity of fluctuations	Statistical procedures may be possible to apply in specific cases, but are beyond the scope of this book	
		Non-replicated	Compare mean curve predicted by model to single curve based on real-system data		
	Deterministic	Replicated	Compare single curve predicted by model to mean curve based on real-system data		
		Non-replicated	Compare single curve predicted by model to single curve based on real-system data		X

Single-value model predictions of interest often are summary statistics from an entire simulation, although they can be any single prediction from a simulation. In our weight fluctuation model the weight of an animal on day 30 of simulated time and total consumption of an animal over a 30-day period both are single-value predictions. We would calculate total consumption by summing daily material transfers from the food source to the weight of animal state variable (Figure 3.2) for each of the 30 days of simulated time. Time-series predictions of interest often consist of sequential values of state variables or material transfers over the course of an entire simulation. The 30 sequential daily weights and the 30 sequential values of daily consumption from the model mentioned above both form time-series predictions.

Comparison of Single-Value Predictions

Replication in Both Model Predictions and Real-System Data

If we are interested in comparing a single-value prediction from a stochastic model to replicated data from the real system, we can use a two-sample *t*-test to test the hypothesis that the mean of model predictions equals the mean of real-system data (Table 4.1). Suppose that we run 14 replicate simulations of our weight fluctuation model (with the sample size of 14 being based on appropriate considerations as presented in Section 3.5) and record the weight of an initially 100 g animal on day 30 of simulated time for each. Assuming that we use a different set of random numbers in each replicated simulation to generate the random environmental temperature driving variable, the 14 predictions of weight on day 30 form a set of independent, identically distributed, random variates. If we also have 14 independent replicate observations of weight on day 30 of initially 100 g animals from the real system, we can compare the 14 replicate model predictions to the 14 replicate estimates from the real system using a *t*-test as shown in Table 4.2.

There are other statistical tests that also can be used to compare model predictions to real system data when we have replication in both. If the values we wish to compare are not normally distributed (normality is an assumption of the *t*-test), we might use a non-parametric procedure such as a Wilcoxon signed rank test (Snedecor and Cochran 1967, p. 130) or a Kolmogorov—Smirnov test (Sokal and Rohlf 1969, p. 571) to compare the frequency distributions of predicted and observed values. However, the *t*-test is quite robust to departures from normality and can be used in most cases.

Replication in Either Model Predictions or Real-System Data

If we are interested in comparing a single-value prediction from a stochastic model to non-replicated data from the real system, or a single-value prediction

TABLE 4.2

Example calculations comparing 14 replicate predictions of weight of an initially 100 g animal on day 30 of simulated time from the weight fluctuation model (Figure 3.2) to 14 replicate estimates of weight on day 30 of initially 100 g animals from the real system using a two-sample *t*-test

	Model Predictions (X_{MP}) of Weight (g) on Day 30	Real-System Estimates (X_{RSD}) of Weight (g) on Day 30
	122.3	117.5
	122.6	116.0
	120.6	122.5
	123.4	117.1
	121.1	126.6
	120.2	116.9
	120.4	118.0
	114.8	119.6
	119.1	121.5
	119.4	118.8
	119.3	114.9
	118.4	121.2
	119.1	117.4
	117.4	120.1
Total	1,678.10 (ΣX_{MP})	1,668.1 (ΣX_{RSD})
Sample size	14 (n_{MP})	14 (n_{RSD})
Mean	119.86 ($\bar{X}_{MP}$)	119.15 ($\bar{X}_{RSD}$)
ΣX^2	201,208.21 (ΣX^2_{MP})	198,876.15 (ΣX^2_{RSD})
Total^2/Sample size	201,144.26 ($(\Sigma X_{MP})^2/n_{MP}$)	198,754.12 ($(\Sigma X_{RSD})^2/n_{RSD}$)
$\Sigma(X - \text{mean})^2$	63.95 ($\Sigma(X_{MP} - \bar{X}_{MP})^2)^A$	122.03 ($\Sigma(X_{RSD} - \bar{X}_{RSD})^2)^A$
Degrees of freedom	13 (df_{MP})	13 (df_{RSD})

Pooled variance (s^2) = (63.95 + 122.03)/(13 + 13) = 7.15

Total degrees of freedom (df) = 13 + 13 = 26

Pooled standard error (s_{MP-RSD}) = $\sqrt{2s^2/n} = \sqrt{2(7.15)/14} = 1.01$

Test statistic (t) = $(\bar{X}_{MP} - \bar{X}_{RSD})/s_{MP-RSD}$ = (119.86 − 119.15)/1.01 = 0.703

Tabulated critical values of the t-distribution indicate that t = 0.703 with 26 degrees of freedom is significant at α = 0.50, but not significant at α = 0.40.

[A] shortcut calculation for $\Sigma(X - \bar{X})^2$ is $\Sigma X^2 - (\Sigma X)^2/n$

from a deterministic model to replicated data from the real system, we can use a modified two-sample t-test to test the hypothesis that the mean of model predictions (or the mean of real system data) equals the single real system datum (or the single deterministic model prediction) (Table 4.1). We assume that the single datum is a sample from a population, but since the population is represented by a single variate the sample does not contribute to the degrees of freedom or to the within group variance calculated as part of the t-test. An example comparing the 14 replicate model predictions of weight of an initially 100 g animal on day 30 of simulated time used in Table 4.2 to a single real-system estimate of weight on day 30 of an initially 100 g animal is shown in Table 4.3. The procedure for comparing a single model prediction to replicated real-system estimates is directly analogous.

No Replication in Model Predictions or Real-System Data

If we are interested in comparing a single-value prediction from a deterministic model to non-replicated data from the real system, our comparison is non-statistical (Table 4.1). We must base our comparison solely on the practical significance of the difference between model prediction and real system estimate.

Comparison of Time-Series Predictions

Comparisons of time-series model predictions to time-series data from the real system follow the same general scheme just presented for comparisons of single-value predictions and data with regard to stochastic versus deterministic models and replicated versus non-replicated real system data (Table 4.1). The question concerning model validation is whether or not the curve based on the time-series of model predictions resembles the curve based on the real system time-series closely enough. Criteria defining ''closely enough'' will vary depending upon model objectives, but in general they will be quantitative versions of the qualitative criteria used to examine qualitative reasonableness of overall model behavior. We quantitatively examine the timing of maximum and minimum values and the amplitude and periodicity of fluctuations within the time-series.

Suppose that validation of our deterministic weight fluctuation model includes comparison of 30 sequential daily predictions of animal weight to 30 sequential daily estimates of animal weight based on non-replicated real-system data. If the curves representing these two time-series are as shown in Figure 4.2A, we see that although the general shapes of the curves are similar, the model underestimates daily weights by 0.5 g to almost 8 g. Upon closer examination we also note that the real system data suggest a decrease in the rate of growth at heavier weights, which is not reflected in model predictions, particu-

TABLE 4.3

Example calculations comparing 14 replicate predictions of weight of an initially 100 g animal on day 30 of simulated time from the weight fluctuation model (Figure 3.2) to a single estimate of weight on day 30 of an initially 100 g animal from the real system using a modified two-sample *t*-test

	Model Predictions (X_{MP}) of Weight (g) on Day 30	Real-System Estimate (X_{RSD}) of Weight (g) on Day 30
	122.3	122.5
	122.6	
	120.6	
	123.4	
	121.1	
	120.2	
	120.4	
	114.8	
	119.1	
	119.4	
	119.3	
	118.4	
	119.1	
	117.4	
Total	1,678.10 (ΣX_{MP})	122.5 (ΣX_{RSD})
Sample size	14 (n_{MP})	1 (n_{RSD})
Mean	119.86 ($\bar{X}_{MP}$)	122.5 ($\bar{X}_{RSD}$)
ΣX^2	201,208.21 (ΣX^2_{MP})	15,006 (ΣX^2_{RSD})
Total^2/Sample size	201,144.26 ($(\Sigma X_{MP})^2/n_{MP}$)	15,006 ($(\Sigma X_{RSD})^2/n_{RSD}$)
$\Sigma(X - \text{mean})^2$	63.95 ($\Sigma(X_{MP} - \bar{X}_{MP})^2)^A$	0 ($\Sigma(X_{RSD} - \bar{X}_{RSD})^2)^A$
Degrees of freedom	13 (df_{MP})	0 (df_{RSD})

Pooled variance (s^2) = (63.95 + 0)/(13 + 0) = 4.92

Total degrees of freedom (df) = 13 + 0 = 13

Pooled standard error (s_{MP-RSD}) $= \sqrt{s^2((n_{MP} + n_{RSD})/n_{MP}n_{RSD})}$

$= \sqrt{4.92((14 + 1)/(14)(1))} = 2.30$

Test statistic (t) $= (\bar{X}_{MP} - \bar{X}_{RSD})/s_{MP-RSD} = (119.86 - 122.5)/2.30 = -1.148$

Tabulated critical values of the t-distribution indicate that $t = -1.148$ with 13 degrees of freedom is significant at $\alpha = 0.40$, but not significant at $\alpha = 0.20$.

^{A}shortcut calculation for $\Sigma(X - \bar{X})^2$ is $\Sigma X^2 - (\Sigma X)^2/n$

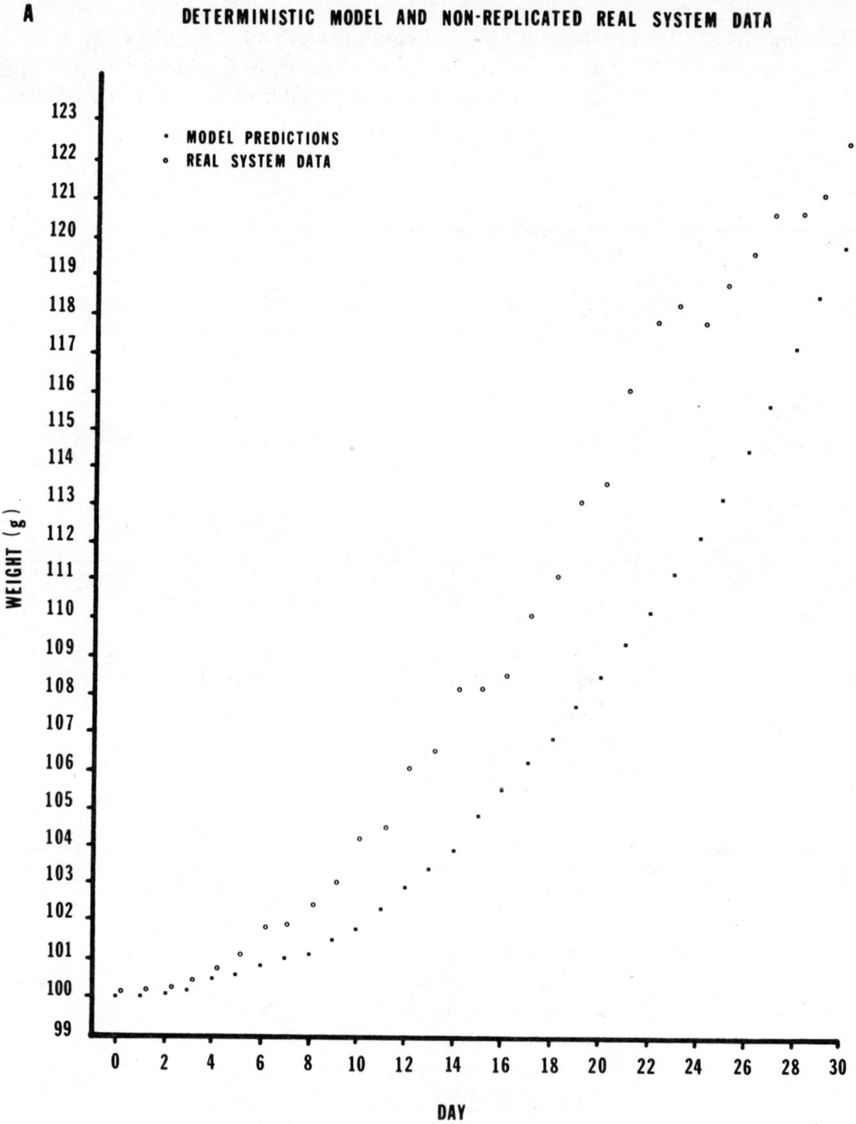

FIGURE 4.2. Comparisons of weights of an animal (A) based on deterministic model predictions and non-replicated real-system data, and (B) based on stochastic model predictions and replicated real-system data. Vertical bars represent ± 1 standard error of the mean.

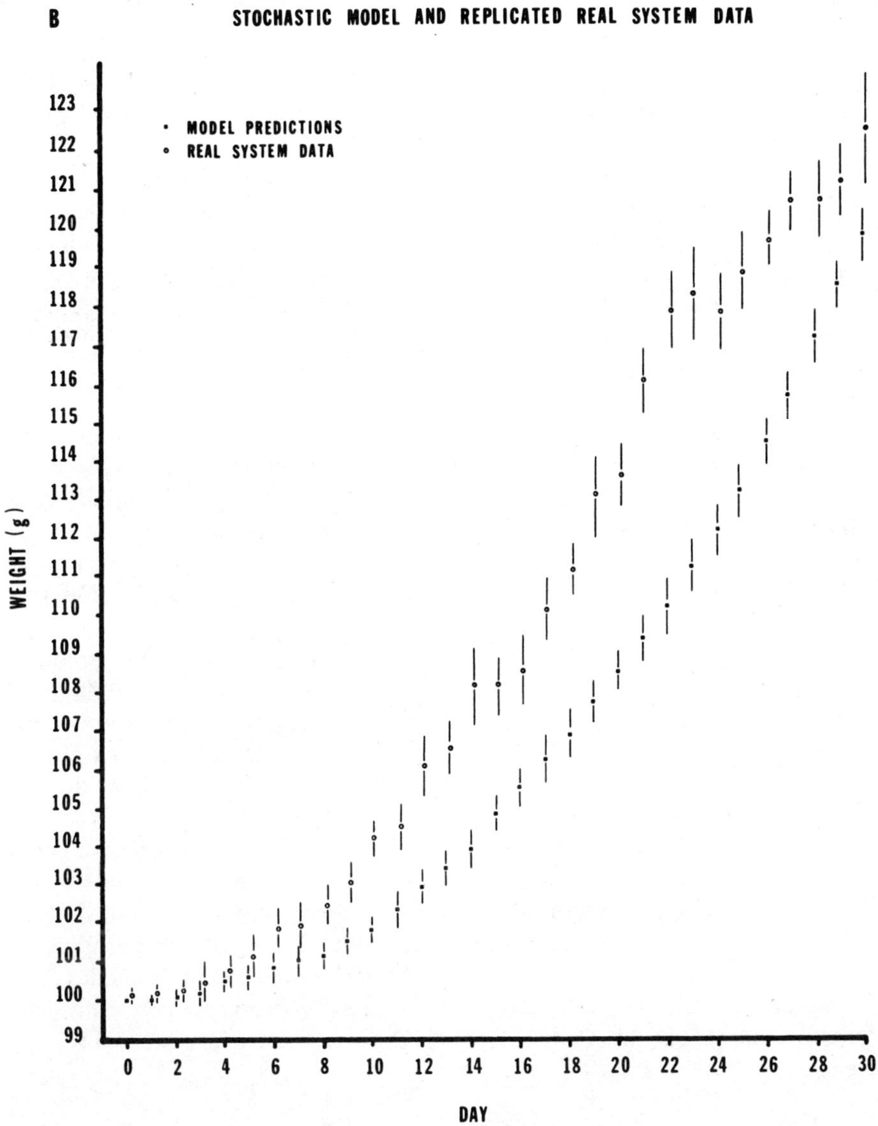

FIGURE 4.2. (*Continued*)

larly during the last 8 days when real system data suggest a 5½ percent increase in weight whereas the model predicts a 9½ percent increase. Whether or not these quantitative differences are sufficient grounds for refuting the model again depends on modeling objectives.

Use of statistical tests to compare time-series model predictions to time-series data from the real system is more complicated than is the case for single-value predictions and data. Complications arise because of autocorrelation between values within each time-series, because often variances increase from beginning to end of simulations, particularly for state variables, and because for most of the situations of interest to us we usually have relatively few (<50) values within each time-series. Exploration of these problems and presentation of possible ways to overcome them in specific situations are beyond the scope of this book. One useful procedure when examining replicated time-series data is to represent the variance (or standard error) around each mean value in the two time-series that we wish to compare, as illustrated for our weight fluctuation model (Figure 4.2B). In Figure 4.2B we note that the variation in real system data on daily weights is noticeably larger than the variation in model predictions. We also can examine the relative distance between mean observed and predicted values compared to the widths of bars representing variability (standard errors of the mean in Figure 4.2B) within observed and predicted values. The general interpretation is that if the bars are broadly overlapping, such as occurs for days 1 through 5 in Figure 4.2B, there is no "significant" difference between observed and predicted values. If the bars do not overlap (days 6 through 30, Figure 4.2B), the difference is "significant." Note that this procedure does not imply a formal statistical comparison of the two curves. We can statistically compare observed and predicted values for any single day on the curves by treating the problem as a single-value comparison. Thus we can use a t-test to compare observed and predicted animal weights on day 10, for example, but, having conducted this single test, we cannot repeat the procedure for other days because our data are not independent from day to day, they are autocorrelated. The weight predicted on day 10 depends upon the weights predicted on preceeding days and affects the weights predicted on subsequent days.

Practical versus Statistical Significance

We should consider for a moment the distinction between practical significance and statistical significance within the framework of model validation. Theoretically there are four possible combinations of practical and statistical significance that we might encounter when comparing model predictions to real system data. Differences may be:

1. both statistically significant and of practical significance to the modeler or resource manager,
2. both statistically non-significant and of no practical significance,
3. statistically significant but of no practical significance, or
4. statistically non-significant but of practical significance (Table 4.4).

Ideally we would encounter only the first two of these cases, in which practical significance and statistical significance are synonymous. If we have designed our validation tests well, if we have selected sample sizes capable of showing that a true difference of a practically significant magnitude between model predictions and real-system data is statistically significant at the desired significance level (α) and probability level (P) (as described in Section 3.5), then this will be the case. Interpretation of such results is clear. If differences are practically/statistically significant we reject the model as invalid. If differences are practically/statistically non-significant we fail to reject the model and the model passes this step in the validation procedure (Table 4.4).

However, if we have overdesigned or underdesigned our validation experiment we may encounter cases in which there is a discrepancy between practical and statistical significance. If our sample size is too large (overdesign) we may have enough statistical power to declare relatively small differences between model predictions and real system data, which are of no practical significance,

TABLE 4.4

Interpretations of the four possible combinations of practical and statistical significance that might be encountered when comparing model predictions to real-system data

	Practically	
Statistically	Significant	Non-significant
Significant	Modeler/manager considers differences important, rejects model as invalid; statistical design of validation appropriate	Modeler/manager considers differences unimportant, fails to reject model as invalid; statistical design of validation overdesigned
Non-significant	Modeler/manager provided with no additional information from statistics; statistical design of validation underdesigned	Modeler/manager considers differences unimportant, fails to reject model as invalid; statistical design of validation appropriate

statistically significant at the chosen α and P levels. But the interpretation of such results also is clear. We do not reject the model, because we have not detected practically significant differences (Table 4.4). Statistical significance alone is not relevant and we have wasted resources by overdesigning the experiment. If our sample size is too small (underdesign) we may not have enough statistical power to declare even practically significant differences statistically significant. Such results are not interpretable, or at least the statistical test provides no useful information. In fact, since we can identify statistical tests that have insufficient power to detect practically significant differences *a priori*, such statistical tests should not be conducted (Table 4.4).

As an example of overdesigned and underdesigned experiments consider the following situation. Suppose that we want to be able to detect a true difference of ten units between the mean of model predictions and the mean of real system data for a particular variable of interest. The level of practical significance is ten units. Further suppose that based on initial estimates of variability of our variable of interest we determine the appropriate sample size needed to detect a true difference of ten units at the desired α and P levels is 50. If we run 500 replicate stochastic simulations, assuming that we also have 500 replicate observations available from the real system, our experiment is overdesigned. Not only can we detect a difference of ten units at the desired α and P levels, we can detect smaller differences if they truely exist. We might declare a difference of one unit statistically significant. But such a small difference has no practical significance for us. We care only that the difference is less than ten units and we could have obtained this information from 50 samples. If we run five replicate stochastic simulations, assuming that we also have five replicate observations available from the real system, our experiment is underdesigned. Even if there is a true difference greater than ten units and our predicted mean differs from the observed mean by more than ten units we still will be unable to detect a statistically significant difference.

Quantitative versus Qualitative Model Validation

At this point we also should consider the relative merits of quantitative as opposed to qualitative methods of model validation. The tendency is to think of quantitative methods of model validation, particularly those involving statistical tests, as being more rigorous and definitive than qualitative methods. We should seek to use quantitative methods of validation whenever appropriate. However, we also should keep in mind that objective quantitative methods ultimately rest on subjective judgements relative to the appropriateness of their use in the validation procedure. Even statistical hypothesis tests rely upon subjective judgement to determine appropriate significance levels for a given problem. Quantitative tests are not inherently better than qualitative tests and both should be viewed within a validation framework that emphasizes evaluation of the ability of the model to meet project objectives.

If the model fails to pass this fourth step in model validation we should reconsider steps 2 and 3 in quantitative specification of the model, choosing functional forms of model equations and choosing the basic time unit for simulations and parameterizing model equations (Figure 4.1).

4.6. SENSITIVITY ANALYSIS OF THE MODEL

The fifth step in model validation is to perform a sensitivity analysis on the model. The objective is to determine the degree of response, or sensitivity, of model behavior to changes in various model variables or transfer rates (Smith 1973, Steinhorst 1979). The basic procedure is to alter the value of one variable or transfer rate at a time by a specified amount throughout an entire simulation and observe the subsequent effect on model behavior. This basic procedure may be modified to include varying sets of parameters at the same time and also may include varying the functional form of relationships within the model. By identifying those parameters or relationships to which model behavior is most responsive, or sensitive, sensitivity analysis provides an indication of the relative accuracy with which each parameter or relationship ideally should be estimated. This information is useful in determining the degree of confidence that should be placed in the model based on the confidence with which the most influential parameters or relationships have been estimated, and also is useful in establishing future research priorities.

Determination of the effect of parameter or relationship changes on model behavior requires specification of procedures for assessing model behavior. These procedures will depend upon objectives of the modeling project, but most commonly will include monitoring selected state variables and material transfer rates and the calculation of selected summary statistics. In our weight fluctuation model (Figure 3.2) we might choose to monitor consumption rate (the material transfer from the food source to the weight of animal state variable) and weight of the animal (state variable), and also calculate total consumption over a 30-day simulation as indicators of model behavior.

The design of the sensitivity analysis, that is, specification of the parameters or relationships to be varied and the manner in which they are to be varied, likewise will depend upon objectives of the modeling project. A possibility with relatively small models is to vary each parameter and each combination of parameters over a selected range of values. However, simulating all possible combinations of even relatively few levels of relatively few parameters can become an overwhelming task. Thus most commonly a sensitivity analysis is designed to address a limited number of questions concerning model sensitivity that are suggested by prior knowledge that the modeler has about system structure. Sensitivity analysis of our weight fluctuation model might consist of noting changes in model behavior, based on the three criteria mentioned above, as we vary the coefficient on the consumption equation ($T_{SO1,1} = 0.05S1$) from 0.04 to 0.06 in increments of 0.005. Our choice of this design for sensitivity analysis might be based on the fact that we lack confidence in our point estimate (0.05) for this coefficient, but feel confident that the true value lies somewhere in the range 0.04 to 0.06.

The manner in which we compare model sensitivity to changes in one parameter with model sensitivity to changes in another parameter follows the same general scheme described for comparison of model predictions to real-system data (Table 4.1). It depends upon the type of model predictions that we are monitoring, single-value or time-series, and whether the model is deterministic or stochastic. If the model is stochastic statistical comparisons are possible; if the model is deterministic statistical comparisons are not possible. In practice, sensitivity analysis of stochastic models of even moderate size is usually conducted using a deterministic version of the model, a version in which all random variables are arbitrarily assigned their mean values. This is done to reduce the number of simulations required and is justified in that initially we are more interested in getting a general feel for the relative magnitude of changes in model behavior caused by various alterations than in statistical comparisons of the changes. If after initial examination of sensitivity analysis results based on the deterministic version of the model we decide that statistical comparisons of certain changes in model behavior are necessary, we then can run the appropriate number of replicate stochastic simulations and conduct the appropriate statistical tests of the specific model alterations in which we are interested.

Interpreting Sensitivity Analysis within Model Validation Framework

Interpretation of sensitivity analysis results differs somewhat from interpretation of results obtained in earlier steps of model validation. Whereas in earlier steps we attempt to refute or invalidate a model, through sensitivity analysis we attempt to evaluate more clearly our confidence in the ability of the validated model to address our questions (Figure 4.1). If we have failed to invalidate a model during the first four steps of the validation procedure, we must consider the model irrefutable based on current knowledge about the real system. However this does not necessarily mean that we have great confidence in the model's ability to answer our questions. Model behavior may be sensitive to changes in some parameters or functional forms of equations that we have estimated based on inadequate information. Our lack of confidence in estimates of these influential parameters or functional forms translates into a lack of confidence in model behavior.

If sensitivity analysis results for our weight fluctuation model (in which we varied the coefficient on the consumption equation from 0.04 to 0.06) indicate that model behavior is not sensitive to changes in the coefficient within this range, then our confidence in the ability of the model to address our questions is unaffected by our lack of confidence in the coefficient estimate. If on the other hand model behavior is changed noticeably by changes in the coefficient, then we are less certain about the model's ability to address our questions, and subsequent interpretations of model behavior should reflect this uncertainty.

4.7. EXAMPLE OF MODEL VALIDATION: HYPOTHETICAL POND MODEL

Examining Capability of the Pond Model to Address the Problem of Interest and Examining Reasonableness of Model Structure and Individual Model Mechanisms

Assuming that our questions of interest about the pond system have not changed, our pond model should pass the first step in model validation. The model can address each of the five specific questions of interest posed in Section 2.6. Questions 1 and 2:

1. What size harvest might be expected in a normal year under the usual scheme of stocking 75 0.227-kg fish on April 15 and harvesting them on November 15, and what profit is associated with this harvest (the baseline situation)?
2. How variable is the harvest (and therefore the profit) from year to year under the baseline stocking and harvesting scheme as the result of the inherent year-to-year variability of the system?

can be addressed by a set of stochastic baseline simulations. The mean harvest and mean profit predicted by the simulations are the harvest and profit expected in a normal year. The year-to-year variability of the harvest and profit are represented by the variances of harvest and profit predicted by the simulations, which might be presented in terms of standard errors of the means or 95 percent confidence intervals.

Questions 3 and 4:

3. Would size of the harvest and the associated profit be changed by harvesting a month earlier or a month later?

4. Would size of the harvest and the associated profit be changed by stocking a month earlier or a month later?

can be addressed by sets of stochastic simulations in which either the beginning (stocking) or the ending (harvesting) date of simulated time (t_{begin} or t_{end}) is either increased or decreased by 30 days relative to baseline.

Question 5:

5. Would size of the harvest and the associated profit be changed by stocking twice as many fish?

can be addressed by a set of stochastic simulations in which the number of fish ($D2$) is doubled relative to baseline. Size of the harvest and the associated profit predicted by the various sets of stochastic simulations can be compared statistically.

Assuming that no new information about the pond system that was not considered during model development has become available, our pond model should also pass the second step in model validation. Model structure and individual model mechanisms are reasonable and cannot be refuted based on the best information available. Further testing of our hypotheses about the structure of individual mechanisms in the model must await availability of new information about the system of interest.

Examining Qualitative Reasonableness of Overall Model Behavior

We begin the third step in model validation by examining results of the set of baseline simulations for obviously impossible system behavior. We already have looked at mean seasonal dynamics of water temperature ($D1$), biomass of plants ($S1$) and fish ($S2$), and weight of individual fish ($A1$) (Figure 3.19) when initially presenting baseline results in Section 3.6. These general dynamics look reasonable. But to take a closer look at model behavior it is convenient to choose one replicate baseline simulation—preferably one in which water temperatures are close to normal—for detailed examination. Table 4.5a summarizes results of one baseline simulation in which water temperatures are 0.3°C warmer than normal. In addition to examining seasonal dynamics of the four variables presented in Figure 3.19, we also examine seasonal dynamics of four material transfers that represent essential biological processes within the system: net plant production ($T_{\text{SO1},1}$), fish consumption ($T_{1,2}$), fish respiration ($T_{2,\text{SI2}}$), and fish excretion ($T_{2,\text{SI3}}$). We need not examine natural mortality rate of plants ($T_{1,\text{SI1}}$) since it is represented as a constant proportion of plant biomass ($S1$). We also need not examine economic variables since they are involved in essentially bookkeeping calculations and should be in order if the biological dynamics are correct. (Actually we need not examine both seasonal dynamics of fish biomass and weight of individual fish, since number of fish is constant and therefore fish biomass always equals a constant times weight of individual fish. However, we choose to present both results because they both are of particular interest to the pond manager.) We note that there are no negative values for state, driving, or auxiliary variables (of course if we examined the profit auxiliary variable ($A4$) a negative value would be possible), or transfer rates. There also are no impossibly high or low values for variables or transfer rates. Maximum values of all variables and transfer rates are within the ranges of real system data presented in Tables 3.3 and 3.4, and information presented elsewhere in Section 3.6.

Next we examine the dynamics of system variables and transfer rates more closely. The general dynamics of each system variable seem reasonable in terms of the relative amplitude and timing of fluctuations (Table 4.5a). Water temperature ($D1$) increases from cooler spring temperatures to warmer summer temperatures and decreases to cooler temperatures again in the fall. Plant biomass ($S1$) decreases monotonically from April through November with the rate of decrease greater during summer, resulting from increased consumption at warmer water temperatures by an increasing biomass of fish. Fish biomass ($S2$) increases monotonically from April through the early fall with rate of increase greater during summer. Weight of individual fish ($A2$) parallels fish biomass, which is expected since the number of fish ($D2$) is constant. Thus both fish biomass and weight of individual fish exhibit general dynamics consistent with the general dynamics of water temperature; the largest rates of increase in biomass and weight resulting from increased consumption are associated with warmer water temperatures.

The general dynamics of each material transfer rate also seem reasonable in terms of the relative amplitude and timing of fluctuations (Table 4.5a). Net production of plants ($T_{SO1,1}$) generally parallels water temperature, increasing from spring to summer and then decreasing into the fall. Plant production is slightly higher at a given water temperature in the spring than it is in the fall because total plant biomass is higher in spring. Consumption of plants by fish ($T_{1,2}$), fish respiration ($T_{2,SI2}$), and fish excretion ($T_{2,SI3}$) also generally parallel water temperature, increasing from spring to summer and decreasing into the fall. In contrast to plant production, fish consumption, respiration, and excretion are noticeably higher at a given water temperature in the fall than they are in the spring because total biomass of fish is higher in fall.

Based on the set of baseline simulations we have no reason to reject the pond model on the grounds that overall model behavior is qualitatively unreasonable. However, at this point we might further examine the qualitative reasonableness of model behavior by running additional simulations in which water temperatures represent either unusually hot or unusually cold years. The pur-

TABLE 4.5

Values of selected variables and transfer rates during (A) one of the replicate baseline simulations in which water temperatures were near normal, (B) a deterministic simulation in which water temperatures were 4°C warmer than normal, and (C) a deterministic simulation in which water temperatures were 4°C cooler than normal

Julian Date (t)	Water Temperature ($D1$) (°C)	Plants ($S1$) (kg/ha)	Fish ($S2$) (kg/ha)	Weight of Individual Fish ($A1$) (kg)	Net Plant Production ($T_{SO1,1}$) (kg/ha-day)	Fish Consumption ($T_{1,2}$) (kg/ha-day)	Fish Respiration ($T_{2,SI2}$) (kg/ha-day)	Fish Excretion ($T_{2,SI3}$) (kg/ha-day)
				A. Normal Water Temperatures				
105 (Apr. 15)	15.4	39,976	17	0.226	21	3	0.1	3
135	20.5	39,302	21	0.280	36	17	0.1	17
165	30.9	38,294	41	0.542	64	78	0.4	77
195	30.8	35,843	85	1.136	59	142	0.8	139
225	29.7	31,600	154	2.048	50	180	1.2	177
255	31.2	26,758	220	2.935	45	189	1.8	185
285	25.1	23,054	259	3.450	28	108	1.4	106
315	18.1	21,246	269	3.582	16	48	0.8	47
319 (Nov. 15)	14.4	21,115	268	3.578	9	21	0.7	20

			B. Water Temperatures 4°C Warmer Than Normal					
105 (Apr. 15)	19.0	40,000	17	0.227	31	10	0.1	11
135	26.3	39,277	26	0.353	52	40	0.2	39
165	33.7	37,642	59	0.788	70	126	0.7	123
195	33.9	33,364	132	1.756	63	216	1.4	211
225	33.9	27,480	214	2.856	52	223	2.1	219
255	33.9	22,380	263	3.508	42	178	2.5	174
285	27.4	18,936	282	3.754	26	94	1.7	92
315	20.0	17,359	280	3.740	15	42	1.0	41
319 (Nov. 15)	19.0	17,228	280	3.730	13	37	0.9	36
			C. Water Temperatures 4°C Cooler Than Normal					
105 (Apr. 15)	11.0	40,000	17	0.227	8	0	0.06	0
135	18.3	39,189	16	0.220	29	9	0.1	9
165	25.7	38,542	25	0.329	49	35	0.2	34
195	25.9	37,409	45	0.601	48	61	0.3	60
225	25.9	35,380	79	1.055	46	96	0.5	94
255	25.9	32,329	128	1.701	42	130	0.8	127
285	19.4	29,439	167	2.233	24	68	0.6	67
315	12.0	28,416	169	2.250	8	0	0.4	0
319 (Nov. 15)	11.0	28,356	167	2.230	6	0	0.4	0

pose of these additional simulations is to expose inadequacies in the model that may be too subtle to identify when the system is operating close to normal conditions. For these simulations we will use a deterministic version of the model in which all random variables are assigned their mean values. For the simulation in which water temperatures represent an unusually hot year we will calculate the water temperature driving variable (*D*1) deterministically, and after each temperature is calculated it will be increased by 4°C. Likewise, for simulations in which water temperatures represent an unusually cold year we will calculate each temperature deterministically and then decrease it by 4°C.

Results of simulations representing an unusually hot and an unusually cold year are summarized in Table 4.5b and Table 4.5c, respectively. We should examine these results in the same manner that we examined the baseline results in Table 4.5a, looking for any signs of unreasonable model behavior. As before there is no obviously impossible system behavior and the general dynamics of each system variable and material transfer rate seem reasonable in terms of the relative amplitude and timing of fluctuations. We note that during the unusually warm year (Table 4.5b) rates of fish consumption, respiration, and excretion all are higher than normal and that fish biomass and weight of individual fish are higher and plant biomass lower than normal on November 15. The opposite is true during the unusually cold year (Table 4.5c) as expected. Additionally we note that during the unusually warm year fish growth slows earlier than normal, although fish grow to a larger size than normal. Further examination shows that this is the result of a restriction in fish consumption due to the fact that plant biomass declines below 20,000 kg/ha during the fall. Thus this important negative feedback on fish growth, which was not operational under normal conditions, is functioning appropriately.

Examining Quantitative Correspondence between Overall Model Behavior and Real System Behavior

The fourth step in validation of the pond model involves quantitative examination of the correspondence between overall behavior of the baseline model and behavior of the real system. Recall that three types of information on the real system are available:

1. information on historical states of the system including data on water temperature, standing crop biomass of plants and fish, and average weight of individual fish at various times of the year for several years (Table 3.3),
2. information on processes occurring within the system including data on rate of plant growth and on rate of consumption of plants by fish (Table 3.4), and

3. information on economic and legal constraints pertaining to management (p. 92).

Data that can be used in this step of model validation include data on standing crop biomass of plants and fish and average weight of individual fish. These data represent the real-world dynamics of state variables *S*1 and *S*2 and auxiliary variable *A*1, respectively, and were not used in the quantitative specification of the model. The remaining data on water temperature, rate of plant growth, rate of consumption of plants by fish, and on the economic and legal constraints were used to quantify various model equations and therefore cannot be used for quantitative model validation.

Since the pond model is stochastic we can make statistical comparisons between attributes of real-system behavior for which we have replicated data and model predictions of these attributes in the set of stochastic baseline simulations. The real-system data on plant biomass, fish biomass, and average weight of individual fish from each replicate sample all are time-series data. We have values for each of these variables at successive points in time. However, we might also think of the data on each variable from each replicate sample as single-value predictions if we concern ourselves only with the value of each variable on the harvest day. Thus we can base our quantitative comparisons of model and real-system behavior on analyses of both time-series and single-value predictions.

Comparison of Single-Value Predictions

Analyses of single-value predictions might consist of three two-sample *t*-tests comparing baseline model predictions of plant biomass, fish biomass, and average weight of individual fish on November 15 (harvest date) (Table 3.10) to real-system data (Table 3.3). We will determine the number of replicate simulations upon which to base our *t*-tests in a manner similar to the way that we determined how many replicate baseline simulations to run in Section 3.6. Recall that the smallest change in harvested fish biomass between management schemes that has a practical significance to the pond manager is 25 kg. Also recall that the pond manager wants to be able to show that a true difference of 25 kg in harvest is statistically significant at the 95 percent confidence level ($\alpha = 0.05$) with an 80 percent probability that the difference will be detected if it exists ($P = 0.80$, or type II error $\beta = 0.20$) (p.119). It follows that a difference of less than 25 kg in harvest between baseline model predictions and real-system data is an appropriate criterion for model validation. If differences in harvest of less than 25 kg between management schemes are not practically significant, then differences of less than 25 kg in harvest between baseline model predictions and real-system data also should be of no practical significance. Thus we wish to know how large a sample is needed to show that a true difference of 25 kg in harvest between baseline model predictions and real-system data is statistically significant at the 95 percent confidence level ($\alpha = 0.05$) with an 80 percent probability that the difference will be detected if it exists ($P = 0.80$).

Recall that we presented a general sample size formula in Section 3.5:

$$n \geq 2\left(\frac{\sigma}{\delta}\right)^2 (t_{\alpha,\gamma} + t_{2(1-P),\gamma})^2$$

where
- n = number of samples
- σ = true standard deviation, which we estimate as the square root of estimated variance within samples
- δ = smallest true difference that we desire to detect
- γ = degrees of freedom of the sample standard deviation with b groups of samples and n samples per group, or $\gamma = b(n - 1)$
- α = significance level
- P = desired probability that a true difference will be found to be significant if it is as small as δ
- $t_{\alpha,\gamma}$ and $t_{2(1-P),\gamma}$ = values from a two-tailed t-table with gamma degrees of freedom and corresponding to probabilities of α and $2(1 - P)$, respectively

However, in the present situation the number of samples that we have from the real system is fixed; we have only 8 years of data available (Table 3.3). We must modify the general sample size formula such that we can set n_2 (real-system sample size) equal to 8 and solve for the required n_1 (simulation sample size).

$$n_1 \geq \frac{\sigma_1^2}{\delta^2/(t_{\alpha,\gamma} + t_{2(1-P),\gamma})^2 - (\sigma_2^2/n_2)}$$

where
- δ = 25 kg
- σ_1^2 = variance of model predictions of harvest = 379
- σ_2^2 = variance of real-system data on harvest (Table 3.3) = 320
- α = 0.05
- P = 0.80

If we assume that $n_1 = 12$, then

$$\gamma = n_1 + n_2 - 2\text{df} = 12 + 8 - 2 = 18\text{df}$$

thus

$$n_1 \geq \frac{379}{(25)^2/(t_{0.05,18} + t_{2(1-0.80),18})^2 - (320/8)}$$

The appropriate number of replicate baseline simulations to use for comparison of model predictions with real-system data in this step of model validation is 12. We will use results of the ten original replicate baseline simulations (Table 3.10) plus results of two additional replicate baseline simulations (Table 4.6) to compare to the real-system data (Table 3.3) in our three two-sample *t*-tests.

Results of the three two-sample *t*-tests indicate no statistically significant differences between model predictions of plant biomass, fish biomass, or weight of individual fish on November 15 (harvest date) and estimates based on real-system data (Table 4.7). Based on these analyses of single-value outputs we have no reason to reject the pond model on the grounds that overall model behavior does not correspond quantitatively to real-system behavior.

Comparison of Time-Series Predictions

Analysis of time-series predictions might consist of quantitatively comparing baseline model predictions of the temporal dynamics of plant biomass, fish biomass, and weight of individual fish to real-system data (Figure 4.3). Means of plant biomass, fish biomass, and weight of individual fish predicted by our model are all within 5 percent of the corresponding means based on real-system data, and confidence intervals (represented in Figure 4.3 as ± 2 standard errors of the mean) about model means overlap broadly with confidence intervals about real-system means. Thus on average model predictions of temporal dynamics parallel real-system dynamics closely, and in view of the variation about both model predictions and real system data we can conclude that there are no important differences between predicted and observed values (although we have not conducted formal statistical comparisons of observed and predicted curves). Likewise, variation about model means is generally the same as variation about real-system means. Based on these analyses of time-series val-

TABLE 4.6

Summary of results of two additional stochastic baseline simulations of the pond model run for use in step 4 of model validation, including biomass of plants and fish, and weight of individual fish on November 15 (harvest date)

Replication	Plants (kg/ha)	Fish (kg/ha)	Weight of Individual Fish (kg)
11	19,316	279	3.72
12	22,587	253	3.38

TABLE 4.7

Calculations and results of *t*-tests comparing model predictions of (A) plant biomass, (B) fish biomass, and (C) weight of individual fish on November 15 (harvest date) (from Tables 3.10 and 4.6) and real-system estimates (from Table 3.3)

A. Plant Biomass

Model Predictions (X_{MP}) of Plant Biomass on November 15 (kg/ha)	Real-System Estimates (X_{RSD}) of Plant Biomass on November 15 (kg/ha)
20,147	24,976
24,482	16,212
22,590	21,774
19,081	21,631
19,511	22,154
21,115	19,104
19,795	20,058
24,032	21,797
19,275	
22,510	
19,316	
22,587	

	Model Predictions	Real-System Estimates
Total	254,441 (ΣX_{MP})	167,706 (ΣX_{RSD})
Sample size	12 (n_{MP})	8 (n_{RSD})
Mean	21,203 ($\bar{X}_{MP}$)	20,963 ($\bar{X}_{RSD}$)
ΣX^2	5,437,070,139 (ΣX^2_{MP})	3,561,831,862 (ΣX^2_{RSD})
Total^2/Sample size	5,395,018,540 ($(\Sigma X_{MP})^2/n_{MP}$)	3,515,662,805 ($(\Sigma X_{RSD})^2/n_{RSD}$)
$\Sigma(X - \text{mean})^2$	42,051,599 ($\Sigma(X_{MP} - \bar{X}_{MP})^2$)	46,169,057 ($\Sigma(X_{RSD} - \bar{X}_{RSD})^2$)
Degrees of freedom	11 (df_{MP})	7 (df_{RSD})

Pooled variance (s^2) = (42,051,599 + 46,169,057)/(11 + 7) = 4,901,148

Total degrees of freedom (df) = 11 + 7 = 18

Pooled standard error (s_{MP-RSD}) = $\sqrt{s^2((n_{MP} + n_{RSD})/n_{MP}n_{RSD})}$

= $\sqrt{4{,}901{,}148((12 + 8)/(12)(8))}$ = 1,010

Test statistic (t) = $(\bar{X}_{MP} - \bar{X}_{RSD})/s_{MP-RSD}$ = (21,203 − 20,963)/1,010 = 0.238

Tabulated critical values of the t-distribution indicate that t = 0.238 with 18 degrees of freedom is significant at α = 0.90, but not significant at α = 0.50, and therefore is not significant at the chosen α = 0.05.

TABLE 4.7 *(Continued)*

B. Fish Biomass

Model Predictions (X_{MP}) of Fish Biomass on November 15 (kg/ha)	Real-System Estimates (X_{RSD}) of Fish Biomass on November 15 (kg/ha)
275	222
229	270
253	262
281	263
279	258
268	281
278	276
236	262
280	
255	
279	
253	

Total	3,166 (ΣX_{MP})	2,094 (ΣX_{RSD})
Sample size	12 (n_{MP})	8 (n_{RSD})
Mean	263.8 ($\bar{X}_{MP}$)	261.8 ($\bar{X}_{RSD}$)
ΣX^2	838,956 (ΣX^2_{MP})	550,342 (ΣX^2_{RSD})
Total²/Sample size	835,296 ($(\Sigma X_{MP})^2/n_{MP}$)	548,104 ($(\Sigma X_{RSD})^2/n_{RSD}$)
$\Sigma(X - \text{mean})^2$	3,660 ($\Sigma(X_{MP} - \bar{X}_{MP})^2$)	2,238 ($\Sigma(X_{RSD} - \bar{X}_{RSD})^2$)
Degrees of freedom	11 (df_{MP})	7 (df_{RSD})

Pooled variance (s^2) = (3,660 + 2,238)/(11 + 7) = 328

Total degrees of freedom (df) = 11 + 7 = 18

Pooled standard error (s_{MP-RSD}) $= \sqrt{s^2((n_{MP} + n_{RSD})/n_{MP}n_{RSD})}$

$= \sqrt{328((12 + 8)/(12)(8))} = 8.3$

Test statistic (t) $= (\bar{X}_{MP} - \bar{X}_{RSD})/s_{MP-RSD} = (263.8 - 261.8)/8.3 = 0.241$

Tabulated critical values of the t-distribution indicate that $t = 0.241$ with 18 degrees of freedom is significant at $\alpha = 0.90$, but not significant at $\alpha = 0.50$, and therefore is not significant at the chosen $\alpha = 0.05$.

TABLE 4.7 *(Continued)*

C. Weight of Individual Fish

Model Predictions (X_{MP}) of Weight of Individual Fish on November 15 (kg/ha)	Real-System Estimates (X_{RSD}) of Weight of Individual Fish on November 15 (kg/ha)
3.67	2.97
3.05	3.60
3.37	3.49
3.74	3.51
3.72	3.44
3.58	3.74
3.70	3.67
3.14	3.49
3.73	
3.40	
3.72	
3.38	

Total	42.2 (ΣX_{MP})	27.9 (ΣX_{RSD})
Sample size	12 (n_{MP})	8 (n_{RSD})
Mean	3.52 ($\bar{X}_{MP}$)	3.49 ($\bar{X}_{RSD}$)
ΣX^2	149 (ΣX^2_{MP})	98 (ΣX^2_{RSD})
Total^2/Sample size	148 ($(\Sigma X_{MP})^2/n_{MP}$)	97 ($(\Sigma X_{RSD})^2/n_{RSD}$)
$\Sigma(X - \text{mean})^2$	1 ($\Sigma(X_{MP} - \bar{X}_{MP})^2$)	1 ($\Sigma(X_{RSD} - \bar{X}_{RSD})^2$)
Degrees of freedom	11 (df_{MP})	7 (df_{RSD})

Pooled variance (s^2) $= (1 + 1)/(11 + 7) = 0.111$

Total degrees of freedom (df) $= 11 + 7 = 18$

Pooled standard error (s_{MP-RSD}) $= \sqrt{s^2((n_{MP} + n_{RSD})/n_{MP}n_{RSD})}$

$= \sqrt{0.111((12 + 8)/(12)(8))} = 0.023$

Test statistic (t) $= (\bar{X}_{MP} - \bar{X}_{RSD})/s_{MP-RSD} = (3.52 - 3.49)/0.023 = 1.304$

Tabulated critical values of the t-distribution indicate that $t = 1.304$ with 18 degrees of freedom is significant at $\alpha = 0.20$, but not significant at $\alpha = 0.10$, and therefore is not significant at the chosen $\alpha = 0.05$.

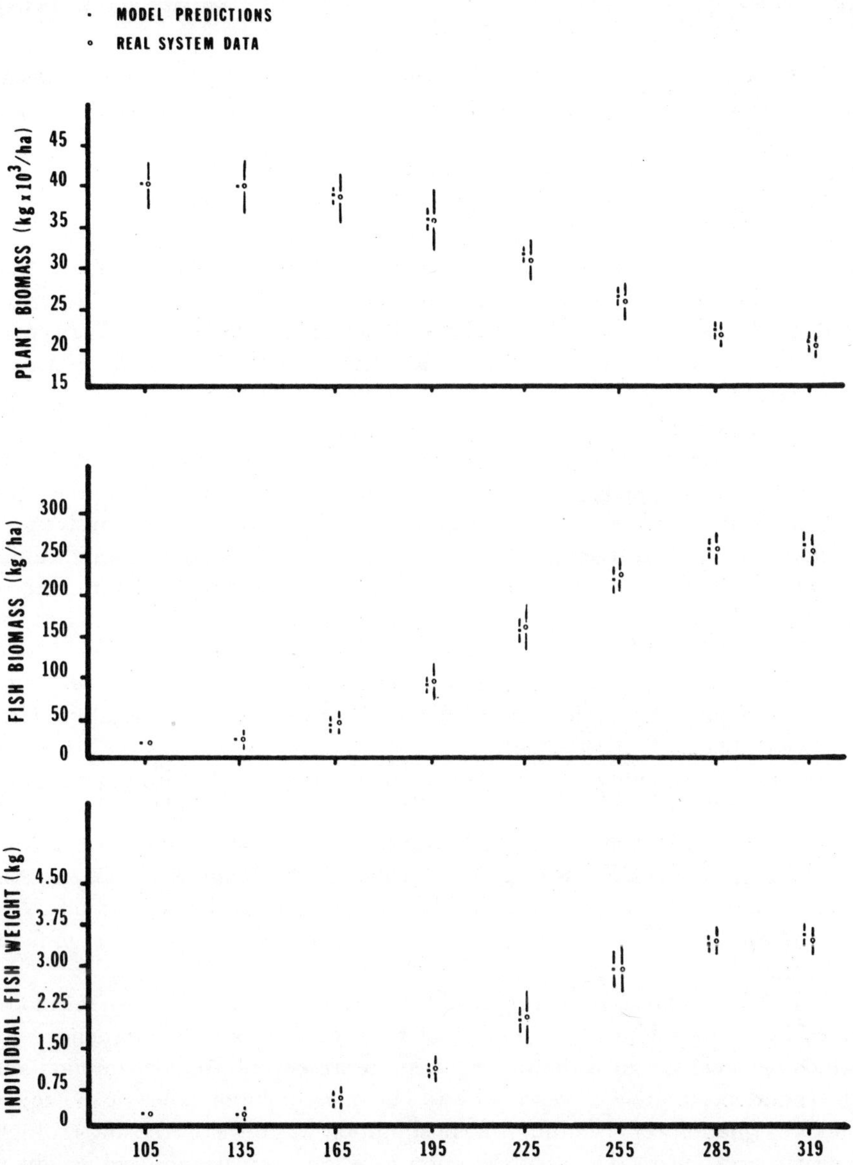

FIGURE 4.3. Comparisons of model predictions of pond dynamics based on 12 replicate stochastic baseline simulations with real-system data from eight different years. Vertical bars represent ± 2 standard errors of the mean.

ues we have no reason to reject the pond model on the grounds that overall model behavior does not correspond quantitatively to real-system behavior.

Sensitivity Analysis of the Model

The fifth and final step in validation of the pond model involves sensitivity analysis of the model. Sensitivity analysis might consist of altering, one at a time, values of each of four material transfer rates ($T_{SO1,1}$, $T_{1,SI1}$, $T_{1,2}$, $T_{2,SI2}$) and one driving variable ($D3$) first by + 10 percent and then by − 10 percent of their baseline values. These four transfer rates and driving variable determine the essential biological dynamics of the pond system. Examination of the sensitivity of model behavior to changes in other variables or transfer rates certainly is legitimate, but as discussed in Section 4.6 we most often confine our attention to selected variables that seem most interesting. With regard to the pond model our main interest is in examining model sensitivity, or responsiveness, to changes in the representation of biologically important processes and subsequent evaluation of the soundness of information used to represent those processes. For sensitivity analysis we will use the deterministic version of the model, in which all random variables are assigned their mean values. Model sensitivity will be assessed by examining effects of the indicated alterations on biomass of plants ($S1$), biomass of fish ($S2$), and weight of individual fish ($A1$) on November 15 (harvest date).

Results of sensitivity analysis indicate that plant biomass, fish biomass, and weight of individual fish are all relatively sensitive to changes in fish consumption ($T_{1,2}$) and that fish biomass and weight of individual fish also are relatively sensitive to changes in assimilation efficiency ($D3$) (Table 4.8). Changes in plant production ($T_{SO1,1}$), plant mortality ($T_{1,SI1}$), and fish respiration ($T_{2,SI2}$) have relatively little effect on plant biomass, fish biomass, or weight of individual fish.

Sensitivity analysis results suggest that confidence in the ability of the model to address our original questions rests in large part on the confidence with which we have specified the rate equation representing fish consumption and the confidence that we have in our estimate of assimilation efficiency. The basic rate equation representing fish consumption was developed using multiple linear regression to relate consumption to water temperature and weight of individual fish based on data from the real system. We have relatively high confidence in our representation of this relationship because 86 percent of variability in observed consumption is accounted for by the regression equation ($r^2 = 0.86$) (p. 100). We also added an additional component to the basic consumption equation to represent decreases in desired consumption resulting from shortages of plant biomass (p. 101), but this additional component affects consumption only at levels of plant biomass (<20,000 kg/ha) lower than those

TABLE 4.8

Results of sensitivity analysis of the deterministic version of the pond model indicating changes relative to baseline in plant biomass, fish biomass, and weight of individual fish on November 15 (harvest date) resulting from changes of + 10 percent and − 10 percent in the indicated material transfer rate or driving variable

Material Transfer Rate or Driving Variable and (Percent Change Relative to Baseline)		Plant Biomass (kg/ha)	Plant Biomass Percent Change Relative to Baseline	Fish Biomass (kg/ha)	Fish Biomass Percent Change Relative to Baseline	Weight of Individual Fish (kg/ha)	Weight of Individual Fish Percent Change Relative to Baseline
Transfer from Source to Plants							
$T_{SO1,1}$	(+ 10)	22,408	+ 4	265	0	3.53	0
	(− 10)	20,684	− 4	265	0	3.53	0
Transfer from Plants to Natural Mortality Sink							
$T_{1,SI1}$	(+ 10)	20,945	− 3	265	0	3.53	0
	(− 10)	22,147	+ 3	265	0	3.53	0
Transfer from Plants to Fish							
$T_{1,2}$	(+ 10)	19,525	− 10	285	+ 8	3.81	+ 8
	(− 10)	24,165	+ 12	234	− 13	3.12	− 13
Transfer from Fish to Respiration Sink							
$T_{2,SI2}$	(+ 10)	21,633	<+ 1	253	− 5	3.38	− 4
	(− 10)	21,533	<− 1	276	+ 4	3.68	+ 4
Assimilation Efficiency							
D3	(+ 10)	21,590	<+ 1	286	+ 8	3.81	+ 8
	(− 10)	22,170	+ 3	234	− 13	3.12	− 13

that usually occur in our simulations. However, our confidence in model behavior under conditions resulting in such low levels of plant biomass for extended periods of simulated time must be evaluated in view of our confidence in the representation of this additional relationship. For our present example we will assume that our confidence in this relationship is high.

With regard to confidence in our estimate of assimilation efficiency, for our present example we have stated simply that assimilation efficiency is known to be 2 percent (p. 89). We will assume that our confidence in this relationship is high.

4.8. EXERCISES

Additional Information

Available information with which to quantitatively compare behavior of our plant–herbivore model with behavior of the real system consists of data on standing crop biomass of live plants and herbivores at selected dates throughout the first year post-introduction of herbivores on those islands where herbivores have been introduced (Table 4.9).

Exercise 4.8.1

Evaluate the qualitative reasonableness of overall plant–herbivore model behavior based on results of the initial baseline simulation (Figure A6.3). Suggest additions to the computer program (Table A6.2) that would prevent material transfer rates from becoming negative. Discuss implications of these additions in terms of our understanding of the processes occurring within the plant–herbivore system and our representation of those processes.

TABLE 4.9

Data on standing crop biomass of live plants and herbivores at selected dates throughout the first year post-introduction of herbivores on islands where herbivores have been introduced

	Standing Crop Biomass (kg/ha)									
	Live Plants on Island Number					Herbivores on Island Number				
Julian Date	1	2	3	4	5	1	2	3	4	5
0	49,761	50,282	49,967	50,140	49,944	1.00	1.00	1.00	1.00	1.00
30	50,885	49,818	50,182	50,518	50,110	1.01	1.00	1.01	1.00	1.00
60	49,926	50,203	50,081	49,921	50,198	1.22	1.22	1.23	1.20	1.23
90	49,673	49,679	49,445	50,120	49,880	1.86	1.83	1.81	1.84	1.84
120	48,169	48,002	48,660	49,650	47,929	3.36	3.32	3.22	3.27	3.32
150	43,552	46,947	42,599	45,385	43,010	6.91	6.62	6.70	6.70	6.85
180	37,999	41,667	36,760	39,441	37,405	15.10	14.36	14.68	14.63	14.98
210	33,088	35,998	31,544	33,944	31,616	22.82	24.84	21.80	23.34	21.88
240	29,834	33,035	27,736	30,251	29,150	20.59	22.70	18.99	20.84	20.04
270	29,861	31,977	27,408	29,087	28,251	20.48	22.02	18.76	19.87	19.33
300	31,688	35,335	30,824	32,642	31,130	21.42	23.84	20.86	22.04	21.00
330	41,770	44,747	39,538	44,278	39,706	24.68	27.85	25.36	26.27	25.00
365	50,718	50,587	50,843	49,974	50,200	25.14	28.61	25.54	26.98	25.70

Exercise 4.8.2

Run a new 1-year baseline simulation with the computer program additions from Exercise 4.8.1 included. Evaluate the qualitative reasonableness of overall model behavior based on results of this new baseline simulation and data in Table 4.9. Run any additional simulations needed to complete examination of the qualitative reasonableness of model behavior.

Exercise 4.8.3

Change the model such that environmental temperature ($D1$) is represented as a random variable

$$D1 = \mathrm{RV1}$$

where RV1 = normal random variate from distribution with

$$\mu = 40 \sin(\pi t/365), \ \sigma^2 = 2$$

Assuming that a practically significant difference in herbivore biomass between model predictions and real-system data is 2.5 kg/ha ($\delta = 2.5$), statistically compare stochastic model predictions of herbivore biomass on the last day of the year ($t = 365$) with real-system data (Table 4.9). Also compare stochastic model predictions of live plant biomass on the last day of the year with real-system data.

Exercise 4.8.4

Conduct an appropriate sensitivity analysis on the deterministic version of the model.

Answers to Chapter 4 exercises are in Appendix 6.

CHAPTER 5

MODEL USE

5.1. INTRODUCTION

The fourth and final phase of systems analysis is use of the model to meet objectives of the modeling project. There is a variety of legitimate modeling objectives as diverse as the objectives of the projects with which models are associated. Models may be useful in educational, research, or management settings. Model use may include:

1. providing a sound conceptual framework for future research,
2. evaluating alternative hypotheses about system structure or function,
3. describing system behavior under normal conditions,
4. predicting system response to specific management schemes or environmental situations, or
5. heuristically exploring the dynamics of the system of interest.

We will focus primarily on the use of models within a management-oriented framework and on the role of models in providing information for the decision-making process. Within this narrower focus we might identify four steps in model use:

1. identifying the management policies or environmental situations to be evaluated and representing them in the model (Section 5.2),
2. developing the experimental design for simulations (Section 5.3),
3. analyzing and interpreting simulation results (Section 5.4), and
4. further examining selected types of management policies or environmental situations (Section 5.5).

This general scheme of model use follows exactly the steps involved in addressing a question through experimentation in the real world.

5.2. IDENTIFYING MANAGEMENT POLICIES OR ENVIRONMENTAL SITUATIONS TO BE EVALUATED AND REPRESENTING THEM IN THE MODEL

The first step in model use is to identify the specific management policies or environmental situations that are to be addressed. The modeler necessarily has identified questions concerning these policies or situations and criteria upon which answers to these questions will be based at the beginning of the modeling project. These questions have guided every phase of model development. Likewise, specific representation of alternative policies or situations in the model has been anticipated by the modeler from the beginning and the model has been constructed in a manner that facilitates representation of each alternative.

Generally modifications to the model required to represent alternative policies or situations fall into one or more of the following categories:

1. changes in values of driving variables,
2. changes in values of parameter estimates for rate equations,
3. changes in functional forms of rate equations, and/or
4. addition of new mechanisms to the model.

The first three, and particularly the first two, of these types of modifications are common in sensitivity analysis of the model, although interpretation of results is somewhat different within the framework of model use.

As an example of model modifications required during model use, recall our model predicting fluctuations in weight of an animal over time (Figure 3.2) (Table 3.1). We used one deterministic version of this model described in Section 3.5 to simulate weight fluctuations of an initially 100 g animal over a 30-day period assuming that environmental temperatures decrease linearly from 20°C to 15°C during the 30 days. The temperature driving variable ($D1$) equation for this version of the model is

$$D1 = 20 - 0.167t$$

where t = day of simulated time (p. 85). Suppose that we wish to compare animal weight fluctuations over two 30-day periods during which temperatures decrease linearly from 21°C to 16°C and from 19°C to 14°C, respectively, with fluctuations under our original (baseline) conditions. Model modifications required to represent these alternative environmental situations include changing only the driving variable equation representing temperature. To represent the warmer situation it becomes

$$D1 = 21 - 0.167t$$

and to represent the cooler situation it becomes

$$D1 = 19 - 0.167t$$

Suppose that we wish to compare weight fluctuations under baseline conditions with weight fluctuations of an animal given a drug that increases consumption per unit of body weight by 50 percent. To represent the drugged animal we reparameterize the rate equation representing consumption from

$$T_{S01,1} = 0.05S1$$

(Table 3.1) to

$$T_{S01,1} = 0.075S1$$

Suppose that the effect of a second drug increases consumption per unit of body weight by 50 percent for a 100 g animal, but the percentage increase in consumption per unit weight decreases linearly as an animal grows larger until, at 200 g, there is no effect. To represent weight fluctuations of an animal given this second drug we must change the functional form of the consumption equation to

$$T_{S01,1} = (0.10 - 0.00025S1)S1$$

This new functional form represents consumption under the influence of the second drug as long as the animal weighs less than or equal to 200 g. But above 200 g the equation predicts that consumption per unit weight is less than baseline (a negative effect rather than no effect), which is not correct. Thus if we wish to simulate conditions under which animal weight might exceed 200 g, we must add a new mechanism that prevents this erroneous decrease in consumption per unit weight. This can be done simply by using two equations to represent consumption, one for animals weighing less than or equal to 200 g,

$$T_{S01,1} = (0.10 - 0.00025S1)S1, \quad \text{if } S1 \leq 200$$

and one for animals weighing more than 200 g,

$$T_{S01,1} = 0.05S1, \quad \text{if } S1 > 200$$

5.3. DEVELOPING THE EXPERIMENTAL DESIGN FOR SIMULATIONS

The second step in model use is to develop the experimental design for simulations. The same principles that apply to the design of experiments conducted in the real world (Cochran and Cox 1957) apply to experiments conducted, or simulated, on the computer. A primary consideration is whether the model is deterministic or stochastic. If the model is deterministic, comparison of system behavior in response to alternative management policies or environmental situations is non-statistical and is based on one simulation of each different alternative.

If the model is stochastic, comparison of system behavior in response to alternative management policies or environmental situations is based on statistical analysis of a set of replicate simulations of each alternative. The question of how many replicate simulations of each alternative to run is directly analogous to the real-world question of how large a sample is needed. We discussed a method for determining the appropriate sample size in Section 3.5 when talking about stochastic baseline simulations. In practice the number of replicate simulations needed for comparisons of alternatives is usually established when running the baseline simulations during the last step of quantitative specification of the model.

If we wish to use the stochastic version of our weight fluctuation model described in Section 3.5 to compare animal weights on day 30 under the baseline, warmer, and cooler conditions mentioned above, we determine the number of replicate simulations to run (n) using the sample size formula presented in Section 3.5:

$$n \geq 2\left(\frac{\sigma}{\delta}\right)^2 (t_{\alpha,\gamma} + t_{2(1-P),\gamma})^2$$

If we want to detect a true difference of $\delta = 3$ g between weights on day 30 resulting from alternative environmental situations at a statistical significance level $\alpha = 0.05$ with probability $P = 0.80$ that the difference will be detected if it exists, we find that 14 replicate simulations are needed:

$$14 \geq 2\left(\frac{2.67}{3}\right)^2 (t_{0.05,39} + t_{2(1-0.80),39})^2$$

$$14 \geq 2\left(\frac{2.67}{3}\right)^2 (2.021 + 0.851)^2$$

$$14 \geq 13.07$$

Recall that the inherent variance (σ^2) of weight on day 30 under baseline conditions is 7.14 (p. 87) and therefore $\sigma = \sqrt{7.14} = 2.67$. Also recall that $\gamma = b(n - 1)$, where b is the number of situations compared (p. 86). Therefore $\gamma = 3(14 - 1) = 39$. Thus for our weight fluctuation model we need the same number of replicate simulations to compare three alternative environmental situations that we need to compare model predictions to real world data during model validation (p. 140). Note however that the sample size needed for model validation may differ from that needed for model use, as is the case with our pond model example (Section 3.6 and Section 4.7).

5.4. ANALYZING AND INTERPRETING SIMULATION RESULTS

The third step in model use is to analyze and interpret simulation results to address the original questions posed at the beginning of the modeling project. The same principles that apply to analysis of real-world experimental data apply to analysis of simulated data. Again a primary consideration is whether the model is deterministic or stochastic. General types of analyses possible for deterministic versus stochastic models are similar to those discussed in Section 4.5 on model validation.

If the model is deterministic, analysis of system behavior in response to management policies or environmental situations is non-statistical. Comparison of single-value model predictions consists of ordinating alternative predictions and assessing the practical significance of differences between them. Comparison of time-series predictions consists of assessing the practical significance of differences with regard to the timing of maximum and minimum values and the amplitude and periodicity of fluctuations within the time-series. Interpretation of results is solely in terms of the magnitude of practically significant differences established by the modeler.

If the model is stochastic, analysis of system behavior in response to alternative management policies or environmental situations is statistical. Comparison of single-value model predictions consists of analysis of variance testing the hypothesis that the mean of model predictions generated under each alternative is the same. If analysis of variance indicates statistically significant differences between alternatives, a multiple comparison (Snedecor and Cochran 1967, p. 271), such as a Duncan's Multiple Range Test (Ott 1984, p. 376), is used to identify statistically significant differences between specific alternatives. This assumes that more than two alternatives are compared. If only two are compared a two-sample t-test is used as described in Section 4.5. Interpretation of results is in terms of both the magnitude and statistical significance of differences in model predictions between alternatives.

Analysis of variance of animal weights on day 30 of simulated time predicted by our stochastic weight fluctuation model under baseline, warmer, and cooler environmental situations is presented in Table 5.1. Mean weights predicted on day 30 under baseline, warmer, and cooler situations are 119.9 g, 110.9 g, and 130.4 g, respectively. Model predictions were generated using the model equations in Table 3.1, except that the temperature driving variable ($D1$) equation was represented as a stochastic function of day of simulated time (t) as described in Section 3.5:

$$D1 = \mathrm{RV}$$

where RV is a normal random variate from a distribution with mean

$$\mu = 20 - 0.167t$$

TABLE 5.1

Analysis of variance of animal weights on day 30 of simulated time predicted by the weight fluctuation model (Figure 3.2) under baseline, warmer, and cooler environmental situations

Model Predictions (X) of Weight (g) on Day 30

	Baseline Situation	Warmer Situation	Cooler Situation
	122.3	107.9	125.6
	122.6	111.4	129.8
	120.6	112.2	132.2
	123.4	106.2	135.8
	121.1	107.2	129.8
	120.2	113.1	133.0
	120.4	110.5	131.8
	114.8	109.1	132.6
	119.1	114.0	127.2
	119.4	110.4	131.6
	119.3	115.2	126.6
	118.4	110.7	128.8
	119.1	114.4	128.6
	117.4	110.4	133.0
Class totals ($T = \Sigma X$)	1678.1	1552.7	1826.4

Grand total ($G = \Sigma T$) = 5057.2

Number of classes (a) = 3

Number of observations per class (n) = 14

Correction for the mean ($C = G^2/an$) = $(5057.2)^2/(3)(14)$ = 608,935.0

Analysis of Variance Table

Source of Variation	Degrees of Freedom (df)	Sum of Squares (SS)	Mean Square (MS)
Between classes	$a - 1 = 2$	$(\Sigma T^2/n) - C = 2681.7$	SS/df = 1340.85
Within classes	$a(n - 1) = 39$	subtract = 268.2	SS/df = 6.88
Total	$an - 1 = 41$	$\Sigma\Sigma X^2 - C = 2949.9$	

$$\text{Test statistic } (F) = \frac{MS_{\text{between classes}}}{MS_{\text{within classes}}} = \frac{1340.85}{6.88} = 194.89$$

Tabulated critical values of the F-distribution indicate that $F = 194.89$ with 2 degrees of freedom for the numerator MS ($MS_{\text{between classes}}$) and 39 degrees of freedom for the denominator MS ($MS_{\text{within classes}}$) is significant at $\alpha = 0.001$.

TABLE 5.2

Duncan's Multiple Range Test indicating specific statistically significant differences in animal weights on day 30 predicted by the weight fluctuation model (Figure 3.2) under baseline, warmer, and cooler environmental situations

Step 1. Rank class means:

Warmer situation (110.9), baseline situation (119.9), cooler situation (130.4)

Step 2. Fill in table:

r	2	3	4 . . . n
$q_{\alpha}(r, \text{df}_{\text{within classes}})$			
W_r			

where r = number of steps by which two means are separated

$q_{\alpha}(r, \text{df}_{\text{within classes}})$ = critical value of the Studentized range required for Duncan's procedure when means being compared are r steps apart

$W_r = q_{\alpha}(r, \text{df}_{\text{within classes}}) \sqrt{\text{MS}_{\text{within classes}}/n}$

For our example from Table 5.1:

$\text{df}_{\text{within classes}} = 39$

$\text{MS}_{\text{within classes}} = 6.88$

$n = 14$

let's choose $\alpha = 0.05$

Therefore the table becomes:

	r = 2	r = 3
$q_{0.5}(r, 39)$	2.86	3.01
W_r	2.00	2.11

since

$$W_2 = q_{0.05}(2, 39)\sqrt{6.88/14} = (2.86)\sqrt{6.88/14} = 2.00$$

$$W_3 = q_{0.05}(3, 39)\sqrt{6.88/14} = (3.01)\sqrt{6.88/14} = 2.11$$

Step 3. Test for significant differences between class means: (two class means are significantly different at the indicated α level if the magnitude of difference between them exceeds W_r):

For our example

Cooler situation − warmer situation = 130.4 − 110.9 = 19.5, W_3 = 2.11, 19.5 > 2.11; therefore means are significantly different at α = 0.05

Cooler situation − baseline situation = 130.4 − 119.9 = 10.5, W_2 = 2.00, 10.5 > 2.00; therefore means are significantly different at α = 0.05

Baseline situation − warmer situation = 119.9 − 110.9 = 9.0, W_2 = 2.00, 9.0 > 2.00; therefore means are significantly different at α = 0.05

and variance

$$\sigma^2 = 2.0006$$

(2.0006 is the MSE of the regression of temperature on day, Figure 3.7) for simulations of the baseline situation, with

$$\mu = 21 - 0.167t \qquad \text{and} \qquad \sigma^2 = 2.0006$$

for simulations of the warmer situation, and

$$\mu = 19 - 0.167t \qquad \text{and} \qquad \sigma^2 = 2.0006$$

for simulations of the cooler situation. Results of analysis of variance suggest statistically significant ($P < 0.001$) differences in animal weights on day 30 between alternative environmental situations. Since there are significant differences between alternatives, we use Duncan's Multiple Range Test to determine significant differences between specific alternatives (Table 5.2). Results of Duncan's Test indicate that animal weights on day 30 predicted under each alternative environmental situation are statistically significantly ($P < 0.05$) different from weights predicted under each of the other alternatives.

Statistical comparison of time-series model predictions is more complicated than is the case for single-value predictions. As discussed in Section 4.5 exploration of problems associated with analysis of time-series and presentation of possible ways to overcome them in specific situations are beyond the scope of this book. The procedure suggested as useful during model validation—graphically representing the variance around each mean value in each time series that we wish to compare—is equally useful during model use. But again such representation does not imply a statistical test.

5.5. FURTHER EXAMINING SELECTED TYPES OF MANAGEMENT POLICIES OR ENVIRONMENTAL SITUATIONS

The fourth step in model use involves further examination of selected types of management policies or environmental situations found most interesting based on results of initial model use simulations. At this point we should have answers to initial questions specified during conceptual model formulation. However, often initial model use simulations raise as many questions as they answer. We may learn something new about the behavior of the system of interest through simulation that suggests a new line of questions relevant to our interests or allows us to refine our original questions to obtain more precise answers. This is exactly the iterative procedure of analysis and redesign upon which all scientific experimentation is based. The new set of questions that we wish to address may necessitate fundamental changes in our model or may even suggest that we must redefine our system of interest. In such cases we return to phase one of systems analysis—conceptual model formulation—and repeat the entire model building procedure. But often our new questions can be addressed using essentially the same model slightly modified to represent a new variant of the management policies or environmental situations that we represented in our original model use simulations.

Results of initial simulations with our weight fluctuation model, in addition to answering our initial question concerning possible differences in weights of animals exposed to three different environmental situations for 30 days (p. 169), may raise a new question concerning identification of temperature regimes required for animals to reach various weights after 30 days. This type of question can be addressed with only minor modification of our present model. We need only modify the temperature driving variable equation to represent the new temperature regimes that we wish to simulate.

Often this step in model use consists of two parts:

1. an exploratory examination of system behavior under relatively many variants of a few original management policies or environmental situations and
2. detailed description and/or formal comparison of system behavior under selected new variants.

During exploratory simulations we are interested primarily in identifying trends in system behavior and not in detailed description or formal comparison of system behavior under all the management or environmental variants simulated. If we have a stochastic model it is more efficient to conduct these exploratory simulations using a deterministic version of the model in which all random variables are assigned their mean values. Trends are clearer without the inherent variability of the stochastic model and fewer simulations are required.

After examining results of exploratory simulations we may identify certain new management policies or environmental situations that we wish to describe in detail and/or compare formally. Formal comparison requires the development of a new experimental design for simulations. For deterministic models we will have run all simulations required by the new experimental design as part of the exploratory analysis. But for stochastic models we must recalculate the sample size, or the number of replicate stochastic simulations, required by the new experimental design, run stochastic simulations, and conduct appropriate statistical analyses as described in Section 5.4.

Suppose that as a result of initial simulations with our weight fluctuation model we become interested in establishing the relationship between temperature on day 0 and animal weight on day 30, assuming that the rate of temperature decline over the 30-day period remains the same as in initial simulations. We establish this relationship by using the deterministic version of our model to simulate a series of temperature regimes. In each of these exploratory simulations we change only the intercept on the temperature driving variable ($D1$) equation, which represents temperature on day 0, and record animal weight on day 30. Results of such simulations indicate that weight on day 30 decreases almost linearly (actually exponentially) as temperature on day 0 increases linearly (Figure 5.1). If we want to describe system behavior under some of the new temperature regimes in more detail or formally compare system behavior

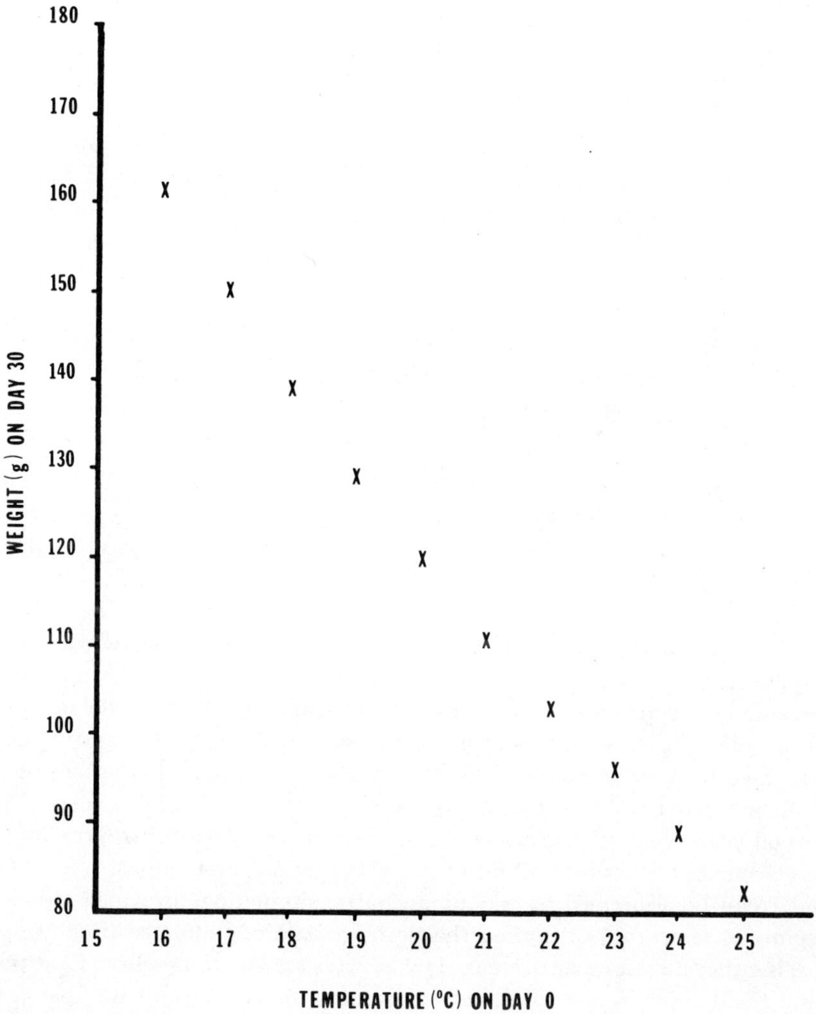

FIGURE 5.1. Results of exploratory simulations using the deterministic version of the weight fluctuation model (Figure 3.2).

under certain new regimes, we develop the appropriate experimental design and run the needed stochastic simulations.

5.6. EXAMPLE OF MODEL USE: HYPOTHETICAL POND MODEL

Identifying Management Policies or Environmental Situations to Be Evaluated and Representing Them in the Model

Five questions to be addressed by the pond model were presented in Section 2.6 when bounding the system of interest during conceptual model formulation:

1. What size harvest might be expected in a normal year under the usual scheme of stocking 75 0.227-kg fish on April 15 and harvesting them on November 15 (the baseline situation) and what profit is associated with this harvest?
2. How variable is the harvest (and therefore profit) from year to year under the baseline stocking and harvesting scheme as the result of inherent year-to-year variability of the system?
3. Would size of the harvest and associated profit be changed by harvesting a month earlier or a month later?
4. Would size of the harvest and associated profit be changed by stocking a month earlier or a month later?
5. Would size of the harvest and associated profit be changed by stocking twice as many fish?

The capability of the model to address these five questions was discussed and the specific representation of each management policy to be evaluated was presented in Section 4.7 when examining the capability of the model to address the problem of interest during model validation. Questions 1 and 2 can be addressed by a set of stochastic baseline simulations. Mean harvest and mean profit predicted by the simulations are the harvest and profit expected in a normal year. Year-to-year variability of harvest and profit are represented by the variances of harvest and profit predicted by the simulations. Questions 3 and 4 can be addressed by sets of stochastic simulations in which either the beginning (stocking) or ending (harvesting) date of simulated time (t_{begin} or t_{end}) is either increased or decreased by 30 days relative to baseline. Question 5 can be addressed by a set of stochastic simulations in which the number of fish ($D2$) is doubled relative to baseline. Thus six versions of the model will be used to address the questions of interest:

1. baseline,
2. early harvest,
3. late harvest,
4. early stocking,
5. late stocking, and
6. double stocking density.

The primary output of interest is profit, although plant and fish biomass and average weight of individual fish are also of interest. All four outputs will be treated as single-value outputs because we are most interested in their values on the harvest day.

Developing the Experimental Design for Simulations

The experimental design for simulations will consist of simulating system behavior under each of the six management policies in which we are interested. Since our model is stochastic we must also specify the number of replicate simulations of each policy to run. Recall from Section 3.6 that the pond manager would like to show that a true difference of 25 kg in harvest between different management policies is statistically significant at the 95 percent significance level ($\alpha = 0.05$) with an 80 percent probability that the difference will be detected if it exists ($P = 0.80$). In Section 3.6 we determined that ten replicate simulations of each policy are necessary to meet this objective. Our experimental design will consist of 60 simulations: ten stochastic replicates of each of the six different versions of the model. For the baseline situation we will use the ten stochastic replicates that we have run already (Table 3.10).

Analyzing and Interpreting Simulation Results

Analyzing Results

Results of simulations of the alternative management policies with regard to outputs of interest—profit, plant and fish biomass, and average weight of individual fish—are summarized in Table 5.3. Results of analyses of variance indicate statistically significant ($P < 0.0001$) management policy effects on biomass of plants, biomass of fish, and weight of individual fish at harvest (Table 5.4). Duncan's Multiple Range Tests indicate that the management policy doubling stocking density results in a significantly ($P < 0.05$) lower biomass of plants, a significantly ($P < 0.05$) higher biomass of fish, and a significantly ($P < 0.05$) lower weight of individual fish than any other policy (Table 5.5). Other policies are not significantly ($P > 0.05$) different from each other with regard to biomass of plants or fish, or weight of individual fish.

TABLE 5.3
Summary of results of ten replicate stochastic simulations of the pond model under each management policy, including biomass of plants and fish, weight of individual fish, and profit on the harvest date

Management Policy	Replication	Plants (kg/ha)	Fish (kg/ha)	Weight of Individual Fish (kg)	Profit (dollars)[a]
Baseline	1	20,147	275	3.67	168
	2	24,482	229	3.05	122
	3	22,590	253	3.37	146
	4	19,081	281	3.74	173
	5	19,511	279	3.72	171
	6	21,115	268	3.58	161
	7	19,795	278	3.70	170
	8	24,032	236	3.14	128
	9	19,275	280	3.73	172
	10	22,510	255	3.40	147
	$\bar{X}(\pm 1SE)$	21,254($\pm$637)	263($\pm$6)	3.51($\pm$0.08)	156($\pm$6)
Early stocking (Mar. 15)	1	21,633	256	3.41	133
	2	18,867	277	3.70	154
	3	25,353	206	2.75	−123
	4	21,163	261	3.48	138
	5	22,914	241	3.22	118
	6	24,084	225	3.00	102
	7	22,719	244	3.26	121
	8	22,717	244	3.26	121
	9	21,999	253	3.37	130
	10	20,377	269	3.58	146
	$\bar{X}(\pm 1SE)$	22,183($\pm$581)	248($\pm$7)	3.30($\pm$0.09)	129($\pm$5)
Late stocking (May 15)	1	23,534	251	3.34	158
	2	22,478	262	3.49	170
	3	21,054	274	3.65	181
	4	23,658	249	3.32	156
	5	22,963	257	3.42	164
	6	24,063	244	3.26	152
	7	21,651	269	3.59	177
	8	23,600	250	3.33	157
	9	21,605	269	3.59	177
	10	20,851	276	3.67	183
	$\bar{X}(\pm 1SE)$	22,546($\pm$374)	260($\pm$4)	3.47($\pm$0.05)	168($\pm$4)

TABLE 5.3 *(Continued)*

Management Policy	Replication	Plants (kg/ha)	Fish (kg/ha)	Weight of Individual Fish (kg)	Profit (dollars)[a]
Early harvest (Oct. 15)	1	24,304	244	3.26	152
	2	22,314	265	3.53	172
	3	24,311	244	3.26	152
	4	19,691	280	3.73	187
	5	27,259	203	2.70	-92
	6	21,728	269	3.59	177
	7	24,029	248	3.30	155
	8	25,522	229	3.05	136
	9	23,178	257	3.42	164
	10	22,727	261	3.48	169
	$\bar{X}(\pm 1SE)$	23,506(±663)	250(±7)	3.33(±0.09)	162(±5)
Late harvest (Dec. 15)	1	17,643	276	3.68	153
	2	20,213	263	3.51	141
	3	25,840	192	2.55	-122
	4	17,672	275	3.67	152
	5	21,432	252	3.36	130
	6	20,674	260	3.46	137
	7	22,917	234	3.12	112
	8	21,531	251	3.34	128
	9	22,878	233	3.11	111
	10	21,380	252	3.36	130
	$\bar{X}(\pm 1SE)$	21,218(±773)	249(±8)	3.32(±0.10)	133(±5)
Double stocking density (150 fish)	1	11,012	360	2.40	-108
	2	8,852	337	2.25	-108
	3	8,337	328	2.19	-108
	4	10,108	353	2.35	-108
	5	8,752	336	2.24	-108
	6	15,299	349	2.32	-108
	7	10,794	359	2.39	-108
	8	10,259	355	2.36	-108
	9	8,163	324	2.16	-108
	10	15,191	350	2.37	-108
	$\bar{X}(\pm 1SE)$	10,677(±824)	345(±4)	2.30(±0.03)	—

[a] $\bar{X}(\pm 1SE)$ for profit does not include negative profits.

TABLE 5.4

Analyses of variance comparing results of the pond model under alternative management policies with regard to (A) plant biomass, (B) fish biomass, and (C) weight of individual fish at harvest. Calculations are based on data in Table 5.3

A. Plant Biomass

Model Predictions (X) of Plant Biomass (kg/ha) at Harvest

	Baseline	Early Stocking	Late Stocking	Early Harvest	Late Harvest	Double Stocking Density
	20,147	21,633	23,534	24,304	17,643	11,012
	24,482	18,867	22,478	22,314	20,213	8,852
	22,590	25,353	21,054	24,311	25,840	8,337
	19,081	21,163	23,658	19,691	17,672	10,108
	19,511	22,914	22,963	27,259	21,432	8,752
	21,115	24,084	24,063	21,728	20,674	15,299
	19,795	22,719	21,651	24,029	22,917	10,794
	24,032	22,717	23,600	25,522	21,531	10,259
	19,275	21,999	21,605	23,178	22,878	8,163
	22,510	20,377	20,851	22,727	21,380	15,191
Class totals ($T = \Sigma X$)	212,538	221,826	225,457	235,063	212,180	106,767

Grand total ($G = \Sigma T$) = 1,213,831
Number of classes (a) = 6
Number of observations per class (n) = 10
Correction for the mean ($C = G^2/an$) = $(1{,}213{,}831)^2/(6)(10) = 2.4556 \times 10^{10}$

Analysis of Variance Table

Source of Variation	Degrees of Freedom (df)	Sum of Squares (SS)	Mean Square (MS)
Between classes	$a - 1 = 5$	$(\Sigma T^2/n) - C = 1.1320 \times 10^9$	SS/df = 2.2640×10^8
Within classes	$a(n - 1) = 54$	subtract = 2.3413×10^8	SS/df = 4.3558×10^6
Total	$an - 1 = 59$	$\Sigma\Sigma X^2 - C = 1.3661 \times 10^9$	

$$\text{Test statistic } (F) = \frac{MS_{\text{between classes}}}{MS_{\text{within classes}}} = \frac{2.2640 \times 10^8}{4.3358 \times 10^6} = 52.22$$

Tabulated critical values of the F-distribution indicate that $F = 52.22$ with 5 df for the numerator MS ($MS_{\text{between classes}}$) and 54 df for the denominator MS ($MS_{\text{within classes}}$) is significant at $\alpha = 0.0001$.

TABLE 5.4 *(Continued)*

B. Fish Biomass

Model Predictions (X) of Fish Biomass (kg/ha) at Harvest

	Baseline	Early Stocking	Late Stocking	Early Harvest	Late Harvest	Double Stocking Density
	275	256	251	244	276	360
	229	277	262	265	263	337
	253	206	274	244	192	328
	281	261	249	280	275	353
	279	241	257	203	252	336
	268	225	244	269	260	349
	278	244	269	248	234	359
	236	244	250	229	251	355
	280	253	269	257	233	324
	255	269	276	261	252	350
Class totals ($T = \Sigma X$)	2634	2476	2601	2500	2488	3451

Grand total ($G = \Sigma T$) = 16,150

Number of classes (a) = 6

Number of observations per class (n) = 10

Correction for the mean ($C = G^2/an$) = $(16{,}150)^2/(6)(10)$ = 4,347,042

Analysis of Variance Table

Source of Variation	Degrees of Freedom (df)	Sum of Squares (SS)	Mean Square (MS)
Between classes	$a - 1 = 5$	$(\Sigma T^2/n) - C = 71{,}287$	SS/df = 14,257
Within classes	$a(n - 1) = 54$	subtract = 19,764	SS/df = 366
Total	$an - 1 = 59$	$\Sigma\Sigma X^2 - C = 91{,}051$	

$$\text{Test statistic } (F) = \frac{MS_{\text{between classes}}}{MS_{\text{within classes}}} = \frac{14{,}257}{366} = 38.95$$

Tabulated critical values of the F-distribution indicate that $F = 38.95$ with 5 df for the numerator MS and 54 df for the denominator MS is significant at $\alpha = 0.0001$.

TABLE 5.4 *(Continued)*

C. Weight of Individual Fish

Model Predictions (X) of Weight of Individual Fish (g) at Harvest

Baseline	Early Stocking	Late Stocking	Early Harvest	Late Harvest	Double Stocking Density
3.67	3.41	3.34	3.26	3.68	2.40
3.05	3.70	3.49	3.53	3.51	2.25
3.37	2.75	3.65	3.26	2.55	2.19
3.74	3.48	3.32	3.73	3.67	2.35
3.72	3.22	3.42	2.70	3.36	2.24
3.58	3.00	3.26	3.59	3.46	2.32
3.70	3.26	3.59	3.30	3.12	2.39
3.14	3.26	3.33	3.05	3.34	2.36
3.73	3.58	3.67	3.48	3.36	2.37
3.40	3.37	3.59	3.42	3.11	2.16

Class totals ($T = \Sigma X$): 35.10, 33.03, 34.66, 33.32, 33.16, 23.03

Grand total ($G = \Sigma T$) = 192.30

Number of classes (a) = 6

Number of observations per class (n) = 10

Correction for the mean ($C = G^2/an$) = $(192.30)^2/(6)(10)$ = 616.32

Analysis of Variance Table

Source of Variation	Degrees of Freedom (df)	Sum of Squares (SS)	Mean Square (MS)
Between classes	$a - 1 = 5$	$(\Sigma T^2/n) - C = 10.128$	SS/df = 2.0256
Within classes	$a(n - 1) = 54$	subtract = 3.329	SS/df = 0.0616
Total	$an - 1 = 59$	$\Sigma\Sigma X^2 - C = 13.457$	

$$\text{Test statistic } (F) = \frac{MS_{\text{between classes}}}{MS_{\text{within classes}}} = \frac{2.0256}{0.0616} = 32.88$$

Tabulated critical values of the F-distribution indicate that $F = 32.88$ with 5 df for the numerator MS and 54 df for the denominator MS is significant at $\alpha = 0.0001$.

TABLE 5.5

Duncan's Multiple Range Tests indicating specific statistically significant differences among pond model results under alternative management policies with (A) plant biomass, (B) fish biomass, and (C) weight of individual fish at harvest. Calculations follow steps outlined in Table 5.2 and are based on data in Table 5.4

A. Plant Biomass

Step 1. Rank class means:

Double stocking density (10,677), late harvest (21,218), baseline (21,254), early stocking (22,183), late stocking (22,546), early harvest (23,506)

Step 2. Fill in table:

$df_{\text{within classes}} = 54$

$MS_{\text{within classes}} = 4.3358 \times 10^6$

$n = 10$

$\alpha = 0.05$

$W_r = q_\alpha(r, df_{\text{within classes}}) \sqrt{MS_{\text{within classes}}/n}$

where r = number of steps by which two means are separated

$q_\alpha(r, df_{\text{within classes}})$ = critical value of the Studentized range required for Duncan's procedure when means being compared are r steps apart

$$W_2 = q_{0.05}(2, 54)\sqrt{4.3358 \times 10^6/10} = (2.83)(2082) = 5892$$
$$W_3 = q_{0.05}(3, 54)\sqrt{4.3358 \times 10^6/10} = (2.98)(2082) = 6204$$
$$W_4 = q_{0.05}(4, 54)\sqrt{4.3358 \times 10^6/10} = (3.08)(2082) = 6412$$
$$W_5 = q_{0.05}(5, 54)\sqrt{4.3358 \times 10^6/10} = (3.14)(2082) = 6537$$
$$W_6 = q_{0.05}(6, 54)\sqrt{4.3358 \times 10^6/10} = (3.20)(2082) = 6662$$

Thus

	$r = 2$	$r = 3$	$r = 4$	$r = 5$	$r = 6$
$q_{0.05}(r, 54)$	2.83	2.98	3.08	3.14	3.20
W_r	5892	6204	6412	6537	6662

Step 3. Test for significant differences between class means (two class means are significantly different at the indicated α level if the magnitude of difference between them exceeds W_r):

Early harvest – double stocking density = 23,506 – 10,677 = 12,829, W_6 = 6662, 12,829 > 6662; therefore means are significantly different at $\alpha = 0.05$

Late stocking – double stocking density = 22,546 – 10,667 = 11,869, W_5 = 6537, 11,869 > 6537; therefore means are significantly different at $\alpha = 0.05$

Early stocking – double stocking density = 22,183 – 10,667 = 1506, W_4 = 6412, 11,506 > 6412; therefore means are significantly different at $\alpha = 0.05$

Baseline – double stocking density = 21,254 – 10,677 = 10,577, W_3 = 6204, 10,577 > 6204; therefore means are significantly different at $\alpha = 0.05$

Late harvest – double stocking density = 21,218 – 10,677 = 10,541, W_2 = 5892, 10,541 > 5892; therefore means are significantly different at $\alpha = 0.05$

TABLE 5.5 *(Continued)*

A. Plant Biomass

Early harvest − late harvest = 23,506 − 21,218 = 2288, W_5 = 6537, 2288 < 6537; therefore means are not significantly different at α = 0.05, thus all remaining pairs of means between early harvest and late harvest must not be significantly different from each other at α = 0.05

B. Fish Biomass

Step 1. Rank class means:

Early stocking (248), late harvest (249), early harvest (250), late stocking (260), baseline (263), double stocking density (345)

Step 2. Fill in table:

$df_{\text{within classes}} = 54$

$MS_{\text{within classes}} = 366$

$n = 10$

$\alpha = 0.05$

$W_2 = q_{0.05}(2, 54)\sqrt{366/10} = (2.83)(6.05) = 17.12$

$W_3 = q_{0.05}(3, 54)\sqrt{366/10} = (2.98)(6.05) = 18.03$

$W_4 = q_{0.05}(4, 54)\sqrt{366/10} = (3.08)(6.05) = 18.63$

$W_5 = q_{0.05}(5, 54)\sqrt{366/10} = (3.14)(6.05) = 19.00$

$W_6 = q_{0.05}(6, 54)\sqrt{366/10} = (3.20)(6.05) = 19.36$

Thus

	$r = 2$	$r = 3$	$r = 4$	$r = 5$	$r = 6$
$q_{0.05}(r, 54)$	2.83	2.98	3.08	3.14	3.20
W_r	17.12	18.03	18.63	19.00	19.36

Step 3. Test for significant differences between class means:

Double stocking density − early stocking = 345 − 248 = 97, W_6 = 19.36, 97 > 19.36; therefore means are significantly different at α = 0.05

Baseline − early stocking = 263 − 248 = 15, W_5 = 19.00, 15 < 19.00; therefore means are not significantly different at α = 0.05, thus all remaining pairs of means between baseline and early stocking must not be significantly different from each other at α = 0.05

Double stocking density − late harvest = 345 − 249 = 96, W_5 = 19.00, 96 > 19.00; therefore means are significantly different at α = 0.05

Double stocking density − early harvest = 345 − 250 = 95, W_4 = 18.63, 95 > 18.63; therefore means are significantly different at α = 0.05

Double stocking density − late stocking = 345 − 260 = 85, W_3 = 18.03, 85 > 18.03; therefore means are significantly different at α = 0.05

Double stocking density − baseline = 345 − 263 = 82, W_2 = 17.12, 82 > 17.12; therefore means are significantly different at α = 0.05

TABLE 5.5 *(Continued)*

C. Weight of Individual Fish

Step 1. Rank class means:

Double stocking density (2.30), early stocking (3.30), late harvest (3.32), early harvest (3.33), late stocking (3.47), baseline (3.51)

Step 2. Fill in table:

$df_{\text{within classes}} = 54$

$MS_{\text{within classes}} = 0.0616$

$n = 10$

$\alpha = 0.05$

$W_2 = q_{0.05}(2, 54)\sqrt{0.0616/10} = (2.83)(0.0785) = 0.2221$

$W_3 = q_{0.05}(3, 54)\sqrt{0.0616/10} = (2.98)(0.0785) = 0.2339$

$W_4 = q_{0.05}(4, 54)\sqrt{0.0616/10} = (3.08)(0.0785) = 0.2417$

$W_5 = q_{0.05}(5, 54)\sqrt{0.0616/10} = (3.14)(0.0785) = 0.2464$

$W_6 = q_{0.05}(6, 54)\sqrt{0.0616/10} = (3.20)(0.0785) = 0.2512$

Thus

	$r = 2$	$r = 3$	$r = 4$	$r = 5$	$r = 6$
$q_{0.05}(r, 54)$	2.83	2.98	3.08	3.14	3.20
W_r	0.2221	0.2339	0.2417	0.2464	0.2512

Step 3. Test for significant differences between class means:

Baseline − double stocking density = 3.51 − 2.30 = 1.21, $W_6 = 0.2512$, 1.21 > 0.2512; therefore means are significantly different at $\alpha = 0.05$

Late stocking − double stocking density = 3.47 − 2.30 = 1.17, $W_5 = 0.2464$, 1.17 > 0.2464; therefore means are significantly different at $\alpha = 0.05$

Early harvest − double stocking density = 3.33 − 2.30 = 1.03, $W_4 = 0.2417$, 1.03 > 0.2417; therefore means are significantly different at $\alpha = 0.05$

Late harvest − double stocking density = 3.32 − 2.30 = 1.02, $W_3 = 0.2339$, 1.02 > 0.2339; therefore means are significantly different at $\alpha = 0.05$

Early stocking − double stocking density = 3.30 − 2.30 = 1.00, $W_2 = 0.2221$, 1.00 > 0.2221; therefore means are significantly different at $\alpha = 0.05$

Baseline − early stocking = 3.51 − 3.30 = 0.21, $W_5 = 0.2464$, 0.21 < 0.2464; therefore means are not significantly different at $\alpha = 0.05$, thus all remaining pairs of means between baseline and early stocking must not be significantly different at $\alpha = 0.05$

We will not conduct analysis of variance on profit because the distribution of profits is extremely non-normal. Extreme lack of normality in the distribution of profits for some management policies is caused by the legal constraint that fish must be at least 3 kg to be sold. If fish weigh less than 3 kg at harvest the pond manager obtains no revenue from harvest and still incurs the full cost of the lease. This means that profit is negative and much less than the smallest profit made if he can sell the fish, resulting in a discontinuous distribution of profits for any set of stochastic simulations containing some replicates in which fish are at least 3 kg and some replicates in which fish are less than 3 kg at harvest.

Analysis of variance is quite robust to violation of the assumption that data are normally distributed and although formal tests for normality exist (Snedecor and Cochran 1967, p. 86) we seldom use them. We often conduct analysis of variance on data that when plotted appear non-normally distributed, as long as the departure from normality is not extreme. But when the departure from normality is extreme, as is the case with profit, we use a different type of analysis. One common procedure is to use nonparametric statistical methods (Snedecor and Cochran 1967, p. 120).

However, examination of the results of our policy evaluation simulations (Table 5.3) suggests an alternative analysis by which we can compare profit of the different management policies. First note that there are two essential components of the comparison of profit between policies:

1. the probability that fish will weigh less than 3 kg at harvest, which determines the probabilities of profit and loss (negative profit) in any given year and
2. size of the average profit and average loss for years when fish are at least 3 kg and less than 3 kg, respectively.

We can estimate the probabilities of profitable ($\geq$ 3-kg fish) and unprofitable ($<$ 3-kg fish) years for each policy by calculating the proportion of the area under the normal curve representing the distribution of fish weights to the left of 3 kg (Figure 5.2). For example, the proportion of the area under the baseline fish weight distribution curve to the left of 3 kg is about 0.02, or 2 percent. Under baseline conditions the pond manager is unable to sell his fish 2 percent of the years. We can estimate the average profit for years when fish are at least 3 kg as we have done in Table 5.3 by averaging all non-negative profits for each policy. We can calculate the average loss for years when fish are less than 3 kg by multiplying unit cost of the lease times length of the lease.

We now can estimate mean annual profit for each management policy by:

1. multiplying the probability of a profitable year by the average profit when fish are at least 3 kg,

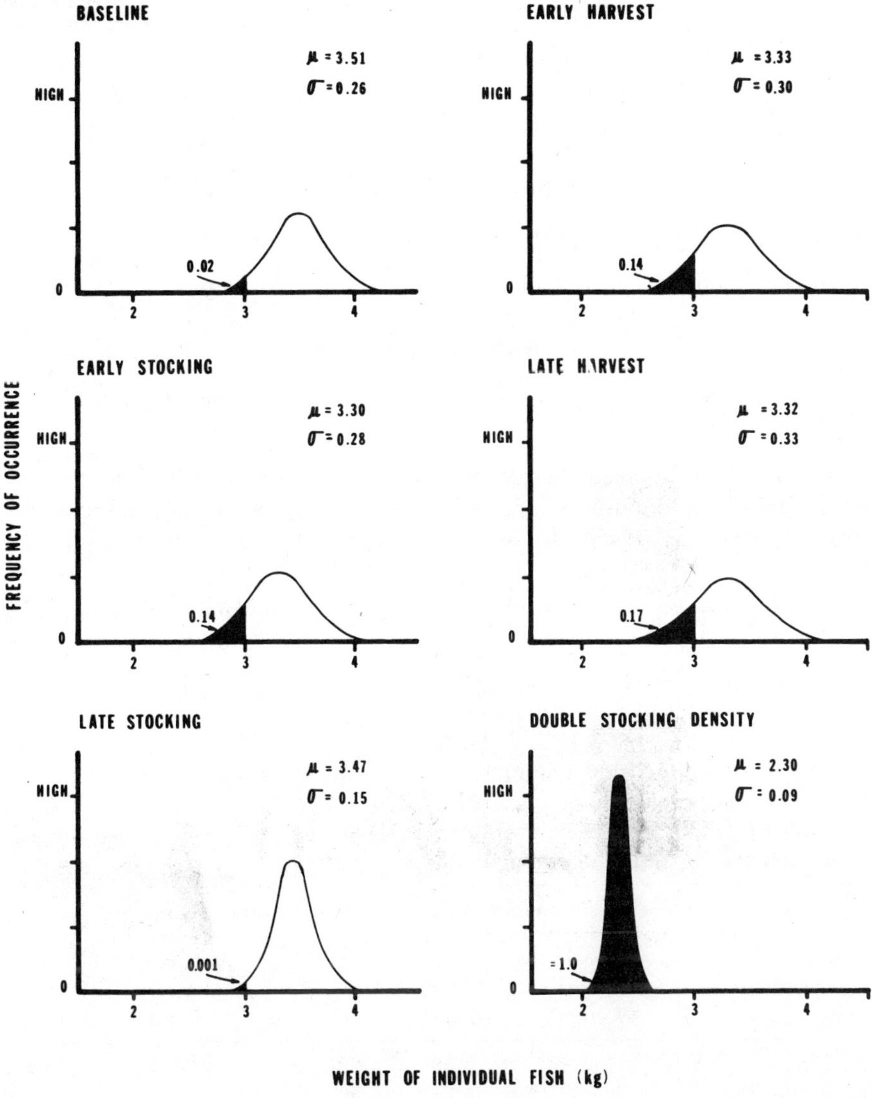

FIGURE 5.2. Distributions of weights of individual fish at harvest resulting from pond model simulations under alternative management policies. Shaded portions represent the proportion of distributions to the left of 3 kg.

2. multiplying the probability of an unprofitable year by average loss when fish are less than 3 kg, and
3. adding the two products.

For example, mean annual profit for the baseline policy is

$$(0.98)(156) + (0.02)(108) = 151 \text{ dollars}$$

Results of profit comparisons are summarized in Table 5.6.

Interpreting Results

We can now answer the five original questions posed at the beginning of our modeling project.

1. The harvest size expected in a normal year under the usual scheme of stocking 75 0.227-kg fish on April 15 and harvesting them on November 15 is 263 kg (Table 5.3). The profit associated with this harvest is 156 dollars (Table 5.3).

2. Year-to-year variability in the harvest under normal conditions due to the inherent variability of the system is ±6 kg, which is the standard error of

TABLE 5.6

Comparison of pond model simulations under alternative management policies with regard to probability of a profitable (≥ 3 kg fish) year, probability of an unprofitable (< 3 kg fish) year, average profit when fish are ≥ 3 kg, average loss when fish are < 3 kg, and mean annual profit

Management Policy	Profitable Years: Probability of Occurrence	Profitable Years: Average Profit (dollars)	Unprofitable Years: Probability of Occurrence	Unprofitable Years: Average Loss (dollars)	Mean Annual Profit (dollars)
Baseline	0.98	156	0.02	108	151
Early stocking (Mar. 15)	0.86	129	0.14	122	94
Late stocking (May 15)	0.999	168	0.001	92	168
Early harvest (Oct. 15)	0.86	163	0.14	92	127
Late harvest (Dec. 15)	0.83	133	0.17	122	90
Double stocking density (150 fish)	≈0	—	≈1	108	- 108

the mean harvest (Table 5.3). We can depict this variability graphically by generating the distribution of harvest sizes expected under baseline conditions (Figure 5.3). Variability in profit is more difficult to describe since expected profits are not distributed continuously because of losses incurred in years when the average weight of individual fish is less than 3 kg. However we can predict that mean annual profit is 151 dollars and that a loss of 108 dollars will be incurred in 2 percent of the years (Table 5.6).

3. Size of the harvest is not changed significantly by harvesting a month earlier or a month later. Mean harvests are 250 kg and 249 kg under the early and late harvest policies, respectively (Table 5.3), but these are not significantly different from baseline because we did not detect a difference greater than 10 percent of the mean baseline harvest (at $\alpha = 0.05$ and $P = 0.80$), which is the level of practical significance set by the pond manager. However, mean annual profit is noticeably lower than baseline under both early (127 dollars) and late (90 dollars) harvest policies because the probability of fish not reaching 3 kg by harvest is quite high under each policy: 14 percent under the early and 17 percent under the late harvest policy (Table 5.6). Early harvest results in a high probability of fish weighing less than 3 kg because of the re-

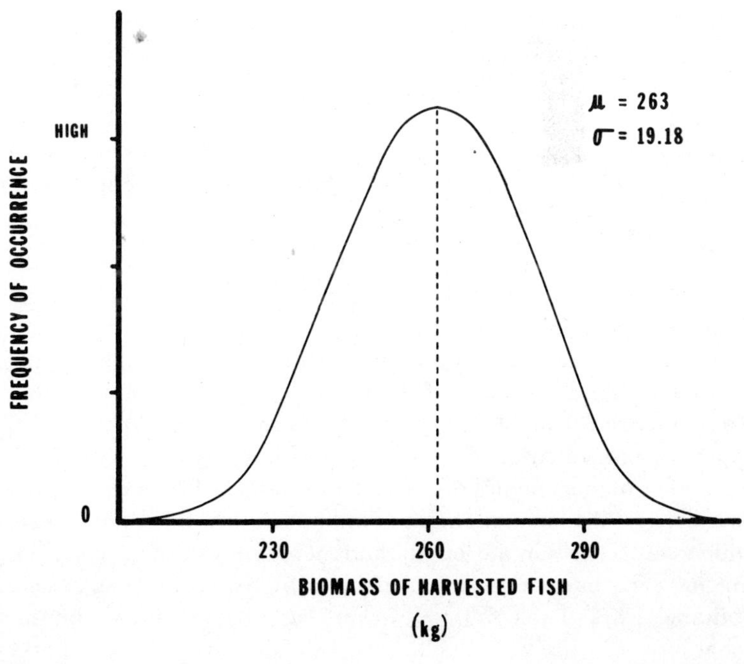

FIGURE 5.3. Expected distribution of harvest sizes under baseline conditions of the pond model.

duced length of time for growth. Late harvest also results in a high probability of gathering fish that weigh less than 3 kg because fish actually lose weight as water temperatures decrease during autumn.

4. Size of the harvest likewise is not changed significantly by stocking a month earlier or a month later. Mean harvests are 248 kg and 260 kg under the early and late stocking policies, respectively (Table 5.3), but again we did not detect a difference greater than 10 percent of the mean baseline harvest at α = 0.05 and P = 0.80. However, mean annual profit is noticeably lower than baseline under the early stocking policy (94 dollars) and is noticeably higher under the late stocking policy (168 dollars) (Table 5.6). Lower profits under the early stocking policy are caused by the 14 percent chance of fish not reaching 3 kg by harvest. Higher profits under the late stocking policy result from the fact that the probability of fish not reaching 3 kg by harvest is low (0.1 percent) even with the shorter, and therefore less costly, length of the lease. Early stocking results in a high probability of fish weighing less than 3 kg because fish actually lose weight during early spring before water temperatures become favorable for growth. Late stocking and the subsequent reduction in the length of time for growth does not result in a high probability of fish weighing less than 3 kg because the time period lost is one during which water temperatures are too low for substantial growth.

5. Size of the harvest is increased significantly by doubling stocking density, but the size of individual fish is decreased significantly. Although size of the harvest is 345 kg, size of individual fish is only 2.30 kg (Table 5.3) which is well below the 3-kg limit for selling fish. Thus mean annual profit is − 108 dollars and the probability of incurring a loss in any given year is essentially 100 percent (Table 5.6). Fish weight is low due to competition for plants (food), which are scarce.

We now summarize results in terms that are of most interest to the pond manager. Doubling stocking density is the only management policy that differs significantly from baseline in terms of size of the harvest or size of individual fish. Unfortunately the larger harvest under this policy is composed of fish that are too small to sell. Presumably fish are small due to lack of food since plant biomass at harvest is roughly half of its usual value and well below the 20,000 kg/ha threshold at which availability of plants limits fish growth (Table 5.3). Although other management policies do not produce differences from baseline in terms of size of the harvest or size of individual fish, there are differences in profit due to differences in the length, and therefore cost, of the lease. Both late stocking and early harvest result in somewhat higher profits than baseline during profitable years (Table 5.3). However, early harvest noticeably increases the probability of incurring a loss because fish fail to reach 3 kg by harvest and hence mean annual profit is less than baseline (Table 5.6). It appears that of the proposed management alternatives the only one worthy of serious consideration by the pond manager is the late stocking policy.

Further Examining Selected Types of Management Policies or Environmental Situations

Exploratory Simulations

Results of initial management policy simulations suggest several interesting possibilities for further consideration. In terms of increasing profit we can (1) reduce the cost of the lease by stocking late or harvesting early, or (2) increase revenue by increasing size of the harvest. The risk we take by reducing the length of the lease is that fish may not have enough time, or enough exposure to warm water temperatures, to grow to 3 kg by harvest. The risk we take by increasing size of the harvest by increasing stocking density is that fish may not have enough food to reach 3 kg. We might examine effects on pond dynamics of smaller changes from baseline in stocking date, harvest date, and stocking density than were represented by our original simulations. The objective is to gain further insight into the response of the pond system to the interplay of these variables and ultimately to see if some new variant of our original management policies is more profitable. We will use the deterministic version of the pond model to conduct these exploratory simulations.

Results of exploratory simulations in which stocking date is varied from March 1 ($t_{begin} = 60$) to June 1 ($t_{begin} = 152$) while harvest date and stocking density remain at baseline values are presented in Figure 5.4. Plant biomass remaining at harvest increases exponentially as stocking date becomes later but remains above 20,000 kg/ha, the point at which availability of plants begins to limit fish consumption, even when fish are stocked on March 1. Fish biomass and the weight of individual fish at harvest both rise slowly from the March 1 stocking date to the April 22 stocking date ($t_{begin} = 112$) but decline sharply thereafter. However, the weight of individual fish remains above 3 kg even when fish are stocked on June 1. Profit rises from the March 1 stocking date to the May 15 stocking date ($t_{begin} = 135$) and then declines.

Results of exploratory simulations in which harvest date is varied from October 1 ($t_{end} = 274$) to December 31 ($t_{end} = 365$) while stocking date and stocking density remain at baseline values are presented in Figure 5.5. Plant biomass remaining at harvest declines exponentially as harvest date becomes later but remains above 20,000 kg/ha even when fish are harvested on December 31. Fish biomass and the weight of individual fish at harvest both rise rapidly from the October 1 harvest date to the November 8 harvest date ($t_{end} = 312$) but decline thereafter. However, the weight of individual fish remains above 3 kg even when fish are harvested on December 31. Profit rises from the October

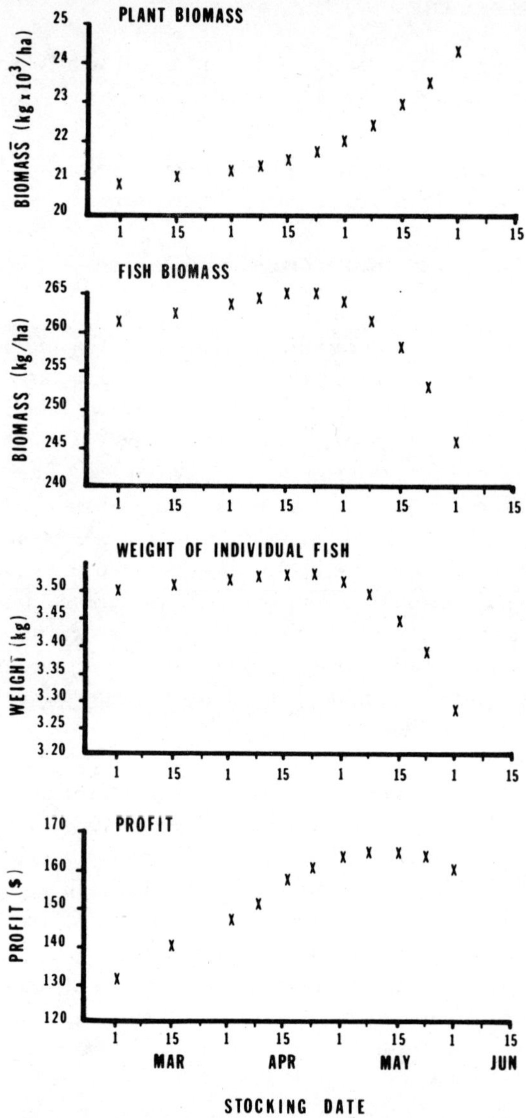

FIGURE 5.4. Results of exploratory simulations using the deterministic version of the pond model indicating the state of the system at harvest under different stocking dates.

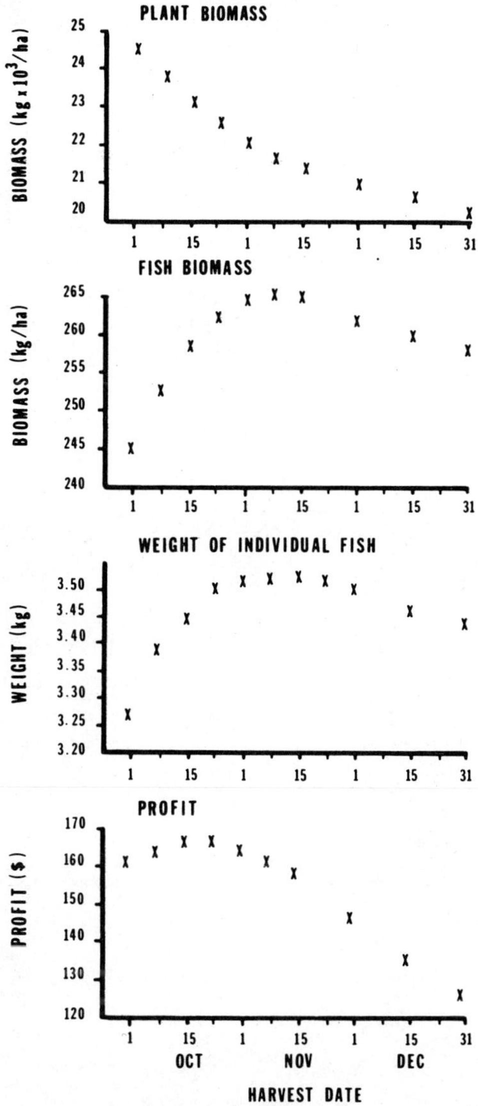

FIGURE 5.5. Results of exploratory simulations using the deterministic version of the pond model indicating the state of the system at harvest under different harvest dates.

1 harvest date to the October 22 harvest date (t_{end} = 295) but declines thereafter.

Results of exploratory simulations in which stocking density is varied from 50 to 150 fish while stocking date and harvest date remain at baseline values are presented in Figure 5.6. Plant biomass remaining at harvest declines exponentially as stocking density increases and becomes less than 20,000 kg/ha when 85 fish are stocked. Fish biomass at harvest increases as stocking density increases but the rate of increase decreases markedly at densities above 85 fish. The weight of individual fish remains essentially constant at stocking densities from 50 to 85 fish but decreases sharply as stocking density increases above 85 fish. Weight declines below 3 kg at a stocking density of 115 fish. Profit increases up to a stocking density of 110 fish but at stocking densities above 115 fish profit is negative.

Formally Comparing New Management Policies

Based on results of these exploratory simulations we might decide to compare formally baseline pond dynamics with dynamics under a new management policy in which we stock 110 fish ($D2$ = 110) on May 15 (t_{begin} = 135) and harvest them on October 22 (t_{end} = 295). Stocking date, harvest date, and stocking density for the new management policy represent conditions under which profit curves peaked in Figures 5.4, 5.5, and 5.6, respectively. We calculate the sample size required by our new experimental design as before using the formula from Section 3.5.

$$n \geq 2\left(\frac{\sigma}{\delta}\right)^2 (t_{\alpha,\gamma} + t_{2(1-P),\gamma})^2$$

Recall that in our case $\sigma^2 = 380$ or $\sigma = 19.49$, $\delta = 25$, $\alpha = 0.05$, and $P = 0.80$ (p. 119). If our initial guess for the correct sample size is 11, then

$$\gamma = 2(11 - 1) = 20$$

Thus

$$11 \geq 2\left(\frac{19.49}{25}\right)^2 (t_{0.05,20} + t_{0.40,20})^2$$

or

$$11 \geq 2\left(\frac{19.49}{25}\right)^2 (2.086 + 0.860)^2$$

and

$$11 \geq 10.55$$

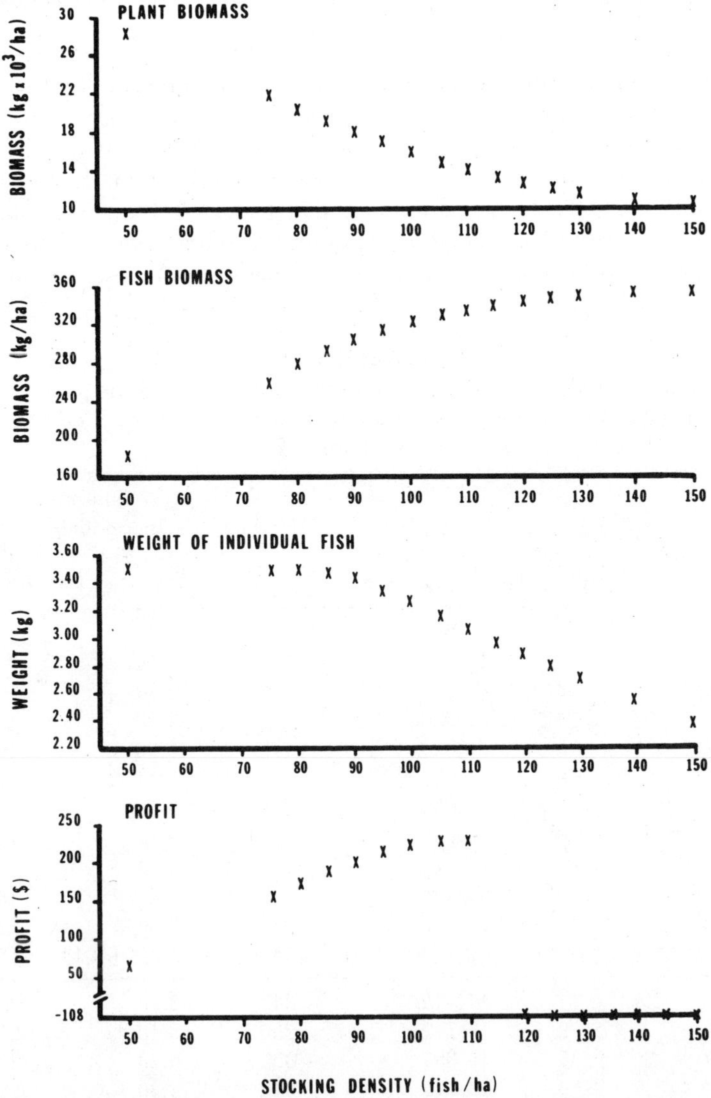

FIGURE 5.6. Results of exploratory simulations using the deterministic version of the pond model indicating the state of the system at harvest under different stocking densities.

and the correct sample size rounded to the nearest whole number is 11. Our new experimental design will consist of 22 simulations—11 stochastic replicates of each of the two versions of the model.

Results of simulations of the baseline and new management policies with regard to outputs of interest—profit, plant and fish biomass, and average weight of individual fish—are summarized in Table 5.7. Biomass of plants and fish and the weight of individual fish under the baseline and new management policies can be compared using three two-sample *t*-tests as described in Section

TABLE 5.7

Summary of results of 11 replicate stochastic simulations of the pond model under baseline conditions and a new management policy under which 110 fish are stocked on May 15 and harvested on October 22

Management Policy	Replication	Plants (kg/ha)	Fish (kg/ha)	Weight of Individual Fish (kg)	Profit (dollars)[a]
Baseline	1	23,608	241	3.22	134
	2	19,480	281	3.74	173
	3	22,298	257	3.42	149
	4	24,326	231	3.08	124
	5	28,029	172	2.30	− 108
	6	19,927	278	3.71	171
	7	19,062	282	3.75	174
	8	25,686	211	2.82	− 108
	9	23,774	239	3.18	131
	10	20,494	274	3.65	166
	11	23,110	247	3.30	140
	$\bar{X}$(± 1SE)	22,708(± 844)	247(± 10)	3.29(± 0.13)	151(± 7)
New	1	17,040	348	3.16	268
	2	16,712	350	3.18	270
	3	14,856	352	3.20	272
	4	14,719	352	3.20	272
	5	12,215	329	2.99	− 80
	6	16,173	352	3.20	272
	7	19,599	322	2.93	− 80
	8	16,289	352	3.20	271
	9	18,568	334	3.04	254
	10	14,240	350	3.18	270
	11	27,357	203	1.85	− 80
	$\bar{X}$(± 1SE)	17,070(± 1198)	331(± 13)	3.01(± 0.12)	269(± 2)

[a]$\bar{X}$(± 1SE) for profit does not include negative profits.

4.5. Profits can be compared using the procedure described earlier in this section. We will not present results of these analyses here, but rather point out that examination of these results may again raise new questions that we might want to explore using our model.

By this point the model has become an integral part of our thinking about the pond system. The manner in which our interests in the pond system change will suggest how we might use the model most effectively to pursue these interests. This may involve only minor changes to the model, or it may involve major changes requiring collection of new data from the real system to improve the representation of important mechanisms within the model. The ideal relationship between modeling and real-world experimentation is iterative and dynamic.

5.7. EXERCISES

Exercise 5.7.1

Use the plant–herbivore model to assess the long-term (5-year) impact of herbivores on the island plant community.

Exercise 5.7.2

Use the model to determine the proportion of herbivore biomass that could be harvested annually on a sustained yield basis assuming that all harvest occurs on the last day of the year.

Answers to Chapter 5 exercises are in Appendix 6.

CHAPTER

6

A FEW WORDS OF CAUTION

6.1. INTRODUCTION

Systems analysis and simulation are powerful tools when used wisely to address questions in wildlife and fisheries sciences. But, as with the use of any tool, there are inherent dangers. In the introduction to Chapter 1 we considered five general areas of potential danger for modelers, all concerned in one way or another with becoming enamored of models themselves. Throughout much of the book we have concerned ourselves with developing, validating, and using models, occasionally noting potential pitfalls in obtaining "valid" models or in interpreting model results. In this final chapter we present a few additional words of caution via two examples demonstrating that sometimes even "valid" models can yield misleading results. The purpose is not to cast doubts on the general utility of modeling through recitation of a lengthy list of potential evils awaiting the unwary modeler. Rather, the purpose is to engender a healthy skepticism regarding model predictions, while maintaining a positive attitude toward wise use of models. We must keep both the virtues and dangers of systems analysis and simulation in proper perspective if we are to realize the full potential of these tools in helping us to deal with complex systems.

6.2. EXAMPLE 1: FISHERY MODEL

Consider the simple fishery model represented in Figure 6.1. Recruitment to the fishable population and natural mortality of fish are solely functions of population size, although recruitment also is confined to the first quarter of the year. Harvest of fish is a function of population size, fishing effort, and the proportion of the population caught by one unit of fishing effort. The objectives of the model are to describe the dynamics of the fish population under

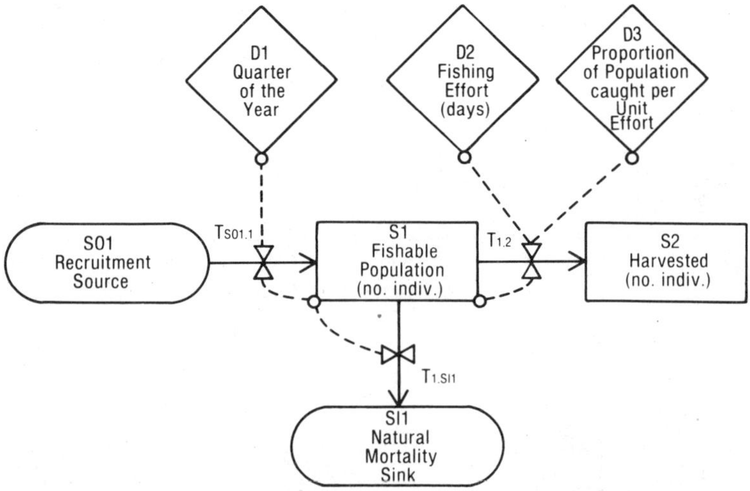

FIGURE 6.1. Conceptual model diagram representing a hypothetical fishery system.

current conditions, in which fishing is prohibited, and to determine the levels of sustainable annual harvest possible under different levels of fishing effort.

Suppose we have data over a rather narrow range of population sizes suggesting that both recruitment and natural mortality are related linearly to population size (Figure 6.2a). However, our biological intuition tells us that the relationships actually are curvilinear (Figure 6.2b), although we have no hard data to support our hypothesis. Further suppose that, using 3 months as the basic time unit for calculations, we can represent the available recruitment and natural mortality data adequately as

$$T_{\mathrm{SO1},1} = 0.7978S1$$

$$T_{1,\mathrm{SI1}} = 0.1525S1$$

or as

$$T_{\mathrm{SO1},1} = 300(1 - e^{-0.003S1})$$

$$T_{1,\mathrm{SI1}} = e^{0.025S1} - 1$$

where symbols are as used in Figure 6.1. (Although the later two equations represent curves shaped like those in Figure 6.2b, it is quite possible that the slope of these curves might be essentially constant over the narrow range of available data, and hence also might fit the data well.) Finally, let's assume the proportion of the population that would be harvested by one day of fishing effort is 0.001, that fishing mortality and natural mortality are completely additive, and that initially there are 82 fish in the system.

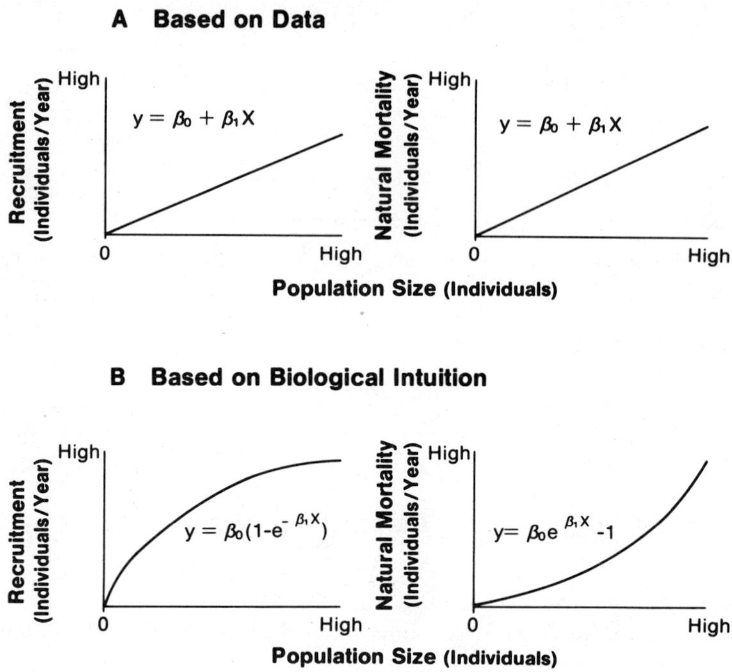

FIGURE 6.2. Hypothesized relationships of recruitment and natural mortality to population size in a simple fishery based (A) on data and (B) on biological intuition.

Based on this information we might develop two different models of the fishery: Model 1 including linear relationships of recruitment and natural mortality (Figure 6.2a), and Model 2 including curvilinear relationships (Figure 6.2b). Equations and BASIC computer programs for Models 1 and 2 are presented in Table 6.1.

If we use each model to simulate dynamics of the fish population over a ten-year period under current conditions (no harvest), results are essentially the same (Figure 6.3a and Table 6.2a). Both models predict regular seasonal fluctuations in population size with a parent stock of 82 to 85 individuals remaining each year just prior to the recruitment period. Let's assume that this accurately reflects what we know about the real fishery and hence that we judge both models to be valid.

If we now use each model to simulate the effects of fishing over a ten-year period at an effort level of, say, 25 days fished per quarter ($D2 = 25$), we notice that predictions of the two models differ markedly (Figure 6.3b and Table 6.2b). Both models predict slight population declines over the first year or two, but the Model 2 population then stabilizes at a pre-recruitment size of

TABLE 6.1

(A) Equations and (B) a BASIC computer program for the baseline version of Models 1 and 2 of the simple fishery system. Symbols for equations are as used in Figure 6.1

A. Model Equations

Initial Conditions of Model

$S1 = 82$
$S2 = 0$
$D2 = 0$
$D3 = 0.001$

Driving Variable Equations

$D1 = 1$ for first quarter of year, 2 for second, etc.

Rate Equations

Model 1

$T_{SO1,1} = 0.7978 S1$ if $D1 = 1$
$T_{SO1,1} = 0$ if $D1 \neq 1$
$T_{1,SI1} = 0.1525 S1$
$T_{1,2} = D2 \times D3 \times S1$

Model 2

$T_{SO1,1} = 300(1 - e^{-0.003S1})$ if $D1 = 1$
$T_{SO1,1} = 0$ if $D1 \neq 1$
$T_{1,SI1} = e^{0.025S1} - 1$
$T_{1,2} = D2 \times D3 \times S1$

State Variable Equations

$S1_{t+1} = S1_t + T_{SO1,1} - T_{1,SI1} - T_{1,2}$
$S2_{t+1} = S2_t + T_{1,2}$

B. BASIC Program

```
10 LPRINTTAB(6); "MODEL 1" [OR "MODEL 2"]
20 S1=82
30 S2=0
40 K1=0.7978   [or K1=300]
50 K2=0        [or K2=-0.003]
60 K3=0.1525   [or K3=0.025]
65 K4=0
70 D1=0
80 LPRINTTAB(6);"QUARTER";TAB(16);"STOCK";TAB(26);
"HARVEST";TAB(36);"RECRUIT";TAB(46);"NAT.MORT.";TAB(56);
"ANN.MORT.";TAB(66);"QUART.HARV.";TAB(76);"ANN.HARV."
81 T=0
82 LPRINTTAB(6);T;TAB(16);S1;TAB(26);S2
83 FOR Y=1 TO 10
84 M=0
85 H=0
90 FOR T=1 TO 4
91 IF T=1 THEN K1=0.7978 ELSE K1=0  [or K1=300 ELSE K1=0]
```

TABLE 6.1 *(Continued)*

B. BASIC Program

```
100 T1=K1*S1   [or T1=K1*(1-EXP(K2*S1))]
110 T2=K3*S1   [or T2=EXP(K3*S1)-1]
120 T3=K4*D1*S1
130 S1=S1+T1-T2-T3
140 S2=S2+T3
141 M=M+T2
142 H=H+T3
143 IF T=4 THEN AM=M ELSE AM=0
144 IF T=4 THEN AH=H ELSE AH=0
150 LPRINTTAB(6);T;TAB(16);S1;TAB(26);S2;TAB(36);T1;TAB
(46);T2;TAB(56);AM;TAB(66);T3;TAB(76);AH
160 NEXT T
161 NEXT Y
170 END
```

about 78 individuals whereas the Model 1 population continues to decline to about 29 individuals at the end of year 10. A closer look indicates that the Model 1 population will continue to decrease indefinitely because each year only 89 percent of total mortality (natural plus fishing) is replaced by recruitment (Table 6.2b). In contrast, recruitment in the Model 2 population completely replaces total mortality during years 2 through 10.

Simulation of a wide range of fishing effort levels confirms that the two models exhibit fundamentally different behavior (Figure 6.4). Model 1 populations decline exponentially with the rate of decline more rapid at higher levels of fishing effort. Model 2 populations decline exponentially at first but then attain a new dynamic (i.e., seasonally varying) equilibrium, with the decline longer and equilibrium population size lower at higher effort levels. Predictions of annual harvest follow the same trend as predictions of population size for each model (Figure 6.5). In terms of determining sustainable annual harvests under different levels of fishing effort—one of our modeling objectives—Model 1 predicts that no harvest is sustainable under any effort level. Model 2 predicts that for practical purposes effort levels from 75 to 100 days fished per quarter result in the highest sustainable annual harvest, or maximum sustainable yield from the fishery (Figure 6.5b).

This example illustrates three important points:

1. A single relationship often can be represented by different equations that exhibit fundamentally different types of behavior, yet fit the available data equally well.

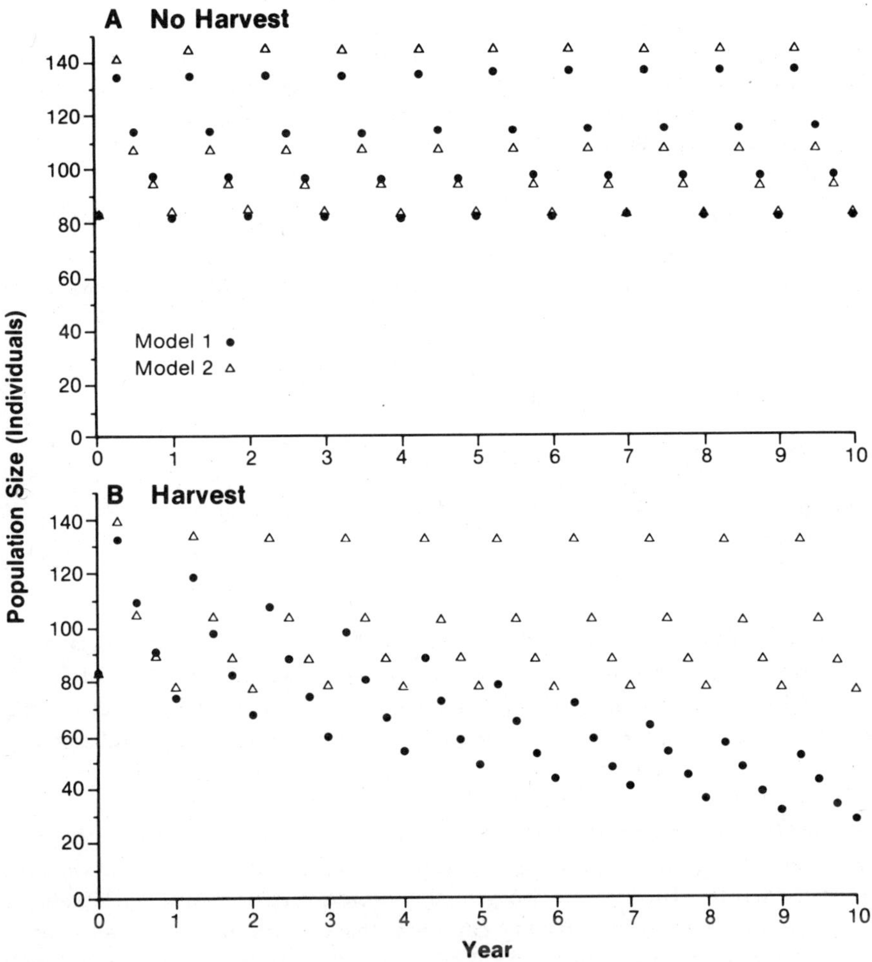

FIGURE 6.3. Population sizes over a ten-year period predicted by different models of the fishery (A) without and (B) with harvest (at a fishing effort of 25 days fished per quarter).

2. Extrapolation of relationships beyond the conditions for which we have data is inherently dangerous.
3. Failure to include self-regulating feedback mechanisms (such as density-dependent control of population growth) in models of living systems, even when we have no data suggesting that these mechanisms exist, makes the behavior of such models under non-equilibrium conditions suspect.

TABLE 6.2

Population size (*N*) at the end of years 2 and 10, and annual recruitment (*R*), natural mortality (NM), and harvest (*H*) during years 2 and 10 predicted by different models of the fishery (A) without and (B) with harvest (at a fishing effort of 25 days fished per quarter). Table entries are in number of individuals except for the ratio *R*/(NM + *H*)

	Year 2					Year 10				
Model	*N*	*R*	NM	*H*	*R*/(NM + *H*)	*N*	*R*	NM	*H*	*R*/(NM + *H*)
A. No Harvest										
Model 1	82	65	65	0	1.0	85	66	66	0	1.0
Model 2	85	67	67	0	1.0	85	67	67	0	1.0
B. Harvest										
Model 1	67	59	57	9	0.89	29	26	25	4	0.89
Model 2	78	63	53	10	1.0	78	62	52	10	1.0

Suppose that recruitment and natural mortality in our simple fishery actually are curvilinear functions of population size as represented in Model 2. Model 1 still appears valid since available data refute neither the linear mechanisms representing recruitment and natural mortality or overall model behavior under baseline conditions. However, when we use Model 1 to evaluate alternative harvesting schemes, which presumably requires extrapolation of both recruitment and natural mortality relationships beyond the range of population sizes over which linear representations are appropriate, our predictions are very wrong. In this case we probably would question Model 1's validity when we observe that the fishery cannot withstand any harvest at all. But in some cases erroneous results from "valid" models can be much more difficult to identify.

Actually, we might have felt uneasy about using Model 1 to assess effects of harvesting even before any harvesting simulations were run. Since per capita rates of recruitment and natural mortality are not represented as density dependent, there is no mechanism allowing the population to compensate for harvest losses. We seldom have good information with which to specify self-regulating, or negative-feedback, mechanisms in ecological models. Failure to include such mechanisms often has no noticeable effect on behavior of models under baseline, or essentially equilibrium, conditions. But their omission can seriously limit the ability of models to respond appropriately if we are simulating new management alternatives that might force the system to operate over a wider range of conditions.

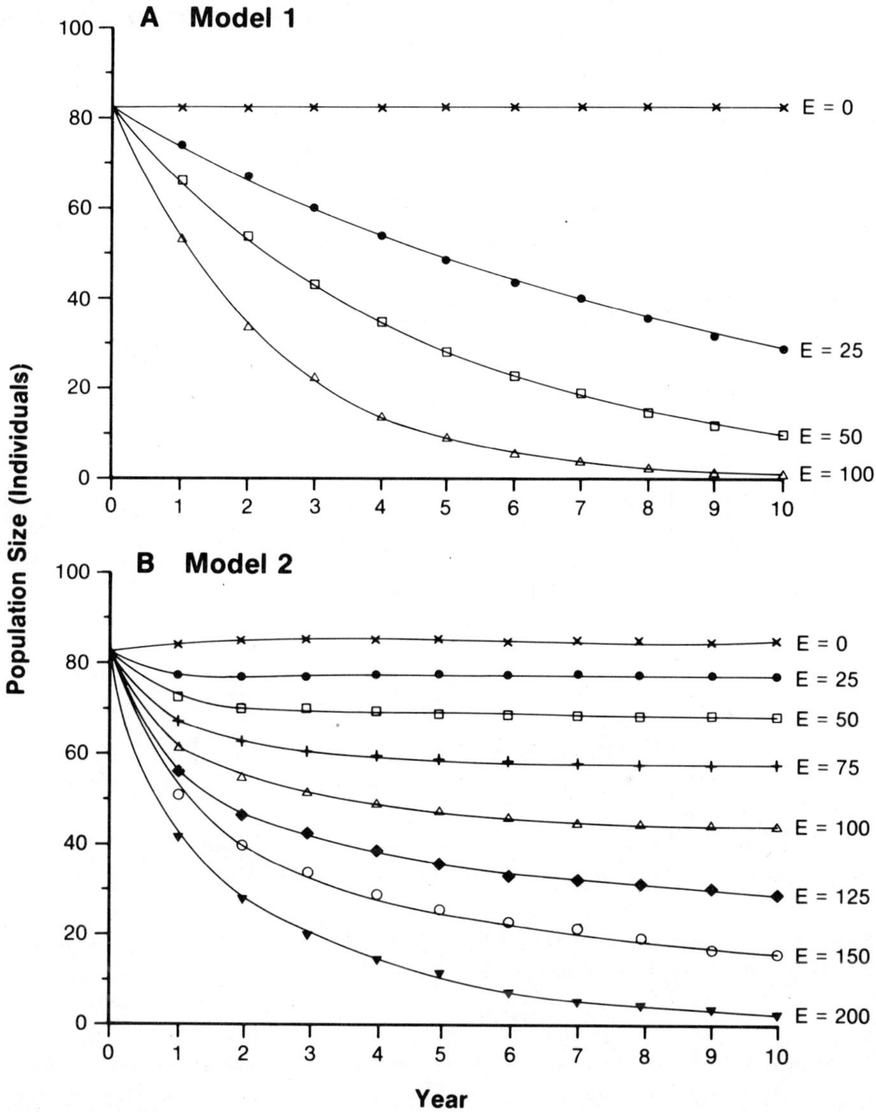

FIGURE 6.4. Population sizes (just prior to annual recruitment) over a ten-year period predicted by different models of the fishery under different levels of fishing effort (E, in days fished per quarter).

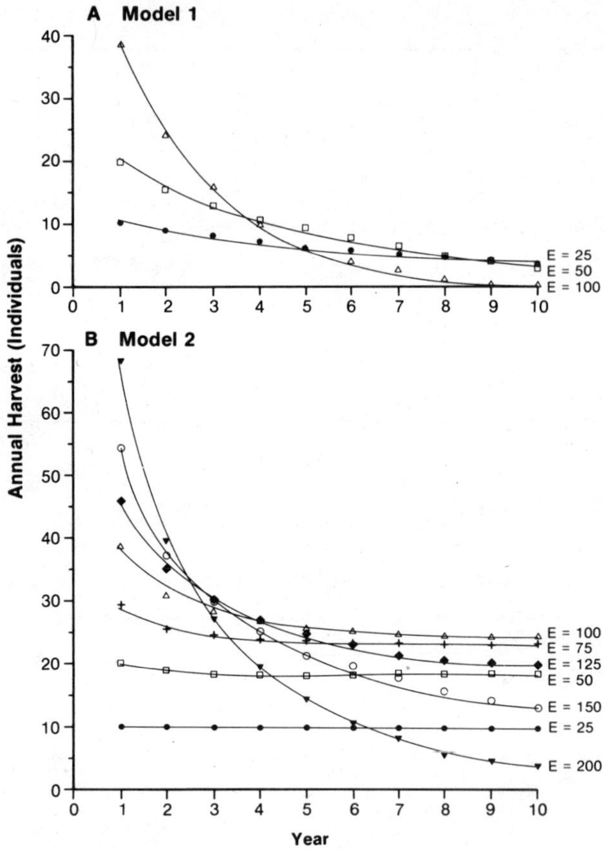

FIGURE 6.5. Annual harvests over a ten-year period predicted by different models of the fishery under different levels of fishing effort (E, in days fished per quarter).

Although both Model 1 and Model 2 assume that fishing and natural mortality are additive, neither model includes a mechanism ensuring that the sum of fishing and natural mortality is less than one. This is not a problem until fishing effort becomes very large, since the proportion of the population harvested by one unit of effort is so low, but it does preclude use of the models to simulate extremely high levels of effort.

6.3. EXAMPLE 2: WILDLIFE MODEL

Consider the simple wildlife population model in Figure 6.6. Natality is a function of total population size and the number of individuals in age classes 2 and

3 (age class 1 individuals are prereproductive). Natural mortality (and survival) of age class 1 and 2 individuals is a function of total population size. No individuals survive past age class 3. The number of individuals harvested from each age class is a function of total population size and the harvest scheme. Material transfers leaving each state variable also are a function of the level of that state variable, although for simplicity the information transfers representing these donor-dependent controls are not shown in Figure 6.6. The objective of the model is to determine the proportion of the population that can be harvested annually without reducing population size below an arbitrarily established limit of 25 individuals. Managers are not concerned about maximizing the number of individuals harvested, but rather they want to provide as many hunters as possible with a chance to hunt.

Suppose data suggest that annual age-specific natality and natural mortality rates are related linearly to population size as indicated in Figure 6.7. (Note that we have related the *per capita* natality rate and the *proportion* of the population dying to population size in Figure 6.7, rather than relating *number* of individuals born and *number* of individuals dying directly to population size as we did in the previous example (Figure 6.2). Thus the self-regulating feedback mechanism that was not present in either the linear recruitment or natural mortality curve in Figure 6.2a is present in both the linear natality and natural mortality curves in Figure 6.7.) We will assume that hunting mortality is completely compensatory with natural mortality. Thus hunting losses do not increase total population mortality until they become greater than the number of individuals that would have died of natural causes in the absence of hunting. We will further assume that age classes are distinguishable in the field and, therefore, that hunting losses can be regulated on an age-specific basis. Finally, let's assume that initially there are 100, 50, and 25 individuals in age classes 1, 2, and 3, respectively. Based on this information we might develop a deterministic population dynamics model consisting of the equations in Table 6.3a and represented by the BASIC computer program in Table 6.3b.

To determine the general response of the population to harvest we might run a series of simulations incrementally increasing harvest levels. But first we would note from the baseline simulation that the population as initially specified is in equilibrium. In the absence of hunting, annual age-specific natural mortality rates for ages 1, 2, and 3 are 0.5, 0.5, and 1.0, respectively. Since hunting mortality is completely compensatory we can harvest at least 50 percent of age classes 1 and 2 and all of the third age class without any noticeable effect on the population. Under any circumstances we can harvest all age 3 individuals without increasing total population mortality because these individuals will die anyway. With this in mind, we might run a series of 50-year simulations in which all age 3 individuals are harvested annually and the annual harvest of age 1 and 2 individuals is increased from 50 toward 100 percent in 5 percent increments.

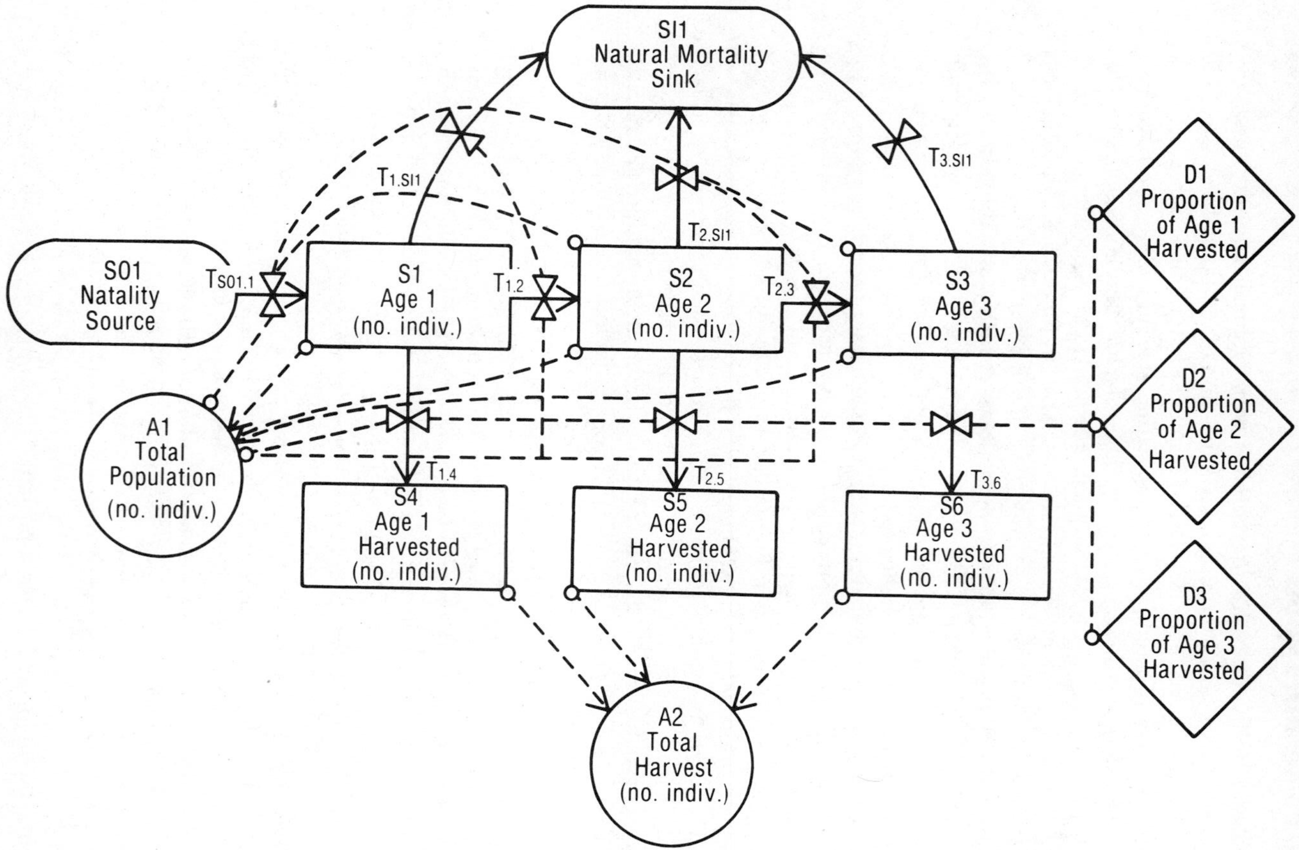

FIGURE 6.6. Conceptual model diagram representing a hypothetical wildlife population system. All material transfers are implicitly donor dependent.

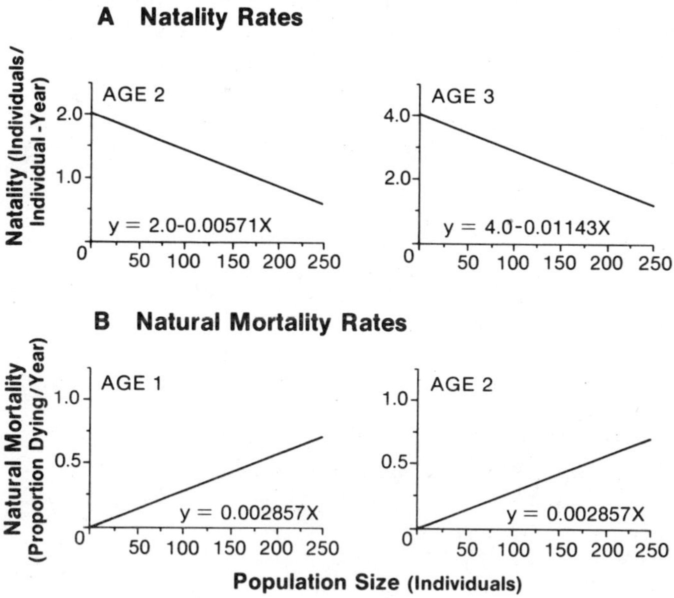

FIGURE 6.7. Relationships of age-specific (A) natality and (B) natural mortality rates to population size in a simple wildlife population system. Age 1 is prereproductive and no individuals survive past age 3.

Results of these simulations suggest that the population can withstand up to roughly 65 percent annual harvest of age 1 and 2 individuals without reducing population size below 25 individuals (Figure 6.8). Sustainable annual harvest, as well as equilibrium population size, is much reduced at higher harvest rates. Sustainable harvests under the 50, 55, 60, and 65 percent schemes are roughly 100, 88, 69, and 38 individuals, respectively. Interpretation of the results of our deterministic population model appears quite straightforward. Of course a tacit assumption upon which our interpretation rests is that the relationships included in our model can be represented adequately in a deterministic manner. But since we have not explicitly represented the effect of varying environmental conditions in the model, we might feel uncomfortable with the deterministic representation of natality and mortality rates, which we know are influenced by environmental factors.

Suppose that we now choose to develop a stochastic model in which we represent the inherent year-to-year variability in natality by specifying natality rates as random variables. Age-specific natality rates will be specified by (1) calculating the rates based deterministically on population size (as we did in the deterministic model) and (2) choosing a random variate from a normal distribution with a mean equal to the deterministically predicted rate and a variance

TABLE 6.3

(A) Equations and (B) a BASIC computer program for the baseline version of a deterministic model of the simple wildlife population system. Symbols for equations are as used in Figure 6.6

A. Model Equations

Initial Conditions of Model

$S1 = 100$
$S2 = 50$
$S3 = 25$
$S4 = 0$
$S5 = 0$
$S6 = 0$
$D1 = 0$
$D2 = 0$
$D3 = 0$

Auxiliary Variable Equations

$A1 = S1 + S2 + S3$
$A2 = S4 + S5 + S6$

Rate Equations

$T_{SO1,1} = (2.0 - 0.00571A1)S2 + (4.0 - 0.01143A1)S3$
$T_{1,SI1} = (0.002857A1 - D1)S1 \text{ if } D1 \leq (0.002857A1)$
$T_{1,SI1} = 0 \text{ if } D1 > (0.002857A1)$
$T_{1,2} = (1.0 - 0.002857A1)S1 \text{ if } D1 \leq (0.002857A1)$
$T_{1,2} = (1.0 - D1)S1 \text{ if } D1 > (0.002857A1)$
$T_{2,SI1} = 0.002857A1 - D2) \text{ if } D2 \leq (0.002857A1)$
$T_{2,SI1} = 0 \text{ if } D2 > (0.002857A1)$
$T_{2,3} = (1.0 - 0.002857A1)S2 \text{ if } D2 \leq (0.002857A1)$
$T_{2,3} = (1.0 - D2)S2 \text{ if } D2 > (0.002857A1)$
$T_{3,SI1} = (1.0 - D3)S3$
$T_{1,4} = D1 \times S1$
$T_{2,5} = D2 \times S2$
$T_{3,6} = D3 \times S3$

State Variable Equations

$S1_{t+1} = S1_t + T_{SO1,1} - T_{1,SI1} - T_{1,2} - T_{1,4}$
$S2_{t+1} = S2_t + T_{1,2} - T_{2,SI1} - T_{2,3} - T_{2,5}$
$S3_{t+1} = S3_t + T_{2,3} - T_{3,SI1} - T_{3,6}$
$S4_{t+1} = S4_t + T_{1,4}$
$S5_{t+1} = S5_t + T_{2,5}$
$S6_{t+1} = S6_t + T_{3,6}$

TABLE 6.3 (*Continued*)

B. BASIC Program

```
10 LPRINTTAB(6);"DET."
20 S1=100
30 S2=50
40 S3=25
50 S4=0
60 S5=0
70 S6=0
80 K1=0
140 K7=0
150 K8=0
160 K9=0
161 LPRINT USING"#.### ";K7;K8;K9
165 T=0
166 A1=S1+S2+S3
170 LPRINT "  YEAR ";"  TPOP ";"  AGE1 ";"  AGE2 ";"  AGE3 ";"
HARV ";"  HAG1 ";" HAG2 ";"  HAG3 "
175 LPRINT USING" ####.  ";T;A1;S1;S2;S3
180 FOR T=1 TO 25
181 K2=2-0.00571*A1
182 K3=4-0.01143*A1
190 T1=K1*S1+K2*S2+K3*S3
191 K4=0.002857*A1
192 K5=0.002857*A1
193 K6=1
197 IF K7>K4 THEN K4=0 ELSE K4=K4-K7
198 IF K8>K5 THEN K5=0 ELSE K5=K5-K8
199 K6=K6-K9
200 T2=(1-K4-K7)*S1
210 T3=(1-K5-K8)*S2
220 T4=K4*S1
230 T5=K5*S2
240 T6=K6*S3
250 T7=K7*S1
260 T8=K8*S2
270 T9=K9*S3
280 S1=S1+T1-T2-T4-T7
290 S2=S2+T2-T3-T5-T8
300 S3=S3+T3-T6-T9
310 S4=S4+T7
320 S5=S5+T8
330 S6=S6+T9
335 A1=S1+S2+S3
340 A2=S4+S5+S6
350 LPRINT USING" ####.  ";T;A1;S1;S2;S3;A2;S4;S5;S6
360 NEXT T
370 END
```

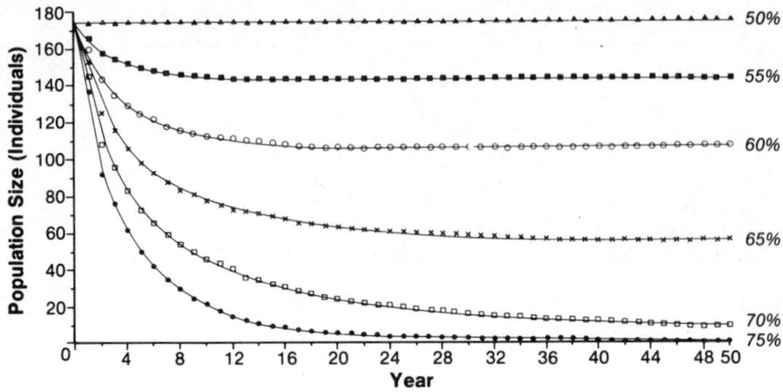

FIGURE 6.8. Population sizes over a 50-year period predicted by the deterministic model under several harvest schemes in which all age 3 individuals are harvested and the percentage of age 1 and 2 individuals harvested is as indicated on the graph.

equal to 1.0. (Assume the choice of a normal distribution and the estimation of variance both were based on information from the real system.) Thus our stochastic model is similar to the deterministic model (Table 6.3a) except that the material transfer from natality source to the age 1 state variable now is calculated as

$$T_{SO1,1} = (\text{RV1} \times S2) + (\text{RV2} \times S3)$$

where RV1 = normal random variate from a distribution with $\mu = 2.0 - 0.00571A1$, $\sigma^2 = 1.0$

RV2 = normal random variate from a distribution with $\mu = 4.0 - 0.01143A1$, $\sigma^2 = 1.0$

To determine response of the population to harvest using our stochastic model, we might run ten replicate stochastic simulations of each harvest scheme that we examined earlier with the deterministic model. (Assume that a sample size of ten was determined in the appropriate fashion.) A BASIC computer program for the stochastic model is presented in Table 6.4.

If we compare mean population sizes predicted by our stochastic model each year under the different harvest schemes (Figure 6.9) with the corresponding curves predicted by the deterministic model (Figure 6.8), we notice that stochastic model predictions generally are lower. These lower predictions result from a downward bias in predicted natality rates that is a mathematical property of the transformation from deterministic to stochastic model. In fact, it can be shown that the mean of a model with stochastic rate equations generally is not the same as the corresponding deterministic model evaluated at the mean rate (Matis and Wehrly 1979). (Note that this same type of bias existed when

TABLE 6.4

BASIC computer program for the baseline version of a stochastic model of the simple wildlife population system

```
1 LPRINTTAB(6);"STOCH."
2 DIM X(50),XS(50),M(50),SD(50)
3 FOR I=1 TO 50
4 X(I)=0
5 XS(I)=0
6 M(I)=0
7 SD(I)=0
8 NEXT I
9 RANDOM
10 FOR REP=1 TO 10
20 S1=100
30 S2=50
40 S3=25
50 S4=0
60 S5=0
70 S6=0
80 K1=0
140 K7=.5
150 K8=.5
160 K9=1
161 LPRINT USING"#.###  ";REP;K7;K8;K9
165 T=0
166 A1=S1+S2+S3
175 LPRINT USING" ####.  ";T;A1;S1;S2;S3
180 FOR T=1 TO 50
181 K2=2-0.00286*A1
182 K3=4-0.00571*A1
183 GOSUB 500
184 K2=K2+Z
185 K3=K3+Z
186 IF K2<0 THEN K2=0 ELSE K2=K2
187 IF K3<0 THEN K3=0 ELSE K3=K3
188 IF K2>2 THEN K2=2 ELSE K2=K2
189 IF K3>4 THEN K3=4 ELSE K3=K3
190 T1=K1*S1+K2*S2+K3*S3
191 K4=0.002857*A1
192 K5=0.002857*A1
193 K6=1
197 IF K7>K4 THEN K4=0 ELSE K4=K4-K7
198 IF K8>K5 THEN K5=0 ELSE K5=K5-K8
199 K6=K6-K9
200 T2=(1-K4-K7)*S1
```

TABLE 6.4 *(Continued)*

```
210 T3=(1-K5-K8)*S2
220 T4=K4*S1
230 T5=K5*S2
240 T6=K6*S3
250 T7=K7*S1
260 T8=K8*S2
270 T9=K9*S3
280 S1=S1+T1-T2-T4-T7
290 S2=S2+T2-T3-T5-T8
300 S3=S3+T3-T6-T9
310 S4=S4+T7
320 S5=S5+T8
330 S6=S6+T9
335 A1=S1+S2+S3
340 A2=S4+S5+S6
341 X(T)=X(T)+A1
342 XS(T)=XS(T)+A1*A1
360 NEXT T
365 NEXT REP
366 GOSUB 600
370 END
500 Z=0
501 FOR I=1 TO 12
502 Z=Z=RND(0)
503 NEXT I
504 Z=Z-6
505 RETURN
600 FOR I=1 TO 50
601 M(I)=X(I)/10
602 SD(I)=(XS(I)-X(I)*X(I)/10)/9
603 SD(I)=SQR(SD(I)/10)
604 LPRINT USING"####.##  ";I;M(I);SD(I)
605 NEXT I
606 RETURN
```

we developed models with stochastic rate equations from deterministic models elsewhere in the text, although for simplicity we chose to ignore the problem in earlier examples.) In the present case mean natality rates from replicate stochastic simulations will be lower than natality rates predicted by the corresponding deterministic simulation. Since the effect on baseline model behavior in this case is quite noticeable, we should adjust the equations representing natality in our stochastic model so that the mean baseline response is similar to

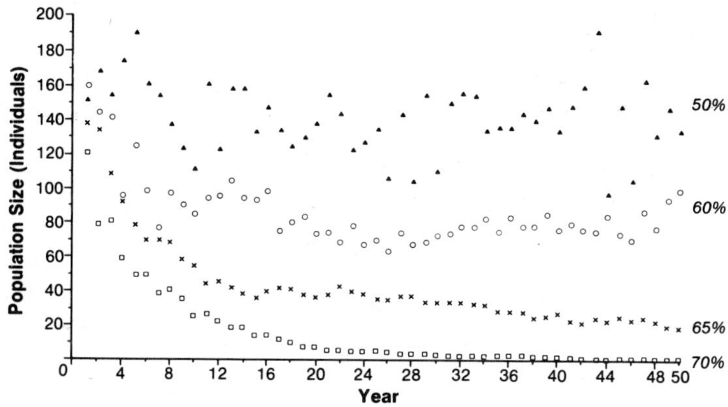

FIGURE 6.9. Population sizes over a 50-year period predicted by the stochastic model under different harvest schemes in which all age 3 individuals are harvested and the percentage of age 1 and 2 individuals harvested is as indicated on the graph. Points are means of ten replicate stochastic simulations.

baseline behavior of the deterministic model. We can do this by decreasing the slopes of the curves relating per capita natality rates to population size (Figure 6.7a) by one half. The material transfer from natality source to the age 1 state variable then is calculated as

$$T_{SO1,1} = (RV1 \times S2) + (RV2 \times S3)$$

where RV1 = normal random variate from a distribution with $\mu = 2.0 - 0.00286A1$, $\sigma^2 = 1.0$

RV2 = normal random variate from a distribution with $\mu = 4.0 - 0.00571A1$, $\sigma^2 = 1.0$

We have chosen to modify only the slopes of these curves because the *y*-intercepts presumably represent physiologically maximum age-specific natality rates.

Results of ten replicate simulations of selected harvest schemes using our modified stochastic model are presented in Figure 6.10. The mean baseline response now is similar to baseline behavior of the deterministic model (Figure 6.8) with population levels fluctuating around 175 individuals for the entire 50-year period. Mean responses of the other harvest schemes also are generally similar to the corresponding deterministic predictions. Note however that for our stochastic model we have plotted mean population sizes $\pm$ 1 standard error of the mean, thus representing the variability in our predictions of population size resulting from the random variability in natality rates.

Based on results of stochastic simulations we can calculate the probability

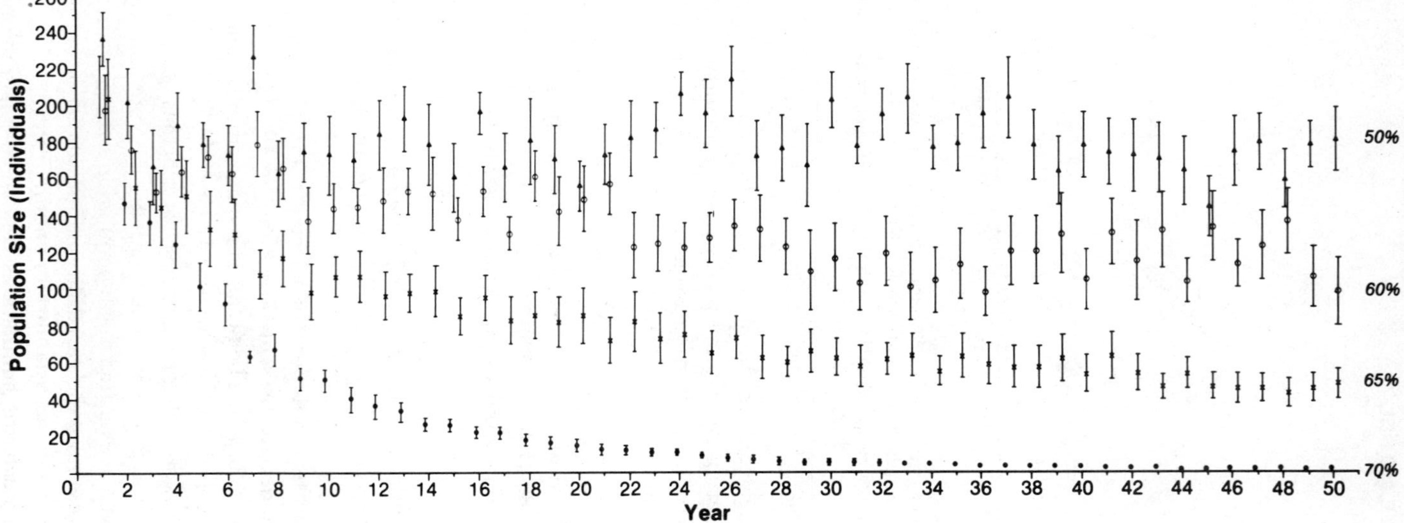

FIGURE 6.10. Population sizes over a 50-year period predicted by the modified stochastic model under different harvest schemes in which all age 3 individuals are harvested and the percentage of age 1 and 2 individuals harvested is as indicated on the graph. Points are means of ten replicate stochastic simulations. Vertical bars represent ± 1 standard error of the mean.

that the population will decline below the arbitrarily established lower limit of 25 individuals during any given year under any given harvest scheme. We can do this by calculating the proportion of the area under the normal curve representing the distribution of population sizes for a given year and harvest scheme to the left of 25 individuals. This is the same general approach that we used to determine probabilities of profitable and unprofitable years under different management schemes for the pond model (Chapter 5, p. 188). For example, mean population size $\pm$ 1 standard error for year 50 under the 64 percent harvest scheme is 49 $\pm$ 8.4 (Figure 6.10). Thus

$$\text{tabulated argument} = \frac{(Y - \bar{X})}{\text{SD}} = \frac{(25-49)}{26.6} = -0.90$$

where

Y = the lower limit on population size

$\bar{X}$ = the mean of the distribution of population sizes for year 50 under the 65 percent harvest scheme

SD = standard deviation of the mean (= standard error of the mean $\times$ $\sqrt{\text{sample size}}$)

tabulated argument = argument in a table containing areas of the normal curve indicating the number of standard deviation units between the mean and Y

If we consult a table containing areas of the normal curve (Rohlf and Sokal 1969, p. 158) we find that 0.3159, or about 32 percent, of the area under the curve lies between the mean (in this case 49) and -0.90 standard deviation units (in this case 25). Since 50 percent of the area under the curve lies to the left of the mean (<49), $0.5000 - 0.3159$, or about 18 percent, lies to the left of -0.90 standard deviation units (or to the left of 25 individuals). Thus the probability that population size will be less than 25 individuals during year 50 under the 65 percent harvest scheme is 18 percent. Similar calculations indicate a 10 percent chance that population size will be less than 25 during year 50 under the 60 percent harvest scheme and a 0.2 percent chance under the 50 percent scheme.

This example illustrates two important points:

1. Certain aspects of model behavior may result directly from the mathematical structure of the model *per se* and interpretation of unexpected behavior in ecological terms must be done cautiously.
2. Failure to explicitly represent inherent randomness in models of living systems, even when we lack adequate data to specify the required random distributions with much confidence, can seriously limit the usefulness of model results within a management framework.

The behavior of all models necessarily results from their mathematical structure. But the degree to which mathematical constraints on model behavior are ecologically interpretable varies considerably. The choice of a particular mathematical format may place certain *a priori* constraints on model behavior regardless of the specific ecological relationships being modeled. Such models still may be quite useful for certain problems as long as we are aware of the mathematical constraints involved. For example, the much used Leslie matrix model of population dynamics, when formulated in the classical manner with time-invariant age-specific natality and survival rates, can exhibit only three types of behavior once the stable age-class distribution has been attained. The simulated population must grow exponentially, remain exactly in equilibrium, or decline exponentially (the Leslie matrix model is described in Appendix 1). In explanatory models mathematical relationships represent ecological mechanisms operating in the real system and, theoretically, the mathematical constraints are ecologically interpretable. But in practice even the behavior of explanatory models is sometimes subtly influenced by mathematical rules operating quite devoid of ecological meaning. The downward bias in predicted natality rates resulting from transformation of our deterministic wildlife model to a stochastic model is an example. The point is simply that we must be cautious when providing ecological interpretations for unexpected model behavior.

Another consideration when interpreting model behavior, particularly within a management framework, is whether the model is deterministic or stochastic. Suppose that resource managers were to use results of our deterministic wildlife model as an aid in establishing harvest levels. They probably would choose to harvest 65 percent of age 1 and 2 individuals (and 100 percent of age 3 individuals) annually. Equilibrium population size under the 65 percent harvest scheme is 57 individuals, which seems comfortably above the lowest acceptable population size of 25 (Figure 6.8). The 70 percent harvest scheme depresses the population to nine individuals, which is unacceptably low, and the 60 percent scheme, with an equilibrium population of 107, seems overly conservative since a management goal is to provide as many hunters as possible with a chance to hunt. However, if resource managers examine results of our stochastic wildlife model, they find there is roughly one chance in five that population size will be less than 25 during year 50 under the 65 percent harvest scheme, even though the most likely population size is 49 (Figure 6.10). Under the 60 percent scheme there is a 10 percent chance that population size will be less than 25 during year 50, and under the 50 percent scheme there is a 0.2 percent chance. Based on this information the managers may or may not want to reconsider their choice of the 65 percent harvest scheme. But certainly explicit representation of the uncertainty associated with our predictions, which is made possible through use of a stochastic model, enhances the interpretation of simulation results considerably.

6.4. CONCLUDING REMARKS

Systems analysis and simulation are still relatively new tools in wildlife and fisheries sciences. As a result my concluding remarks regarding the current "state of the art" may seem somewhat contradictory. On the one hand there are numerous examples of wise and effective use of models in both wildlife and fisheries. We also can begin to identify a common scheme underlying the diverse ways in which wildlife and fisheries professionals have applied the systems approach to different problems. Thus a viable protocol for developing and using effective models is becoming established. On the other hand there is still considerable resistance among some workers to any use of models. Such unqualified mistrust of models results primarily from a misunderstanding of the basic principles of systems analysis and simulation. This situation can and should be remedied in the near future. But even the most experienced ecological modelers retain considerable skepticism about the models they develop and use. This is a healthy situation given the current state of the art, and one that should persist for some time. Adherence to the modeling guidelines set forth in this (or any other) book does not ensure that the resulting model will provide correct answers to our questions. Ecological modeling is not yet a well enough defined science that we can set forth universally applicable procedures and standards. Although we can guide beginning modelers past many potential pitfalls, we cannot yet provide a comprehensive list of modeling problems, much less a corresponding list of guaranteed solutions. Personal modeling experience and a questioning mind remain among the most valuable assets of the modeler.

APPENDIX 1

MATRIX ALGEBRA

A1.1. INTRODUCTION

This appendix presents an overview of matrix algebra and describes some uses of matrix algebra in wildlife and fisheries sciences. Only basic matrix terminology and calculations, and a few more involved calculations used in specific applications of interest, are presented. There are several excellent matrix algebra texts available that provide detailed treatment of the subject (e.g., Searle 1966).

One of the most common uses of matrix algebra in wildlife or fisheries ecology is to simulate population dynamics (Usher 1972). Using what is known as a Leslie matrix (Leslie 1945, 1948), which contains information about age-specific natality and survival rates of a population, we can predict via simulation the future size and age structure of a population. Under certain assumptions we also can treat the Leslie matrix analytically to determine the rate at which a population is increasing or decreasing and the stable age-class distribution.

Another interesting use of matrix algebra is to simulate ecological succession (Bledsoe and Van Dyne 1971). If the probabilities of succession proceeding from one seral stage to another are arranged appropriately in a probability transition matrix, the probability of a community being in each seral stage at specific future times can be predicted via simulation. We also can treat the probability transition matrix analytically to determine the mean length of time that a community spends in each seral stage and some other interesting attributes of the successional process.

We will examine both the Leslie matrix model for simulating population dynamics and the probability transition matrix model for simulating ecological succession later, but first we present some basic matrix terminology and calculations.

A1.2. MATRIX TERMINOLOGY AND BASIC MATRIX CALCULATIONS

Terminology

A matrix is a rectangular array of numbers or functions arranged in m rows and n columns in which the arrangement is meaningful. The elements of a matrix are individual entries in a matrix that are uniquely identified by the row and the column in which they occur. The order or dimension of a matrix is defined by the number of rows and columns. For example, matrix **A** below is a 3×4 matrix (3 rows by 4 columns) and the number 6 is the 2,1 element of that matrix (element in the second row, first column).

$$\mathbf{A} = \begin{bmatrix} 0 & 7 & 1 & 10 \\ 6 & -8 & 15 & -3 \\ 8 & 2 & 4 & 12 \end{bmatrix}$$

A vector is a special matrix that has either one column (column vector) or one row (row vector). A scalar is a 1×1 matrix, or a single number. For example, vector **x** below is a column vector and vector **z** is a row vector.

$$\mathbf{x} = \begin{bmatrix} 6 \\ 7 \\ 2 \\ -1 \end{bmatrix}, \qquad \mathbf{z} = \begin{bmatrix} -5 & 16 & 3 & 7 \end{bmatrix}$$

Most commonly the array of numbers representing a matrix is enclosed in brackets. Symbolically matrices are usually represented by the beginning capital letters of the alphabet (A,B,C, . . .), vectors by the ending lower case letters (. . . ,x,y,z), and scalars by the beginning lower case letters (a,b,c, . . .). Elements of matrices and vectors are usually represented by doubly subscripting lower case letters corresponding to the letter designating the entire matrix or vector, with the first and second subscripts representing the row and column, respectively, in which the element occurs. Matrix **A** below is a $m \times n$ matrix (m rows $\times$ n columns) and each element a_{ij} represents a number in the ith row, jth column.

$$\mathbf{A} = \begin{bmatrix} a_{11} & a_{12} & a_{13} & . & . & . & a_{1n} \\ a_{21} & a_{22} & a_{23} & . & . & . & a_{2n} \\ a_{31} & a_{32} & a_{33} & . & . & . & a_{3n} \\ . & . & . & . & . & . & . \\ . & . & . & . & . & . & . \\ . & . & . & . & . & . & . \\ a_{m1} & a_{m2} & a_{m3} & . & . & . & a_{mn} \end{bmatrix}$$

Matrix Addition and Subtraction

Matrix addition and subtraction are defined only for matrices of the same order. If two matrices are of the same order they can be added or subtracted element by element.

$$\begin{bmatrix} 5 & 3 \\ 1 & -4 \\ 1 & 0 \end{bmatrix} + \begin{bmatrix} 11 & 0 \\ 1 & 6 \\ 7 & -3 \end{bmatrix} = \begin{bmatrix} 16 & 3 \\ 2 & 2 \\ 8 & -3 \end{bmatrix}$$

$$\begin{bmatrix} 5 & 3 \\ 1 & -4 \\ 1 & 0 \end{bmatrix} - \begin{bmatrix} 11 & 0 \\ 1 & 6 \\ 7 & -3 \end{bmatrix} = \begin{bmatrix} -6 & 3 \\ 0 & -10 \\ -6 & 3 \end{bmatrix}$$

Matrix Multiplication

A matrix may be multiplied by a scalar (scalar multiplication) by multiplying each element of the matrix by the scalar.

$$5 \begin{bmatrix} 5 & 3 \\ 1 & -4 \\ 1 & 0 \end{bmatrix} = \begin{bmatrix} 25 & 15 \\ 5 & -20 \\ 5 & 0 \end{bmatrix}$$

Matrix multiplication is defined only for pairs of matrices in which the number of columns of the first matrix (called the pre-multiplier) is equal to the number of rows of the second matrix (called the post-multiplier). When two matrices that meet this dimensional constraint are multiplied the result is a matrix that has the same number of rows as the pre-multiplier and the same number of columns as the post-multiplier. The element in the ith row and jth column of the resulting matrix is the sum of products of elements in the ith row of the pre-multiplier and the jth column of the post-multiplier.

$$\begin{bmatrix} 3 \\ 3 \\ 1 \end{bmatrix} \begin{bmatrix} 2 & 1 \end{bmatrix} = \begin{bmatrix} 6 & 3 \\ 6 & 3 \\ 2 & 1 \end{bmatrix} \text{ or } \begin{bmatrix} (3 \times 2) & (3 \times 1) \\ (3 \times 2) & (3 \times 1) \\ (1 \times 2) & (1 \times 1) \end{bmatrix}$$

$$\begin{bmatrix} 1 & 5 \\ 2 & 4 \end{bmatrix} \begin{bmatrix} 3 \\ 6 \end{bmatrix} = \begin{bmatrix} 33 \\ 30 \end{bmatrix} \text{ or } \begin{bmatrix} (1 \times 3) + (5 \times 6) \\ (2 \times 3) + (4 \times 6) \end{bmatrix}$$

$$\begin{bmatrix} 1 & 5 \\ 2 & 4 \end{bmatrix} \begin{bmatrix} 3 & 2 \\ 6 & 1 \end{bmatrix} = \begin{bmatrix} 33 & 7 \\ 30 & 8 \end{bmatrix} \text{ or } \begin{bmatrix} (1 \times 3) + (5 \times 6) & (1 \times 2) + (5 \times 1) \\ (2 \times 3) + (4 \times 6) & (2 \times 2) + (4 \times 1) \end{bmatrix}$$

Matrix Inversion

There is no such thing as matrix division. However pre-multiplication of a matrix by its inverse is the matrix algebra analog to division. Calculating the inverse of a matrix requires use of several additional matrix operations including matrix transposition and calculation of the determinant and cofactors of a matrix.

Matrix Transposition

A matrix is transposed by interchanging its rows and columns such that each element a_{ij} of the original matrix **A** becomes the a_{ji} element of the transposed matrix **A′** (called **A** transpose).

If

$$\mathbf{A} = \begin{bmatrix} 3 & 4 & 2 \\ 6 & 7 & 9 \\ 1 & 3 & 5 \end{bmatrix}$$

then

$$\mathbf{A}' = \begin{bmatrix} 3 & 6 & 1 \\ 4 & 7 & 3 \\ 2 & 9 & 5 \end{bmatrix}$$

or if

$$\mathbf{A} = \begin{bmatrix} 6 & 4 & 9 \\ 1 & 7 & 0 \end{bmatrix}$$

then

$$\mathbf{A}' = \begin{bmatrix} 6 & 1 \\ 4 & 7 \\ 9 & 0 \end{bmatrix}$$

Determinant of a Matrix

A determinant is the sum of certain products of the elements of the matrix from which it is derived. Determinants are defined only for square matrices. The determinant of matrix **A** usually is indicated as $|\mathbf{A}|$. The determinant of a 1×1 matrix (scalar) is the value of its sole element. The determinant of a 2×2 matrix (second-order matrix) is calculated by subtracting the product of off-diagonal elements from the product of diagonal elements.

If

$$\mathbf{A} = \begin{bmatrix} 7 & 3 \\ 4 & 6 \end{bmatrix}$$

then

$$|\mathbf{A}| = \begin{vmatrix} 7 & 3 \\ 4 & 6 \end{vmatrix} = (7 \times 6) - (3 \times 4) = 30$$

The determinant of a 3×3 (third-order) matrix is calculated as a linear function of three second-order determinants derived from the original third-order determinant. The determinant of the 3×3 matrix below can be calculated by expansion of elements of the first row (1,2,3).

If

$$\mathbf{A} = \begin{bmatrix} 1 & 2 & 3 \\ 4 & 5 & 6 \\ 7 & 8 & 10 \end{bmatrix}$$

then

$$|\mathbf{A}| = \begin{vmatrix} 1 & 2 & 3 \\ 4 & 5 & 6 \\ 7 & 8 & 10 \end{vmatrix}$$

$$= 1(+1)\begin{vmatrix} 5 & 6 \\ 8 & 10 \end{vmatrix} + 2(-1)\begin{vmatrix} 4 & 6 \\ 7 & 10 \end{vmatrix} + 3(+1)\begin{vmatrix} 4 & 5 \\ 7 & 8 \end{vmatrix}$$

$$= 1(50 - 48) - 2(40 - 42) + 3(32 - 35) = -3$$

Each of the three second-order determinants in the expansion is formed by elements of the third-order determinant remaining after deleting the row and column in which one of the expanded elements occurs. Each second-order determinant is multiplied by the expanded element with which it is associated and also by either $+1$ or -1. If the sum of the row number and column number in which the expanded element occurs is even (such as for the 1,1 element) the multiplier is $+1$. If the sum is odd (such as for the 2,1 element) the multiplier is -1. A matrix can be evaluated by expansion of elements of any row or column. We might evaluate the 3×3 matrix above by elements of the first column (1,4,7) rather than by elements of the first row (1,2,3).

$$|\mathbf{A}| = \begin{vmatrix} 1 & 2 & 3 \\ 4 & 5 & 6 \\ 7 & 8 & 10 \end{vmatrix}$$

$$= 1(+1)\begin{vmatrix} 5 & 6 \\ 8 & 10 \end{vmatrix} + 4(-1)\begin{vmatrix} 2 & 3 \\ 8 & 10 \end{vmatrix} + 7(+1)\begin{vmatrix} 2 & 3 \\ 5 & 6 \end{vmatrix}$$

$$= 1(50 - 48) - 4(20 - 24) + 7(12 - 15) = -3$$

Determinants of matrices of orders higher than 3 can be calculated in the same manner by successive expansions.

Cofactors of a Matrix

Each element of a matrix has an associated cofactor that can be calculated as illustrated below.

If

$$\mathbf{A} = \begin{bmatrix} 1 & 2 & 3 \\ 4 & 5 & 6 \\ 7 & 8 & 10 \end{bmatrix}$$

then the cofactor of

$$1 = (+1) \begin{vmatrix} 5 & 6 \\ 8 & 10 \end{vmatrix} = 2$$

the cofactor of

$$4 = (-1) \begin{vmatrix} 2 & 3 \\ 8 & 10 \end{vmatrix} = 4$$

the cofactor of

$$7 = (+1) \begin{vmatrix} 2 & 3 \\ 5 & 6 \end{vmatrix} = -3$$

and so on. Note that calculation of the cofactor associated with a given element is identical to calculation of the term in the determinant expansion associated with that element, except that the element itself does not enter into the calculation.

Identity Matrix

Finally, before defining the inverse of a matrix and presenting the steps by which it is calculated, we must define the identity matrix (usually symbolized by $\mathbf{I}$). An identity matrix has diagonal elements equal to one and all other elements equal to zero. For example a 3×3 identity matrix has the form

$$\mathbf{I} = \begin{bmatrix} 1 & 0 & 0 \\ 0 & 1 & 0 \\ 0 & 0 & 1 \end{bmatrix}$$

Calculating the Inverse

We can now define the matrix inverse. Pre-multiplication of a matrix ($\mathbf{A}$) by its inverse (usually denoted as $\mathbf{A}^{-1}$) results in the identity matrix ($\mathbf{I}$). This operation is the matrix algebra analog to division. To calculate the inverse of a matrix we:

1. calculate the determinant of the matrix,
2. calculate the cofactor for each element of the matrix and form a new matrix (cofactor matrix) by replacing each element of the original matrix by its cofactor,
3. transpose the cofactor matrix to form the adjugate matrix,
4. multiply the adjugate matrix by the inverse of the determinant to obtain the matrix inverse, and
5. pre-multiply the matrix by the matrix inverse to verify that the result is the identity matrix.

If

$$\mathbf{A} = \begin{bmatrix} 1 & 2 & 3 \\ 4 & 5 & 6 \\ 7 & 8 & 10 \end{bmatrix}$$

then

$$|\mathbf{A}| = -3$$

$$\text{the cofactor matrix} = \begin{bmatrix} 2 & 2 & -3 \\ 4 & -11 & 6 \\ -3 & 6 & -3 \end{bmatrix}$$

$$\text{the adjugate matrix} = \begin{bmatrix} 2 & 4 & -3 \\ 2 & -11 & 6 \\ -3 & 6 & -3 \end{bmatrix}$$

and

$$\mathbf{A}^{-1} = \frac{1}{-3}\begin{bmatrix} 2 & 4 & -3 \\ 2 & -11 & 6 \\ -3 & 6 & -3 \end{bmatrix} = \begin{bmatrix} -2/3 & -4/3 & 3/3 \\ -2/3 & 11/3 & -6/3 \\ 3/3 & -6/3 & 3/3 \end{bmatrix}$$

and as a check we verify that $\mathbf{A}^{-1}\mathbf{A} = \mathbf{I}$

$$\begin{bmatrix} -2/3 & -4/3 & 3/3 \\ -2/3 & 11/3 & -6/3 \\ 3/3 & -6/3 & 3/3 \end{bmatrix}\begin{bmatrix} 1 & 2 & 3 \\ 4 & 5 & 6 \\ 7 & 8 & 10 \end{bmatrix} = \begin{bmatrix} 1 & 0 & 0 \\ 0 & 1 & 0 \\ 0 & 0 & 1 \end{bmatrix}$$

For a 2 × 2 matrix we can obtain the adjugate matrix directly from the original matrix, and therefore skip steps 2 and 3 above, by exchanging the diagonal elements and changing the signs on the off-diagonal elements.

If

$$\mathbf{A} = \begin{bmatrix} 2 & 5 \\ 3 & 8 \end{bmatrix}$$

then

$$\text{the adjugate matrix} = \begin{bmatrix} 8 & -5 \\ -3 & 2 \end{bmatrix}$$

A1.3. EXAMPLES OF USE OF MATRIX ALGEBRA IN WILDLIFE AND FISHERIES SCIENCES

Leslie Matrix

Simulating Population Dynamics

A common use of matrix algebra in wildlife or fisheries ecology is to simulate population dynamics using a Leslie matrix format. A Leslie matrix is a square matrix in which elements in the first row represent age-specific natality rates of a population and sub-diagonal elements (elements directly below diagonal elements) represent age-specific survival rates.

$$\mathbf{M} = \begin{bmatrix} f_0 & f_1 & f_2 & . & . & . & f_k \\ p_0 & 0 & 0 & . & . & . & 0 \\ 0 & p_1 & 0 & . & . & . & 0 \\ 0 & 0 & p_2 & . & . & . & . \\ . & . & . & . & . & . & . \\ . & . & . & . & . & . & . \\ . & . & . & . & . & . & . \\ 0 & 0 & 0 & . & . & p_{k-1} & 0 \end{bmatrix}$$

where $\mathbf{M}$ = Leslie matrix

f_i = number of females alive at time $t+1$ that were born during time t to $t+1$ to a female aged i at time t

p_i = probability that a female aged i at time t survives to time $t + 1$
$k + 1$ = number of age classes

Note that natality rates (f_i) of the Leslie matrix are defined exactly like m_x values of a standard life table and that survivorship rates (p_i) are defined such that they equal l_{x+1}/l_x in terms of l_x values of a standard life table. If we also define a vector that contains as its elements the numbers of individuals in each age class in the population at time t,

$$\mathbf{v}_t = \begin{bmatrix} v_{0,t} \\ v_{1,t} \\ v_{2,t} \\ v_{3,t} \\ \cdot \\ \cdot \\ \cdot \\ v_{k,t} \end{bmatrix}$$

where $\mathbf{v}_t$ = age–class vector at time t
$v_{i,t}$ = number of individuals in age class i at time t

then the result of pre-multiplying the age–class vector at time t by the Leslie matrix is another age-class vector that represents the number of individuals in each age class at time $t + 1$.

$$\mathbf{Mv}_t = \mathbf{v}_{t+1}$$

or

$$\begin{bmatrix} f_0 & f_1 & f_2 & \cdot & \cdot & \cdot & f_k \\ p_0 & 0 & 0 & \cdot & \cdot & \cdot & 0 \\ 0 & p_1 & 0 & \cdot & \cdot & \cdot & 0 \\ 0 & 0 & p_2 & \cdot & \cdot & \cdot & \cdot \\ \cdot & \cdot & \cdot & \cdot & \cdot & \cdot & \cdot \\ \cdot & \cdot & \cdot & \cdot & \cdot & \cdot & \cdot \\ \cdot & \cdot & \cdot & \cdot & \cdot & \cdot & \cdot \\ 0 & 0 & 0 & \cdot & \cdot & p_{k-1} & 0 \end{bmatrix} \begin{bmatrix} v_{0,t} \\ v_{1,t} \\ v_{2,t} \\ v_{3,t} \\ \cdot \\ \cdot \\ \cdot \\ v_{k,t} \end{bmatrix} = \begin{bmatrix} v_{0,t+1} \\ v_{1,t+1} \\ v_{2,t+1} \\ v_{3,t+1} \\ \cdot \\ \cdot \\ \cdot \\ v_{k,t+1} \end{bmatrix}$$

or

$$\mathbf{v}_{0,t+1} = f_0 v_{0,t} + f_1 v_{1,t} + f_2 v_{2,t} + \ldots + f_k v_{k,t}$$

$$\mathbf{v}_{1,t+1} = p_0 v_{0,t}$$

$$\mathbf{v}_{2,t+1} = p_1 v_{1,t}$$

$$\mathbf{v}_{3,t+1} = p_2 v_{2,t}$$

$$\vdots$$

$$\mathbf{v}_{k,t+1} = p_{k-1} v_{k-1,t}$$

By iteratively pre-multiplying the present age-class vector by the Leslie matrix we can project both total population size and age-class distribution into the future.

$$\mathbf{M}\mathbf{v}_{t+1} = \mathbf{v}_{t+2}, \quad \mathbf{M}\mathbf{v}_{t+2} = \mathbf{v}_{t+3}, \quad \text{etc.}$$

Consider a population represented by the following age-specific natality (f_i) and survival (p_i) rates and initial age-class distribution ($\mathbf{v}_{i,t}$).

$$f_0 = 0 \qquad p_0 = 0.33 \qquad v_{0,t} = 0$$

$$f_1 = 9 \qquad p_1 = 0.5 \qquad v_{1,t} = 0$$

$$f_2 = 12 \qquad \qquad v_{2,t} = 1$$

Thus $\mathbf{M}\mathbf{v}_t = \mathbf{v}_{t+1}$ becomes

$$\begin{bmatrix} 0 & 9 & 12 \\ 0.33 & 0 & 0 \\ 0 & 0.5 & 0 \end{bmatrix} \begin{bmatrix} 0 \\ 0 \\ 1 \end{bmatrix} = \begin{bmatrix} 12 \\ 0 \\ 0 \end{bmatrix}$$

$\mathbf{M}\mathbf{v}_{t+1} = \mathbf{v}_{t+2}$ becomes

$$\begin{bmatrix} 0 & 9 & 12 \\ 0.33 & 0 & 0 \\ 0 & 0.5 & 0 \end{bmatrix} \begin{bmatrix} 12 \\ 0 \\ 0 \end{bmatrix} = \begin{bmatrix} 0 \\ 3.96 \\ 0 \end{bmatrix}$$

$\mathbf{M}\mathbf{v}_{t+2} = \mathbf{v}_{t+3}$ becomes

$$\begin{bmatrix} 0 & 9 & 12 \\ 0.33 & 0 & 0 \\ 0 & 0.5 & 0 \end{bmatrix} \begin{bmatrix} 0 \\ 3.96 \\ 0 \end{bmatrix} = \begin{bmatrix} 35.64 \\ 0 \\ 1.98 \end{bmatrix}$$

and so on.

If population dynamics are simulated in this manner the population will attain a stable age-class distribution (unless the population becomes extinct first). A point will be reached when the proportion of individuals in each age class will remain constant over time. After the stable age-class distribution is reached the population will either increase exponentially, decrease exponentially, or remain exactly stable. If we continue to simulate dynamics of the population above we will reach a point when the proportion of individuals in age classes 0, 1, and 2 will remain at 0.83, 0.14, and 0.03, respectively. After this stable age-class distribution is reached the population will double in size each time step. Thus the finite rate of increase of the population is 2.

Analytical Treatment of Leslie Matrix

The Leslie matrix also can be analyzed directly to obtain information on rates of population increase or decrease and to determine the stable age-class distribution. We may represent conditions under which a stable age-class distribution exists as

$$\mathbf{M}\mathbf{v}_s = \lambda\mathbf{v}_s$$

where $\mathbf{M}$ and $\mathbf{v}_s$ are the Leslie matrix and the stable age-class vector, respectively, and λ is a scalar.

If the proportion of individuals in each age class is not changing over time there is a scalar (λ) that when multiplied times the age-class vector at any time t yields the age-class vector at time $t + 1$. λ represents the finite rate of increase (or decrease) of the population. Calculating the finite rate of increase and stable age-class distribution of a population from a Leslie matrix involves solving the equation $\mathbf{M}\mathbf{v}_s = \lambda\mathbf{v}_s$. This is done analytically by calculating eigenvalues and eigenvectors (or latent roots and latent vectors) of the Leslie matrix. λ is the dominant (largest) positive real eigenvalue of the Leslie matrix and $\mathbf{v}_s$ is the eigenvector associated with this eigenvalue. A detailed description of this procedure is presented by Searle (1966).

Two other characteristics of a population can be calculated directly from the finite rate of population increase (λ). One is the instantaneous rate of population increase (r) which may be calculated as $r = \ln \lambda$. The other is the percentage of the population that can be harvested on a sustained-yield basis (H) which may be calculated as $H = 100((\lambda - 1)/\lambda)$.

Leslie Matrix with Time-Varying Rates

We have examined Leslie matrices in which elements representing age-specific natality and survival rates are constant. Although Leslie matrices are classically represented in this way it is possible to relax the constraint that age-specific rates are constant and allow them to be time-varying functions. For example we may want to represent either natality or survival rates, or both, as functions of population density. This poses no problem in the simulation of population dynamics. Appropriate elements of the Leslie matrix are recalculated each time step according to appropriate rules prior to updating the age-class vector through pre-multiplication by the Leslie matrix. However, the analytical treatment of the Leslie matrix presented above is very difficult if age-specific rates are time-varying.

Probability Transition Matrix

Simulating Ecological Succession

Another interesting use of matrix algebra is to simulate the process of ecological succession. This is done through use of a probability transition matrix. A probability transition matrix is a square matrix whose elements represent the probabilities of a community changing from seral stage j to seral stage i during the next time step.

$$\mathbf{P} = \begin{bmatrix} p_{11} & p_{12} & p_{13} & \cdot & \cdot & \cdot & p_{1k} \\ p_{21} & p_{22} & p_{23} & \cdot & \cdot & \cdot & p_{2k} \\ p_{31} & p_{32} & p_{33} & \cdot & \cdot & \cdot & p_{3k} \\ \cdot & \cdot & \cdot & \cdot & \cdot & \cdot & \cdot \\ \cdot & \cdot & \cdot & \cdot & \cdot & \cdot & \cdot \\ \cdot & \cdot & \cdot & \cdot & \cdot & \cdot & \cdot \\ p_{k1} & p_{k2} & p_{k3} & \cdot & \cdot & \cdot & p_{kk} \end{bmatrix}$$

where $\mathbf{P}$ = probability transition matrix
p_{ij} = probability of the community changing from seral stage j to seral stage i during the next time step

Note that with the p_{ij}s defined in this manner, each column of **P** must sum to 1.0. We also define a state vector that contains as its elements the probabilities of the community being in each seral stage at time t,

$$\mathbf{z}_t = \begin{bmatrix} z_{1,t} \\ z_{2,t} \\ z_{3,t} \\ \cdot \\ \cdot \\ \cdot \\ z_{k,t} \end{bmatrix}$$

where $\mathbf{z}_t$ = state vector at time t

$z_{i,t}$ = probability of the community being in seral stage i at time t

If we pre-multiply the state vector at time t by the probability transition matrix the result is another state vector that represents the probabilities of the community being in each seral stage at time $t + 1$.

$$\mathbf{P}\mathbf{z}_t = \mathbf{z}_{t+1}$$

By iteratively pre-multiplying the present state vector by the probability transition matrix we can simulate succession in a manner similar to the simulation of population dynamics using the Leslie matrix.

$$\mathbf{P}\mathbf{z}_{t+1} = \mathbf{z}_{t+2}, \qquad \mathbf{P}\mathbf{z}_{t+2} = \mathbf{z}_{t+3}, \quad \text{etc.}$$

If we simulate succession in this manner the state vector will eventually reach steady state. Probabilities of the community being in each seral stage will no longer change with time and the stable state vector will represent the probabilities that various climax communities will be attained.

Consider the following probability transition matrix (**P**), which represents a successional process including a bare dunes (BD) community, a grass (G) community, and an oak (O) community.

$$\mathbf{P} = \begin{matrix} & \text{BD} & \text{G} & \text{O} \\ \text{BD} & 0.3 & 0.2 & 0.1 \\ \text{G} & 0.6 & 0.5 & 0.1 \\ \text{O} & 0.1 & 0.3 & 0.8 \end{matrix}$$

where $p_{1,1} = 0.3 =$ probability of succession remaining in the bare dunes stage for the next time unit

$p_{2,1} = 0.6 =$ probability of succession proceeding from bare dunes to the grass stage during the next time unit

$p_{3,1} = 0.1 =$ probability of succession proceeding from bare dunes to the oak stage during the next time unit, and so on.

If we define the initial state vector ($\mathbf{z}_t$) as

$$\mathbf{z}_t = \begin{bmatrix} 0.1 \\ 0.1 \\ 0.8 \end{bmatrix}$$

then $\mathbf{Pz}_t = \mathbf{z}_{t+1}$ becomes

$$\begin{bmatrix} 0.3 & 0.2 & 0.1 \\ 0.6 & 0.5 & 0.1 \\ 0.1 & 0.3 & 0.8 \end{bmatrix} \begin{bmatrix} 0.1 \\ 0.1 \\ 0.8 \end{bmatrix} = \begin{bmatrix} 0.13 \\ 0.19 \\ 0.68 \end{bmatrix}$$

$\mathbf{Pz}_{t+1} = \mathbf{z}_{t+2}$ becomes

$$\begin{bmatrix} 0.3 & 0.2 & 0.1 \\ 0.6 & 0.5 & 0.1 \\ 0.1 & 0.3 & 0.8 \end{bmatrix} \begin{bmatrix} 0.13 \\ 0.19 \\ 0.68 \end{bmatrix} = \begin{bmatrix} 0.145 \\ 0.241 \\ 0.614 \end{bmatrix}$$

$\mathbf{Pz}_{t+2} = \mathbf{z}_{t+3}$ becomes

$$\begin{bmatrix} 0.3 & 0.2 & 0.1 \\ 0.6 & 0.5 & 0.1 \\ 0.1 & 0.3 & 0.8 \end{bmatrix} \begin{bmatrix} 0.145 \\ 0.241 \\ 0.614 \end{bmatrix} = \begin{bmatrix} 0.153 \\ 0.269 \\ 0.578 \end{bmatrix}$$

and so on. If we continue this process eventually we will obtain the steady state vector

$$\mathbf{z}_\infty = \begin{bmatrix} 0.163 \\ 0.302 \\ 0.535 \end{bmatrix}$$

Thus the probabilities of the climax community being bare dunes, grass, and oak are about 0.163, 0.302, and 0.535, respectively.

Analytical Treatment of Probability Transition Matrix

The probability transition matrix also can be analyzed directly to obtain additional information about the successional process. We can determine the mean length of time that a community spends in any given seral stage before first reaching any other seral stage, given that succession began from any specified stage. We also can determine the mean length of time it takes to first reach any given seral stage from any other stage.

Consider a probability transition matrix (**P**) whose elements (p_{ij}) are as defined above

$$\mathbf{P} = \begin{bmatrix} p_{11} & p_{12} & p_{13} \\ p_{21} & p_{22} & p_{23} \\ p_{31} & p_{32} & p_{33} \end{bmatrix}$$

Suppose that we want to calculate the mean length of time that the community spends in seral stage 2 before reaching seral stage 3, given that succession began in stage 1. We first create a new matrix (**Q**) that consists of elements of **P** remaining after the row and column representing transition probabilities to and from stage 3 are eliminated.

$$\mathbf{Q} = \begin{bmatrix} p_{11} & p_{12} \\ p_{21} & p_{22} \end{bmatrix}$$

We then calculate $(\mathbf{I} - \mathbf{Q})^{-1}$ to form another new matrix (**R**).

$$(\mathbf{I} - \mathbf{Q})^{-1} = \mathbf{R} = \begin{bmatrix} r_{11} & r_{12} \\ r_{21} & r_{22} \end{bmatrix}$$

R is a matrix of mean residence times. Each r_{ij} represents the mean length of time spent in seral stage i (mean residence time) before first reaching the seral stage that has been dropped from **P** to obtain **Q**, given that succession began in stage j. In terms of our original question r_{21} represents mean length of time spent in seral stage 2 before first reaching seral stage 3, given that succession began in stage 1.

If we also are interested in the mean length of time that it takes to first reach seral stage 3 from seral stage 1, we can calculate this as the sum of elements in the first column of **R**. Likewise if we want to know the mean length of time that it takes to first reach seral stage 3 from seral stage 2, we can sum the second column of **R**. In general Σr_{ij} for a given j represents the mean length of time that it takes to first reach the seral stage eliminated from **P** to obtain **Q**, given that succession began in stage j.

Suppose that **P** is the probability transition matrix for our bare dunes (BD) – grass (G) – oak (O) successional process

$$\mathbf{P} = \begin{array}{c} \\ BD \\ G \\ O \end{array} \begin{array}{c} \begin{array}{ccc} BD & G & O \end{array} \\ \begin{bmatrix} 0.3 & 0.2 & 0.1 \\ 0.6 & 0.5 & 0.1 \\ 0.1 & 0.3 & 0.8 \end{bmatrix} \end{array}$$

If we want to calculate the mean length of time spent in grass before first succeeding to oak, given that succession began in bare dunes, we first create a new matrix (**Q**) that consists of elements of **P** remaining after we eliminate the row and column representing transition probabilities to and from oak.

$$\mathbf{Q} = \begin{array}{c} \\ BD \\ G \end{array} \begin{array}{c} \begin{array}{cc} BD & G \end{array} \\ \begin{bmatrix} 0.3 & 0.2 \\ 0.6 & 0.5 \end{bmatrix} \end{array}$$

Then we calculate $(\mathbf{I} - \mathbf{Q})^{-1}$.

$$(\mathbf{I} - \mathbf{Q}) = \begin{bmatrix} 1 & 0 \\ 0 & 1 \end{bmatrix} - \begin{bmatrix} 0.3 & 0.2 \\ 0.6 & 0.5 \end{bmatrix} = \begin{bmatrix} 0.7 & -0.2 \\ -0.6 & 0.5 \end{bmatrix}$$

$$(\mathbf{I} - \mathbf{Q})^{-1} = \frac{1}{(0.7 \times 0.5) - (-0.6 \times -0.2)} \begin{bmatrix} 0.5 & 0.2 \\ 0.6 & 0.7 \end{bmatrix}$$

$$(\mathbf{I} - \mathbf{Q})^{-1} = \frac{1}{0.23} \begin{bmatrix} 0.5 & 0.2 \\ 0.6 & 0.7 \end{bmatrix} = \begin{bmatrix} 0.5/0.23 & 0.2/0.23 \\ 0.6/0.23 & 0.7/0.23 \end{bmatrix}$$

$$(\mathbf{I} - \mathbf{Q})^{-1} = \begin{array}{c} \\ BD \\ G \end{array} \begin{array}{c} \begin{array}{cc} BD & G \end{array} \\ \begin{bmatrix} 2.17 & 0.87 \\ 2.61 & 3.04 \end{bmatrix} \end{array} = \mathbf{R}$$

The 2,1 element of **R** ($r_{21} = 2.61$) represents the mean length of time spent in grass (or mean residence time) before first succeeding to oak, given that succession began in bare dunes. Likewise, r_{11} represents the mean residence time in bare dunes before first passage to oak, given that succession began in bare dunes; r_{12} represents the mean residence time in bare dunes before first passage to oak, given that succession began in grass; and r_{22} represents the mean residence time in grass before first passage to oak, given that succession began in grass.

If we want to calculate the mean length of time that it takes to first succeed from bare dunes to oak, we sum the elements of the first column of **R**:

$$\sum_{i=1}^{i=2} r_{i1} = r_{11} + r_{21} = 2.17 + 2.61 = 4.78$$

Likewise the mean residence time in the system before first succeeding to oak from grass is the sum of elements of the second column of **R**:

$$\sum_{i=1}^{i=2} r_{i2} = r_{12} + r_{22} = 0.87 + 3.04 = 3.91$$

Probability Transition Matrix with Time-Varying Probabilities

We have examined probability transition matrices in which the elements representing transition probabilities are constant. As with the Leslie matrix it is possible to relax the constraint that transition probabilities are constant and allow them to be functions. For example we may want to represent successional probabilities as functions of current values in the state vector. This poses no problem in simulation but analytical treatment of the probability transition matrix presented above is very difficult if probabilities are time-varying.

APPENDIX 2

DIFFERENTIAL AND DIFFERENCE EQUATIONS

A2.1. INTRODUCTION

This appendix presents an overview of differential and difference equations and describes some uses of these equations in wildlife and fisheries sciences. Only the most basic concepts and terminology from calculus are presented. There are various excellent calculus texts available that provide detailed treatment of the subject (e.g., Salas and Hille 1982).

Differential and difference equations are used widely in wildlife and fisheries modeling to represent the rate of change in one variable with respect to the rate of change of another variable. Strictly speaking, if the variables involved are thought of as continuous—if they can be represented by a continuous set of variate values—differential equations are used. If the variables involved are thought of as discrete—if they can be represented only by discrete variate values—difference equations are used. However, this strict distinction is not always observed in practice. In any case differential equations represent instantaneous rates of change whereas difference equations represent finite rates of change.

Familiar ecological examples of differential equations include the Lotka–Volterra competition equations (Smith 1980, p. 505) of the form

$$\frac{dN_1}{dt} = r_1 N_1 \left(\frac{K_1 - N_1 - \alpha N_2}{K_1} \right)$$

$$\frac{dN_2}{dt} = r_2 N_2 \left(\frac{K_2 - N_2 - \beta N_1}{K_2} \right)$$

where dN_1/dt = instantaneous rate of change in size of population N_1
r_1 = intrinsic rate of increase of population N_1

K_1 = carrying capacity of the environment for population N_1
α = competition coefficient representing the negative effect on growth of population N_1 caused by the addition of one individual of population N_2.

dN_2/dt, r_2, K_2 and β are defined similarly for population N_2.

Familiar ecological examples of difference equations include Lindeman's trophic dynamic model of energy transfer (Smith 1980, p. 141) which has the form

$$\frac{\Delta \Lambda_n}{\Delta t} = \lambda_n - \lambda_n'$$

where $\Delta\Lambda_n/\Delta t$ = finite rate of change of energy content of trophic level n over time
λ_n = energy received from trophic level $n - 1$
λ_n' = energy passed on to trophic level $n + 1$

Population growth often is represented in both differential and difference equation format. Population growth in an unlimited environment, or exponential growth, can be represented by the differential equation

$$\frac{dN}{dt} = rN$$

where dN/dt = instantaneous rate of change in size of population N
r = intrinsic rate of population increase (an instantaneous rate)

or by the difference equation

$$\frac{\Delta N}{\Delta t} = N_t - N_{t-1}$$

where $\Delta N/\Delta t$ = finite rate of change in size of population N
N_t = population size at time t

A2.2. DIFFERENTIAL AND DIFFERENCE EQUATION TERMINOLOGY AND BASIC CALCULUS CONCEPTS

The goal of this section is to develop an intuitive sense of what differential and difference equations are and to become familiar with some terminology associated with their use. We begin with informal definitions of dx, $\int$, and dy/dx. "dx" means an element of X, or a little bit of X, where the little bit is infinitely small. "$\int$" means the integral, or sum of, or whole. Thus $\int dx = X$, or the sum of all the elements of X, is X. "dy/dx" means the derivative of Y with respect to X, or the small change in Y per small change in X. It is the proportion of dy to dx when both are infinitely small. It is an instantaneous rate of change. Of course we can find this ratio dy/dx only if Y is related to X in some way.

Differential Equations

A differential equation is an equation that contains one or more derivatives. Differential equations are classified with regard to (1) type, (2) order, (3) degree, and (4) linearity.

Type

Type refers to whether or not the differential equation contains ordinary or partial derivatives. If the dependent variable is a function of only one independent variable, such as in the relationship

$$Y = 3X^2 + 2X + 5$$

then differentiating will result in an ordinary differential equation,

$$\frac{dy}{dx} = 6X + 2$$

If the dependent variable is a function of more than one independent variable, such as in the relationship

$$Y = 3X^2 + 4Z^3 + 2$$

then differentiating will result in partial differential equations,

$$\frac{\partial y}{\partial x} = 6X$$

and

$$\frac{\partial y}{\partial z} = 12Z^2$$

"$\partial y/\partial x$" means the partial derivative of Y with respect to X. It is the proportion of ∂y to ∂x (or dy to dx) when both are infinitely small and all other independent variables affecting Y are held constant.

Order and Degree

Order refers to the highest order derivative in the equation and degree refers to the exponent of the highest power of the highest order derivative, after the equation has been cleared of fractions and radicals in the dependent variable and its derivatives.

$$\left(\frac{d^2y}{dx^2}\right)^3 + \frac{dy}{dx} + \frac{y}{x^2} + 1 = e^x$$

is an ordinary differential equation of order 2 and degree 3.

Linearity

Linearity refers to whether a differential equation is linear or nonlinear. To determine if a differential equation is linear, we compute the degree of each of its terms by adding the exponents of the dependent variable and any of its derivatives that occur in the term. If every term is of degree 0 or 1 the equation is linear. An ordinary differential equation of the first order always can be put into the standard form

$$\frac{dy}{dx} + Py = Q$$

where P and Q are functions of X.

Concern about whether or not a differential equation is linear results from the fact that there is a well-developed theory of linear differential equations that allows analytical (as opposed to numerical) solutions to some sets of these equations. The theory also allows us to determine certain properties of a system of linear differential equations, such as whether or not the system is stable. Such a body of theory does not exist for non-linear differential equations. In most cases we must solve sets of non-linear differential equations numerically rather than analytically. Analytical solutions are general solutions that are applicable to all situations whereas numerical solutions are specific to a particular situation.

Difference Equations

Difference equations are similar to differential equations in that they represent the rate of change of one variable with respect to another variable. However, in difference equations the ratio $\Delta y/\Delta x$ represents a finite rate of change whereas in differential equations the derivative dy/dx represents an instantaneous rate of change. In difference equations we are concerned with the finite, measurable change in Y per finite, measurable change in X.

We have little concern about linearity in difference equations because all solutions are calculated numerically. To gain an intuitive feel for the relationship between Δy and dy we will compare the differential equation and difference equation forms of the simple function

$$Y = e^{0.25X}$$

Differentiating this equation we obtain

$$\frac{dy}{dx} = 0.25e^{0.25X}$$

(We will not concern ourselves with rules for differentiation.) The analogous difference equation is

$$\frac{\Delta y}{\Delta x} = e^{0.25X_2} - e^{0.25X_1}$$

where X_1 and X_2 are values of X at points that are $1\Delta x$ unit apart. We can calculate dy/dx at any value of X. It represents the instantaneous rate of change at that point or, graphically, the tangent to the curve

$$Y = e^{0.25X}$$

at that point (Figure A2.1a). We can calculate $\Delta y/\Delta x$ only between points that are $1\Delta x$ unit apart, such as between X_1 and X_2, and we assume that the rate of change of Y is constant over that Δx (Figure A2.1b). In Figure A2.1 we see that $\Delta y/\Delta x$ for the interval from X_3 to X_4 is intermediate between dy/dx at X_3 and dy/dx at X_4. But since the function

$$Y = e^{0.25X}$$

is not linear between X_3 and X_4 the constant slope assumed by the difference equation is steeper than the instantaneous slopes of the differential equation close to X_3 and flatter than the instantaneous slopes near X_4. The smaller Δx becomes the smaller the difference between the difference equation and differential equation curves becomes. This relationship between Δx and dx is important in understanding numerical approximation procedures for solving differential equations (that cannot be solved analytically), which we discuss next. It also is important in understanding criteria for selecting the length of the time-step in simulation models, which we discuss in Chapter 3.

Numerical Approximation of Differential Equations

Simple differential equations that can be solved analytically, such as the example in Figure A2.1, are not representative of the types of differential equations that we require to adequately represent most wildlife and fisheries systems. More often models of wildlife and fisheries systems are composed of more com-

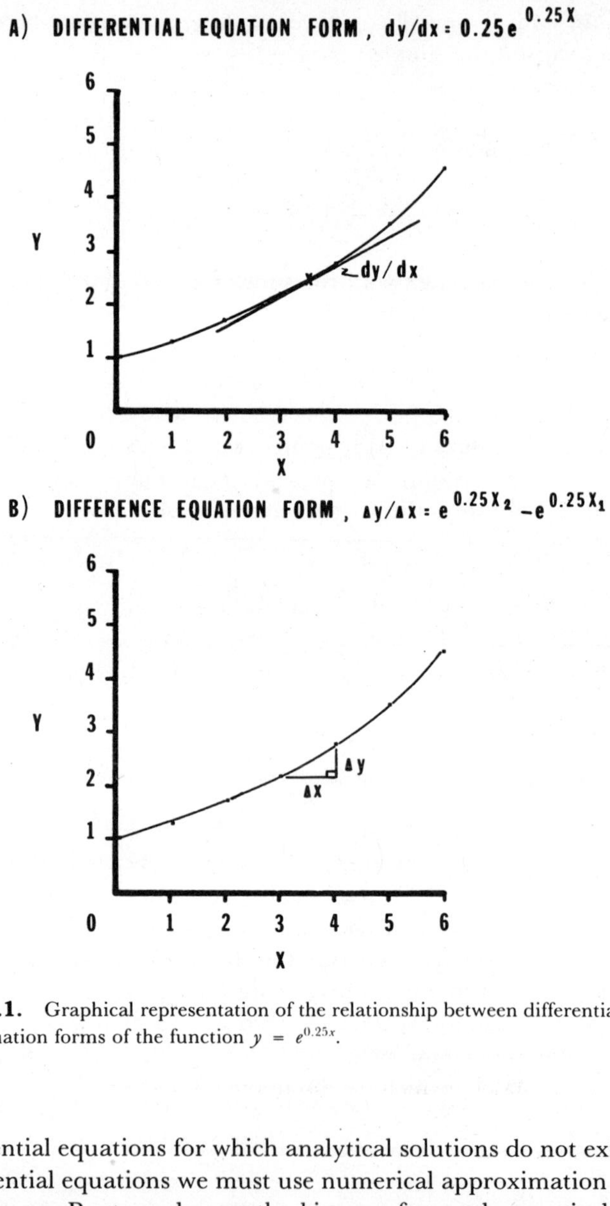

FIGURE A2.1. Graphical representation of the relationship between differential equation and difference equation forms of the function $y = e^{0.25x}$.

plex differential equations for which analytical solutions do not exist. To solve such differential equations we must use numerical approximation procedures.

The Euler, or Rectangular, method is one of several numerical approximation procedures for solving differential equations (see Olinick 1978 or Giordano and Weir 1985 for other methods). Consider the analytically unsolvable function $X(t)$ graphed in Figure A2.2. Assuming that we know an initial value

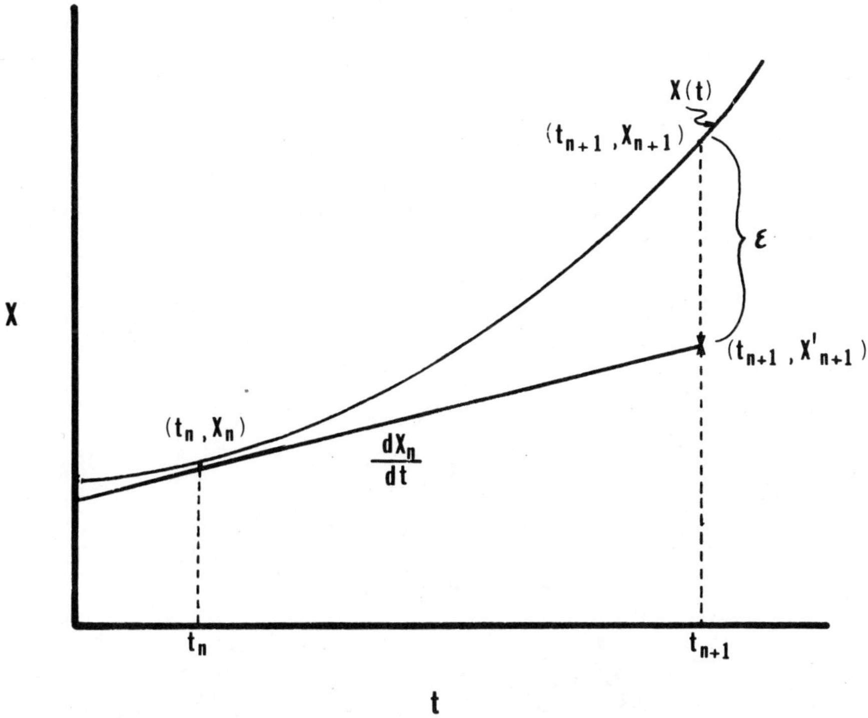

FIGURE A2.2. Graphical representation of the Euler (Rectangular) method of numerical approximation for differential equations. See text for symbol definitions.

of the function $X(t)$ at time n (X_n) and the instantaneous rate of change at time n (dx_n/dt), the objective of numerical approximation is to estimate the value of $X(t)$ at some future time $n + 1$ (X_{n+1}). This estimate is made by assuming that the instantaneous rate of change at time n is representative of the rate of change over the entire finite time interval (Δt) from t_n to t_{n+1}. Graphically this means that we project the tangent to $X(t)$ at point (t_n, X_n) Δt into the future to t_{n+1}, thus obtaining an estimate of X_{n+1}—call it X'_{n+1} (Figure A2.2). If dx/dt actually is changing from t_n to t_{n+1}, as is the case in Figure A2.2, our estimate of X_{n+1} is wrong. However, we can control size of the error (ε) by adjusting the size of Δt. The smaller Δt is the smaller ε will be. This procedure of projecting the instantaneous rate of change over a finite time interval to estimate the value of the function in the future is repeated. However, errors associated with successive estimates are cumulative.

A2.3. EXAMPLES OF USE OF DIFFERENTIAL AND DIFFERENCE EQUATIONS IN WILDLIFE AND FISHERIES SCIENCES

As mentioned above population growth is often represented in both differential and difference equation format. We will present a differential equation model and a difference equation model of population growth in a limited environment, or sigmoidal growth, as simple examples of the use of these two equation types.

Differential Equations

The differential equation representing sigmoidal population growth is

$$\frac{dN}{dt} = rN - \left(\frac{r}{K}\right)N^2$$

where dN/dt = instantaneous rate of change in size of population N
r = intrinsic rate of increase of population N
K = carrying capacity of the environment for population N

The integral form of this equation is

$$N_t = \frac{K}{1 + e^{-r(t - t_0)}}$$

where N_t = population size at time t
t_0 = parameter of the model which is calculated such that N_0 equals the desired initial population size

Suppose that we want to predict growth of a population that initially contains 10 individuals ($N_0 = 10$) over the next 10 time units (from $t = 0$ to $t = 10$). Further suppose that for this population $r = 0.5$ and $K = 50$.

First we must complete parameterization of the integral form of the equation by calculating t_0. We want to calculate t_0 such that $N_0 = 10$ when $t = 0$, given that $r = 0.5$ and $K = 50$. Substituting these values into our equation and solving for t_0, we have:

$$N_t = \frac{K}{1 + e^{-r(t - t_0)}}$$

$$10 = \frac{50}{1 + e^{-0.5(0 - t_0)}}$$

$$10 = \frac{50}{1 + e^{0.5t_0}}$$

$$10(1 + e^{0.5t_0}) = 50$$

$$10 + 10e^{0.5t_0} = 50$$

$$10e^{0.5t_0} = 40$$

$$e^{0.5t_0} = 4$$

$$0.5t_0 = \ln 4$$

$$t_0 = \frac{\ln 4}{0.5}$$

$$t_0 = \frac{1.39}{0.5}$$

$$t_0 = 2.77$$

Thus our two equations in parameterized form are

$$\frac{dN}{dt} = 0.5N - \left(\frac{0.5}{50}\right)N^2$$

$$N_t = \frac{50}{1 + e^{-0.5(t-2.77)}}$$

We can predict population growth over the next 10 time units by substituting values for t into the integral form of the equation. We can predict instantaneous rate of change in population size by substituting values for N (predicted by the integral form of the equation) into the differential form of the equation. The predictions are presented in Table A2.1a.

Note that we need not simulate population growth to predict population size at $t = 10$. We can substitute $t = 10$ into our equation and solve for N_{10} directly. We can obtain an analytical solution. We can solve for dN/dt for any N_t directly.

Difference Equations

The difference equation representing population growth is

$$\frac{\Delta N}{\Delta t} = N_{t+1} - N_t$$

TABLE A2.1

Predictions of population growth using (A) differential equations and (B) difference equations

A. Differential Equations

$$\frac{dN}{dt} = rN - \left(\frac{r}{K}\right) N^2$$

$$N_t = \frac{K}{1 + e^{-r(t-t_0)}}$$

$N_0 = 10,\ r = 0.5,\ K = 50,\ t_0 = 2.77$

t	N_t	dN/dt
0	10.00	4.00
1	14.61	5.17
2	20.25	6.02
3	26.44	6.23
4	32.45	5.69
5	37.65	4.65
6	41.70	3.46
7	44.62	2.40
8	46.59	1.59
9	47.88	1.02
10	48.69	0.64

B. Difference Equations

$$\Delta N/\Delta t = N_{t+1} - N_t$$

$$N_{t+1} = \frac{\lambda N_t}{1 + N_t(\lambda - 1)/K}$$

$N_0 = 10,\ \lambda = e^r = e^{0.5} = 1.649,\ K = 50$

t	N_t	$\Delta N/\Delta t$
0	10.00	4.60
1	14.60	5.64
2	20.24	6.19
3	26.43	6.02
4	32.45	5.20
5	37.65	4.05
6	41.70	2.91
7	44.61	1.98
8	46.59	1.29
9	47.88	0.81
10	48.69	

where $\Delta N/\Delta t$ = finite rate of change in size of population N

N_t = population size at time t

The "integral" form of this equation for sigmoidal growth is

$$N_{t+1} = \frac{\lambda N_t}{1 + N_t(\lambda - 1)/K}$$

where λ = finite rate of increase of population N

K = carrying capacity of the environment for population N

Suppose that we want to predict growth of the same population in the example above over the next 10 time units (from $t = 0$ to $t = 10$). Recall that for this population $N_0 = 10$, $r = 0.5$, and $K = 50$. The general relationship between an instantaneous rate of increase (r) and a finite rate of increase (λ) is

$$r = \ln \lambda$$

or

$$\lambda = e^r$$

For our population

$$\lambda = e^r = e^{0.5} = 1.649$$

Thus our two equations in parameterized form are

$$\frac{\Delta N}{\Delta t} = N_{t+1} - N_t$$

$$N_{t+1} = \frac{1.649 N_t}{1 + N_t(1.649 - 1)/50}$$

We can predict population growth over the next 10 time units by substituting the initial value $N_0 = 10$ into the "integral" form of the equation and solving for N_1, then substituting the value of N_1 and solving for N_2, etc. We can predict the finite rate of change in population size by substituting consecutive values for N (predicted by the "integral" form of the equation) into the difference equation. These predictions are presented in Table A2.1b.

Note that we must simulate population growth to predict population size at $t = 10$. We cannot calculate N_{10} directly. We cannot obtain an analytical solution. We also must calculate $\Delta N/\Delta t$ for any Δt by simulation.

APPENDIX 3

REGRESSION ANALYSIS

A3.1. INTRODUCTION

This appendix presents an overview of regression analysis and describes some uses of regression in wildlife and fisheries sciences. Only the most basic regression terminology and calculations are presented. There are various excellent regression texts available that provide detailed treatment of the subject (e.g., Draper and Smith 1966).

Regression analysis is used widely in wildlife and fisheries ecology to determine the degree of influence of one or several independent, or regressor, variables on one dependent variable. We can use regression techniques to address questions such as:

1. How much does the dependent variable change for a given change in an independent variable?
2. What is the shape of the curve relating the dependent to an independent variable?
3. How accurately can the dependent variable be predicted from knowledge of an independent variable?

We should note that although the term dependent variable implies a causal relationship between the dependent variable and those variables used to predict it, this need not be the case. In fact evidence of causality must come from sources other than the regression analysis itself.

We often use regression analysis in exploratory type analyses to search for correlation between a particular variable of interest and a wide variety of other variables. This is a valid and useful procedure but one that has some inherent dangers. Use of too many independent variables (over-parameterization of the regression model) can lead to meaningless results. Inclusion of as many independent variables as there are data points guarantees that all variability in the dependent variable will be accounted for by the independent variables, no matter which independent variables are chosen. Indiscriminate use of ecologically uninterpretable independent variables also can lead to meaningless results.

A3.2. REGRESSION TERMINOLOGY AND SIMPLE LINEAR REGRESSION

Terminology

The simplest type of linear regression model contains one dependent and one independent variable. This is referred to as a "simple linear regression" and the model has the general form

$$Y = \beta_0 + \beta_1 X + \varepsilon$$

where

Y = dependent variable
X = independent variable
β_0 and β_1 = parameters of the model
ε = error term

Of course $Y = \beta_0 + \beta_1 X$ is simply the equation for a straight line. It is the straight line that we wish to estimate from data to represent the relationship between Y and X. Using data (in a manner described below) we obtain an estimate of β_0 (b_0) and an estimate of β_1 (b_1) such that the sum of the squared deviations ($\Sigma\varepsilon_i^2$) of data points from the line $Y = b_0 + b_1 X$ is minimized. b_0 is our estimate of the Y-intercept of the regression line and b_1 is our estimate of the slope of the regression line (Figure A3.1).

Three of the most widely used statistics associated with linear regression analysis are:

1. the slope of the regression line (b_1),
2. the square of the correlation coefficient (r^2), and
3. the error mean square, or mean square about regression (s^2).

Examination of these three statistics yields much information about the relationship of Y to X. The slope (b_1) represents the amount of change in Y for a given change in X. The square of the correlation coefficient (r^2) represents the proportion of variability in the dependent variable (Y) that can be accounted for by variability in the independent variable (X). The error mean square (s^2) represents the degree of scatter of data points about the regression line.

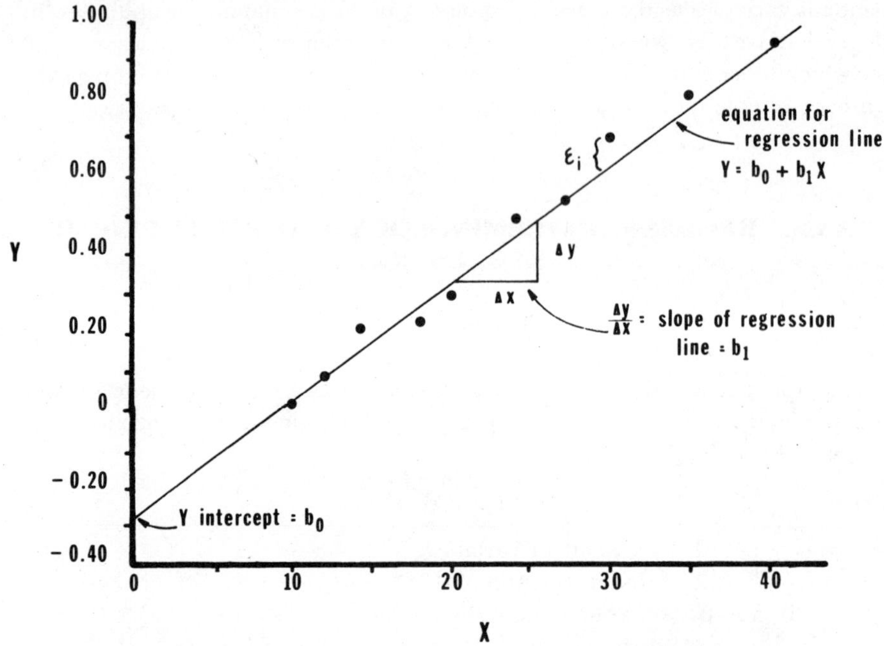

FIGURE A3.1. Graphical representation of a simple linear regression.

The relationship between these three statistics is presented in Figure A3.2. The slope is greater in cases A and B than in cases C and D and therefore b_1 is greater in the former two. The scatter about the regression lines is small in cases A and C and large in cases B and D and therefore s^2 is small in the former two and large in the latter two. The square of the correlation coefficient is large in case A, where the slope is large and the scatter is small, and is small in the other three cases, unless s^2 is very small.

Calculations

Having defined parameters of a simple linear regression model and some useful statistics associated with regression analysis, we now turn our attention to the manner in which these items are calculated.

Regression Equation

Consider the data set presented in Table A3.1. Examination of a plot of these data (Figure A3.1) confirms that the relationship is essentially linear and that

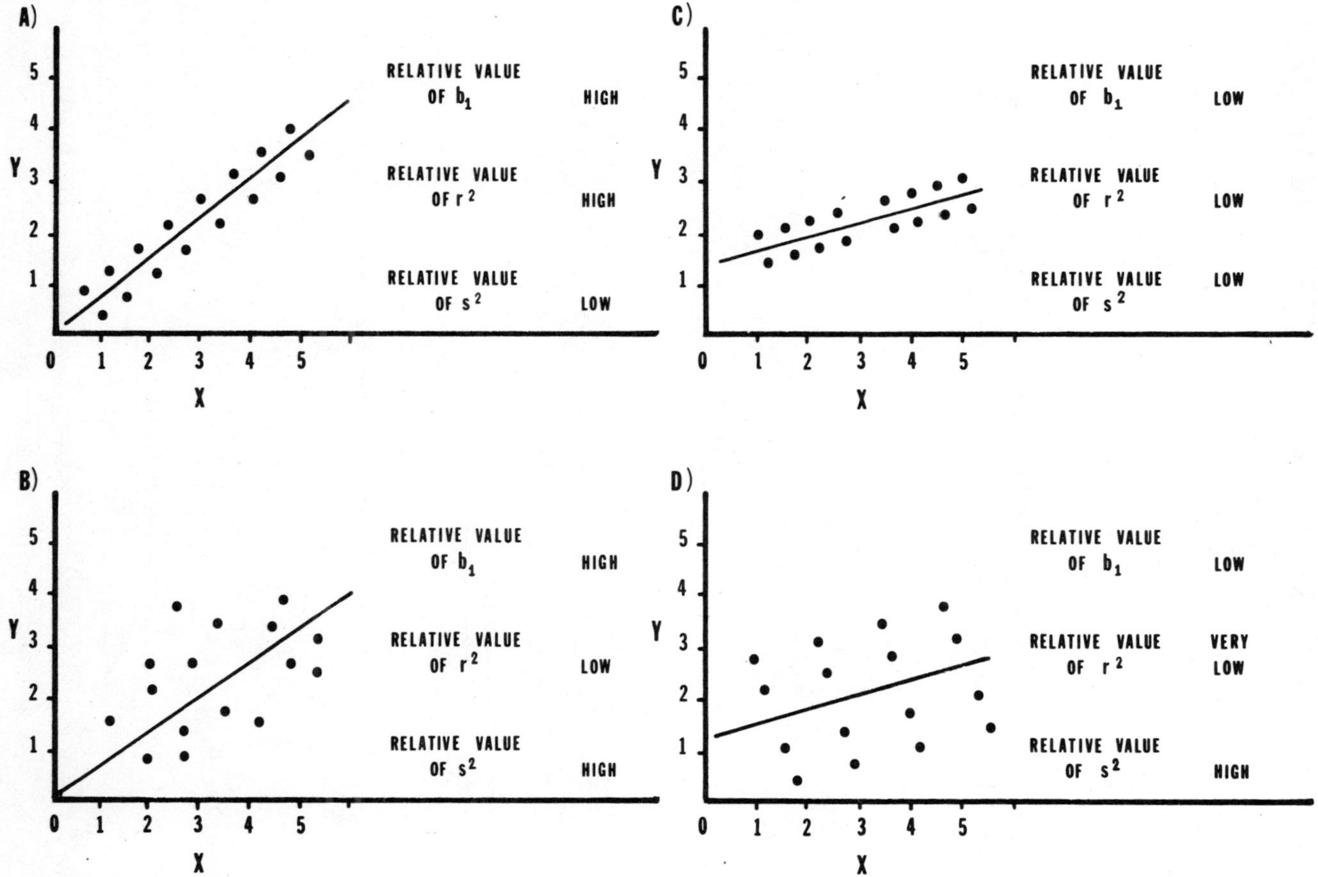

FIGURE A3.2. Relationships between the slope of the regression line (b_1), the square of the correlation coefficient (r^2), and the error mean square (s^2).

TABLE A3.1

Hypothetical data set relating consumption of an ectotherm (in g food consumed/g live body wt.-day) to temperature (°C) and the associated linear regression analysis

Observation Number (i)	Consumption (Y_i)	Temperature (X_i)	$X_i Y_i$	X_i^2	Y_i^2
1	0.01	10	0.1	100	0.0001
2	0.09	12	1.08	144	0.0081
3	0.21	14	2.94	196	0.0441
4	0.22	18	3.96	324	0.0484
5	0.30	20	6.0	400	0.09
6	0.49	24	11.76	576	0.2401
7	0.54	27	14.58	729	0.2916
8	0.70	30	21.0	900	0.49
9	0.80	35	28.0	1225	0.64
10	0.95	40	38.0	1600	0.9025

$\Sigma Y_i = 4.31 \quad \Sigma X_i = 230 \quad \Sigma X_i Y_i = 127.42 \quad \Sigma X_i^2 = 6194 \quad \Sigma Y_i^2 = 2.7549$

$(\Sigma Y_i)^2 = 18.5761 \quad (\Sigma X_i)^2 = 52{,}900$

$\bar{Y} = 0.431 \quad \bar{X} = 23$

$$b_1 = \frac{\Sigma X_i Y_i - (\Sigma X_i \Sigma Y_i)/n}{\Sigma X_i^2 - (\Sigma X_i)^2/n} = \frac{127.42 - (230)(4.31)/10}{6194 - (230)^2/10} = 0.0313$$

$$b_0 = \bar{Y} - b_1 \bar{X} = 0.431 - (0.0313)(23) = -0.2889$$

linear regression analysis therefore is appropriate. We can calculate the linear regression

$$Y = \beta_0 + \beta_1 X + \varepsilon$$

by estimating β_1 as

$$b_1 = \frac{\Sigma X_i Y_i - (\Sigma X_i \Sigma Y_i)/n}{\Sigma X_i^2 - (\Sigma X_i)^2/n}$$

where Y_i and X_i = value of the ith observation on the dependent and independent variables, respectively

n = sample size

Σ = "the sum of"

and estimating β_0 as

$$b_0 = \bar{Y} - b_1\bar{X}$$

where $\bar{Y}$ = mean of the Y values
$\bar{X}$ = mean of the X values

These calculations are illustrated in Table A3.1. The resulting regression equation is $Y = -0.2889 + 0.0313X$.

r^2, s^2, *and Test for Significant Regression*

We also can calculate r^2 and s^2 and test to see if the regression is statistically significant (test the null hypothesis that $\beta_1 = 0$). These calculations are based on the fact that the total sum of squared deviations of values of the dependent variable (Y) from the mean of the dependent variable ($\bar{Y}$), or $\Sigma(\bar{Y} - Y_i)^2$, which is called the total sum of squares (SS_y), can be partitioned into a sum of squares due to regression (SS_r) and an error sum of squares (SS_e). Then $r^2 = SS_r/SS_y$. SS_r and SS_e divided by the appropriate number of degrees of freedom yield the mean square for regression (MS_r) and the error mean square (s^2), respectively. We then can test statistical significance of the regression by calculating a test statistic (F) as $F = MS_r/s^2$. We compare the calculated F-value with the critical value of the F-distribution associated with the appropriate degrees of freedom and significance level (α). If our calculated F-value is greater than the tabulated critical F-value the regression is statistically significant at the given α level.

TABLE A3.2

Calculation of the total sum of squares (SS_y), sum of squares due to regression (SS_r), sum of squares due to error (SS_e), mean square for regression (MS_r), error mean square (s^2), square of the correlation coefficient (r^2), and F-statistic based on data in Table A3.1

$$SS_y = SS_r + SS_e$$

$$SS_y = \Sigma Y_i^2 - (\Sigma Y_i)^2/n = 2.7549 - 18.5761/10 = 0.8973$$

$$SS_r = b_1(\Sigma X_i Y_i - (\Sigma X_i \Sigma Y_i)/n) = 0.0313(127.42 - (230)(4.31)/10) = 0.8855$$

$$SS_e = SS_y - SS_r = 0.8973 - 0.8855 = 0.0118$$

$$r^2 = SS_r/SS_y = 0.8855/0.8973 = 0.9868$$

$$s^2 = SS_e/(n - 2) = 0.0118/(10 - 2) = 0.0015$$

$$MS_r = SS_r/1 = 0.9868/1 = 0.9868$$

$$F = MS_r/s^2 = 0.9868/0.0015 = 658$$

Calculations for r^2, s^2, and F are illustrated using data from Table A3.1 in Table A3.2. $r^2 = 0.99$, which means that we can account for 99 percent of the variability in Y by variability in X. s^2 value is quite low (0.0015), which means that there is little scatter of data points about the regression line (Figure A3.1 suggested this). The calculated F-value is 658. We compare this value with critical F-values that we look up in a statistical table (Rohlf and Sokal 1969). A portion of such a table (usually called "critical values of the F-distribution" or simply "F-table") is reproduced in Table A3.3. Each F-value in the table is associated with a number of degrees of freedom of the numerator mean square (V_1), a number of degrees of freedom of the denominator mean square (V_2), and a significance level (α). In our example we calculated $F = MS_r/s^2$ with 1 degree of freedom associated with the numerator mean square (MS_r) and $n - 2$ or 8 degrees of freedom associated with the denominator mean square (s^2). If we wish to see if the regression is significant at $\alpha = 0.05$, we find the tabulated F-value associated with $V_1 = 1$, $V_2 = 8$, $\alpha = 0.05$, which is 5.32. Since our calculated F-value is larger than 5.32 the regression is significant at the $\alpha = 0.05$ level. In fact the regression is significant at the $\alpha = 0.001$ level since 658 is larger than 25.4.

There are various additional calculations associated with linear regression analysis that are of interest. We can calculate standard errors and confidence limits for the estimates of β_0, β_1, and for estimates of Y values obtained from the regression equation. These calculations are explained in standard regression texts (Draper and Smith 1966).

TABLE A3.3
Selected critical values of the *F*-distribution (Rohlf and Sokal 1969)

Degrees of Freedom of Denominator Mean Square	α (significance level)	Degrees of Freedom of Numerator Mean Square 1
	0.75	0.109
	0.50	0.499
	0.25	1.54
	0.10	3.46
8	0.05	5.32
	0.025	7.57
	0.01	11.3
	0.005	14.7
	0.001	25.4

Assumptions

Simple linear regression is based on two basic assumptions about the model

$$Y = \beta_0 + \beta_1 X + \varepsilon$$

1. The Xs are measured without error, and
2. the εs are random, independently and normally distributed about a mean of zero, and have a common variance.

One additional assumption required for multiple linear regression is that the effects of independent variables are additive. These assumptions are presented formally and discussed in regression texts (Draper and Smith 1966).

A3.3. MULTIPLE LINEAR REGRESSION

Calculations

The concept of simple linear regression can be extended to multiple linear regression in which two or more independent variables are related to the single dependent variable. The general model for multiple linear regression has the form

$$Y = \beta_0 + \beta_1 X_1 + \beta_2 X_2 + \ldots + \beta_k X_k + \varepsilon$$

where

$$\begin{aligned} Y &= \text{dependent variable} \\ X_1, X_2, \ldots, X_k &= k \text{ independent variables} \\ \beta_0, \beta_1, \beta_2, \ldots, \beta_k &= \text{parameters of the model} \\ \varepsilon &= \text{error term} \end{aligned}$$

Calculations of simple linear regression can be extended to multiple linear regression by using a matrix algebra format where $\mathbf{Y}$ is a vector of observations of the dependent variable, $\mathbf{X}$ is a matrix of observations of the independent variables, $\boldsymbol{\beta}$ is a vector of parameters to be estimated, and $\boldsymbol{\varepsilon}$ is a vector of error terms. The regression equation in matrix form is

$$\mathbf{Y} = \mathbf{X}\boldsymbol{\beta} + \boldsymbol{\varepsilon}$$

Estimates of the βs (the bs), r^2, s^2, and F can be calculated as illustrated in Table A3.4. For purposes of simplicity and comparison the example in Table A3.4 deals with the simple linear regression data set analyzed above (Table A3.1). However, the matrix calculations are identical for any number of independent variables.

TABLE A3.4

Calculation of regression coefficients (*b*s), total sum of squares (SS_y), sum of squares due to regression (SS_r), and sum of squares due to error (SS_e) using a matrix algebra approach. Calculations are based on data in Table A3.1

$\mathbf{X'Xb} = \mathbf{X'Y}$, solve for $\mathbf{b}$

$$\mathbf{X'X} = \begin{bmatrix} 1 & 1 & \cdots & 1 \\ X_1 & X_2 & \cdots & X_k \end{bmatrix} \begin{bmatrix} 1 & X_1 \\ 1 & X_2 \\ \vdots & \vdots \\ 1 & X_k \end{bmatrix} = \begin{bmatrix} n & \Sigma X_i \\ \Sigma X_i & \Sigma X_i^2 \end{bmatrix}$$

$$\mathbf{X'Y} = \begin{bmatrix} 1 & 1 & \cdots & 1 \\ X_1 & X_2 & \cdots & X_k \end{bmatrix} \begin{bmatrix} Y_1 \\ Y_2 \\ \vdots \\ Y_k \end{bmatrix} = \begin{bmatrix} \Sigma Y_i \\ \Sigma X_i Y_i \end{bmatrix}$$

$$\mathbf{X'X} = \begin{bmatrix} 1 & 1 & 1 & 1 & 1 & 1 & 1 & 1 & 1 & 1 \\ 10 & 12 & 14 & 18 & 20 & 24 & 27 & 30 & 35 & 40 \end{bmatrix} \begin{bmatrix} 1 & 10 \\ 1 & 12 \\ 1 & 14 \\ 1 & 18 \\ 1 & 20 \\ 1 & 24 \\ 1 & 27 \\ 1 & 30 \\ 1 & 35 \\ 1 & 40 \end{bmatrix} = \begin{bmatrix} 10 & 230 \\ 230 & 6194 \end{bmatrix}$$

$$\mathbf{X'Y} = \begin{bmatrix} 1 & 1 & 1 & 1 & 1 & 1 & 1 & 1 & 1 & 1 \\ 10 & 12 & 14 & 18 & 20 & 24 & 27 & 30 & 35 & 40 \end{bmatrix} \begin{bmatrix} 0.01 \\ 0.09 \\ 0.21 \\ 0.22 \\ 0.30 \\ 0.49 \\ 0.54 \\ 0.70 \\ 0.80 \\ 0.95 \end{bmatrix} = \begin{bmatrix} 4.31 \\ 127.42 \end{bmatrix}$$

$$(\mathbf{X'X})^{-1} = \begin{bmatrix} \dfrac{\Sigma X_i^2}{n\Sigma(X_i - \bar{X})^2} & \dfrac{-\bar{X}}{\Sigma(X_i - \bar{X})^2} \\ \dfrac{-\bar{X}}{\Sigma(X_i - \bar{X})^2} & \dfrac{1}{\Sigma(X_i - \bar{X})^2} \end{bmatrix}$$

TABLE A3.4 *(Continued)*

$$(\mathbf{X'X})^{-1} = \begin{bmatrix} 6194/(10)(904) & -23/904 \\ -23/904 & 1/904 \end{bmatrix} = \begin{bmatrix} 0.6852 & -0.0254 \\ -0.0254 & 0.0011 \end{bmatrix}$$

$$(\mathbf{X'X})^{-1}(\mathbf{X'X})\mathbf{b} = (\mathbf{X'X})^{-1}\mathbf{X'Y}$$

$$\mathbf{Ib} = (\mathbf{X'X})^{-1}\mathbf{X'Y}$$

$$\mathbf{b} = (\mathbf{X'X})^{-1}\mathbf{X'Y}$$

$$\begin{bmatrix} 0.6852 & -0.0254 \\ -0.0254 & 0.0011 \end{bmatrix} \begin{bmatrix} 10 & 230 \\ 230 & 6194 \end{bmatrix} \begin{bmatrix} b_0 \\ b_1 \end{bmatrix} = \begin{bmatrix} 0.6852 & -0.0254 \\ -0.0254 & 0.0011 \end{bmatrix} \begin{bmatrix} 4.31 \\ 127.42 \end{bmatrix}$$

$$\begin{bmatrix} 1 & 0 \\ 0 & 1 \end{bmatrix} \begin{bmatrix} b_0 \\ b_1 \end{bmatrix} \begin{matrix} = \\ = \end{matrix} \begin{bmatrix} -0.28 \\ 0.03 \end{bmatrix}$$

$$\begin{bmatrix} b_0 \\ b_1 \end{bmatrix} \begin{matrix} = \\ = \end{matrix} \begin{bmatrix} -0.28 \\ 0.03 \end{bmatrix}$$

$$SS_y = \mathbf{Y'Y} - (\Sigma Y_i)^2/n$$

$$\mathbf{Y'Y} = [0.01 \quad 0.09 \quad 0.21 \quad 0.22 \quad 0.30 \quad 0.49 \quad 0.54 \quad 0.70 \quad 0.80 \quad 0.95] \begin{bmatrix} 0.01 \\ 0.09 \\ 0.21 \\ 0.22 \\ 0.30 \\ 0.49 \\ 0.54 \\ 0.70 \\ 0.80 \\ 0.95 \end{bmatrix}$$

$$= 2.7549$$

$$SS_y = 2.7549 - 18.5761/10 = 0.8973$$

$$SS_r = \mathbf{b'X'Y} - (\Sigma Y_i)^2/n$$

$$\mathbf{b'X'Y} = [-0.28 \quad 0.03] \begin{bmatrix} 4.31 \\ 127.42 \end{bmatrix} = 2.7431$$

$$SS_r = 2.7431 - 18.5761/10 = 0.8855$$

Calculations for SS_e, MS_r, s^2, r^2, and F are identical to those presented in Table A3.2.

Selection of Independent Variables

In multiple linear regression analysis we often wish to know if it is worthwhile to include additional independent variables in the model. The question of whether or not to include an additional term is basically the same as the question in simple linear regression of whether or not there is a regression relationship in the first place. Recall that in simple linear regression we test the null hypothesis that the regression coefficient (β_1) is equal to zero. Stated another way we ask if the model

$$Y = \beta_0 + \beta_1 X + \varepsilon$$

is a better predictor of Y than the model

$$Y = \beta_0 + \varepsilon$$

where $\beta_0 = \bar{Y}$. In multiple linear regressions we ask if the model

$$Y = \beta_0 + \beta_1 X_1 + \beta_2 X_2 + \varepsilon$$

is a better predictor of Y than the model

$$Y = \beta_0 + \beta_1 X_1 + \varepsilon$$

or if the model

$$Y = \beta_0 + \beta_1 X_1 + \beta_2 X_2 + \beta_3 X_3 + \varepsilon$$

is a better predictor than the model

$$Y = \beta_0 + \beta_1 X_1 + \beta_2 X_2 + \varepsilon \text{ etc.}$$

We can evaluate the worth of adding another independent variable to a multiple linear regression model in several ways:

1. We can examine r^2 with and without a given variable in the model and decide if the improvement in r^2 is great enough to justify adding the variable.
2. We can examine s^2 with and without the variable in the model and decide if the reduction in s^2 is great enough to justify adding the variable.
3. We can examine sequential F-values to see if the added variable accounts for a significant additional portion of the total sum of squares over and above the portion of the total sum of squares accounted for by the variables already in the model.

However, it is important to note that the order in which additional independent variables are added to the model will influence the apparent worth of the variables as judged by any of the criteria above. A variable added early when only a few variables have preceded it may appear to improve the model substantially. Yet that same variable added late when many variables have preceded it may appear to improve the model little. The practicality of examining all possible regressions containing one or more of a given set of potential independent variables decreases sharply as the number of potential variables increases. We should rely on ecological intuition as well as the various statistical selection criteria to identify the "best" regression model.

A3.4. LINEAR TRANSFORMATIONS

Thus far we have discussed only linear regression models. Linearity refers to linearity of the parameters, the βs, not linearity of the independent variables. For instance

$$Y = \beta_0 + \beta_1 X + \beta_2 X^2 + \varepsilon$$

is a linear model, but

$$Y = \beta_0 X^{\beta_1} + \varepsilon$$

is not. The significance of the distinction between linear and non-linear regression models is that estimation procedures and calculations for linear regression are well-established whereas estimation procedures and calculations for non-linear models are not.

However, the relationship between dependent and independent variables need not necessarily be linear to use linear regression analysis. As long as the relationship can be transformed into a linear relationship, linear regression can be used. For example

$$Y = \beta_0 X_1^{\beta_1} X_2^{\beta_2} \varepsilon$$

can be linearized by the natural log transformation

$$\ln Y = \ln \beta_0 + \beta_1 \ln X_1 + \beta_2 \ln X_2 + \varepsilon$$

and

$$Y = \beta_0 e^{\beta_1 X_1 + \beta_2 X_2 + \varepsilon}$$

also can be linearized by the natural log transformation

$$\ln Y = \ln \beta_0 + \beta_1 X_1 + \beta_2 X_2 + \varepsilon$$

In both cases the log transform is used to establish a linear relationship between ln Y and the independent variables, linear regression is performed, and the relationship is then detransformed to regain the relationship between Y and the independent variables. A variety of other linear transformations are described in standard regression texts (Draper and Smith 1966).

Often we do not know the correct model relating dependent and independent variables. We might first hypothesize that the relationship is linear and run a linear regression. One way to explore the possibility that a linear transformation might improve the model is to examine the distribution of differences between model predictions of the dependent variable and observed values. These differences are called residuals. The form of the distribution of residuals may suggest an appropriate transformation, or may suggest that an additional term of a specific form should be added to the model. Consider the scatter of data points and the linear regression line estimated from the data presented in Figure A3.3. The regression model underestimates Y-values for low values of X, overestimates for middle values of X, and again underestimates for high values of X. This suggests that a log transformation of Y might result in a more nearly linear relationship,

$$\ln Y = \beta_0 + \beta_1 X + \varepsilon$$

and thus improve the model. As mentioned above, when performing linear regression analysis it is assumed that prediction errors, or residuals, are independent, have a mean of zero, a constant variance, and follow a normal distri-

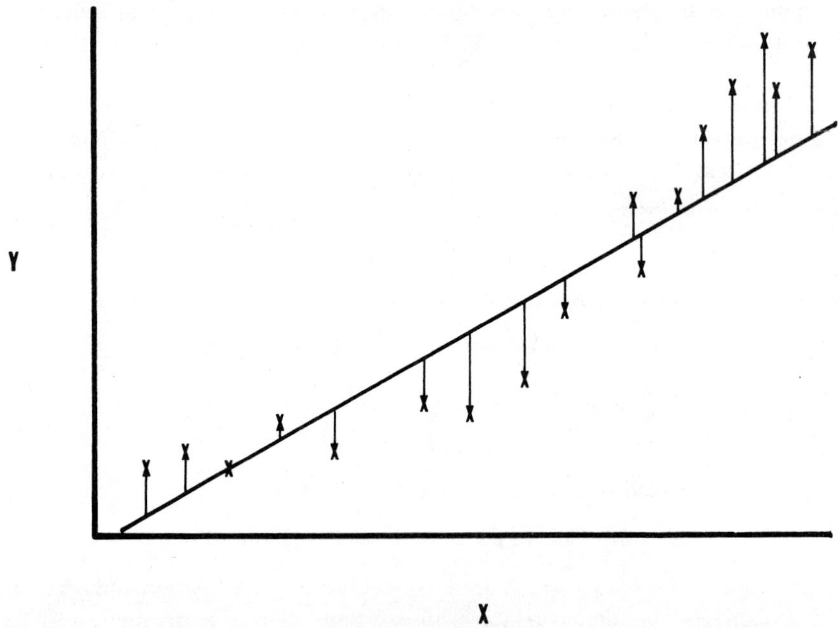

FIGURE A3.3. Hypothetical relationship between data points and the estimated linear regression line. Vertical lines represent residuals. The distribution of residuals suggests that a log transformation is appropriate.

bution. If examination of the residuals suggests that any of these assumptions have been violated the regression model may be inappropriate. In Figure A3.3 residuals do not appear to be independent since underestimates tend to occur together over certain ranges in X and overestimates tend to occur together over other ranges in X.

A3.5. EXAMPLES OF USE OF REGRESSION ANALYSIS IN WILDLIFE AND FISHERIES SCIENCES

Regression analysis is used in wildlife and fisheries sciences to identify and quantify relationships between variables in exactly the same manner as it is used in other fields. Some of the most widely used empirical relationships in both wildlife and fisheries have been established by simple linear regression. For example the relationship relating average daily metabolic rate to body size of the form

$$\text{ADMR} = aW^b$$

where ADMR = average daily metabolic rate in kcal/day
W = live weight in kg
a and b = parameters of the equation

was established by simple linear regression of the form

$$\ln \text{ADMR} = \ln a + b \ln W$$

In the von Bertalanffy equation predicting growth of the form

$$l_t = L_\infty(1 - e^{-k(t-t_0)})$$

where l_t = length at time t
L_∞ = asymptotic length measured in the same units as l_t (a parameter of the equation)
t_0 = theoretical age at which length is zero (a parameter of the equation)

the parameter k commonly is estimated from a simple linear regression relating length at time $t + 1$ to length at time t. The slope of this regression equals e^{-k}. In short, regression analysis is a tool commonly used in wildlife and fisheries research.

Regression analysis also is an important tool used in systems analysis. Within this context regression analysis represents one of the methods that may be useful in identifying and quantifying equations that represent the transfer of material from one system component to another. An example of such usage is presented in Chapter 3.

APPENDIX 4

COMPUTERS AND PROGRAMMING

A4.1. INTRODUCTION

Today the use of computers is widespread in many fields. They are useful tools for processing large amounts of information rapidly. They are used for word processing, information storage and retrieval, and numerical calculations. The algorithmic languages in which many computer programs are written also serve as effective means of communicating scientific methodology. Wildlife and fisheries scientists use computers for all of these purposes. This appendix presents a brief overview of the nature of computers and programming as they are applied to perform the numerical calculations associated with systems modeling. More detailed overviews of the use of computers for modeling in the general field of systems ecology can be found in Patten (1971b) and Kitching (1983).

In Chapter 1 we mentioned the role of models in wildlife and fisheries research, education, and the decision-making or management process. The fact that so many models in use today require computers for efficient solutions emphasizes the need for computer literacy on the part of wildlife and fisheries scientists, educators, and managers (Ziesenis and Adams 1980). Computer literacy implies the ability to communicate efficiently with computers. This communication can be direct by writing our own computer programs, or indirect by instructing trained programmers to write programs for us. Because of the growing diversity of computing equipment and programming languages a good argument can be made that the use of trained programmers is cost-efficient in most research or management programs where much programming is required. Nonetheless some basic knowledge about computers and programming and some hands-on experience writing computer programs will serve wildlife and fisheries professionals well because (1) it will greatly enhance our ability to communicate effectively with trained programmers and (2) it will provide the flexibility to do our own programming if appropriate.

There are two basic types of computers, analog and digital, which are quite different in general nature. Analog computers process continuous information. Digital computers process discrete information. They are more versatile and much more widely used than analog computers.

A4.2. ANALOG COMPUTERS

Analog computers are essentially single-purpose devices designed to solve differential equations. Consequently we can use them to analyze behavior of models expressed as sets of differential equations. They are particularly useful in experimenting with differential equation models for which we need to determine coefficients by trial and error, because the reaction of the model to coefficient changes is instantaneous.

An analog computer consists of a set of basic components that can be interconnected so that an electrical current, governed by the same differential equations as those representing the model, flows through them. On analog computers there is only one independent variable (time) and only one dependent variable (voltage). The general strategy in programming an analog computer consists of:

1. diagramming or ''patching'' an electrical circuit connecting appropriate analog components such that current flow is directed by the same relationships that direct flow of the quantity of interest in the model,
2. scaling voltage such that all model behavior is compressed into voltage range available on the analog computer being used (usually $-10\text{v} \leq E \leq +10\text{v}$, or for ecological systems this becomes $0\text{v} \leq E \leq +10\text{v}$, since negative quantities are generally inappropriate), and
3. scaling time such that simulations representing desired length of problem time are run in a reasonable amount of computer time or real time.

Patten (1971b) provides a primer for ecologists on the use of analog computers.

A4.3. DIGITAL COMPUTERS

There are many types of digital computers representing a wide variety of computing capabilities. Most commonly digital computers are classified by size as mainframes, minis, or micros. Although the dividing lines between these size categories are somewhat arbitrary, we can conveniently think of central computers at large universities as mainframes, personal computers as micros, and anything in between as minicomputers.

Regardless of size digital computers all have five basic functions:

1. input,
2. memory,
3. arithmetic-logic,
4. control, and
5. output.

These functions are accomplished through interaction of the computer's physical structure, called hardware, and a variety of programs and programming systems, collectively called software. For all practical purposes we purchase a computer's hardware intact from the manufacturer. We may purchase various hardware modules that have been designed to function together at different times, but the modules themselves usually are not altered by the user. We also purchase some types of software as intact packages, but we may write other types of software ourselves to meet specific needs. What follows is a brief discussion of various types of software or computer programs.

Computer Programs

A computer program is a sequence of instructions that a computer is to carry out in solving a problem or in handling data. Each instruction calls upon the computer to perform one of the operations for which it is designed. A computer automatically performs each instruction in the order indicated by the program.

Types of Languages

We write instructions for a computer in a code that a particular computer is designed to accept. The basic code that a computer can accept, interpret, and execute is called machine language. However machine languages for different types of computers are different and all are highly abstract. We rarely write programs directly in machine language. Usually we write programs in one of several other languages that are more convenient to use and these programs are then translated into machine language by manufacturer-supplied software packages.

The other languages fall into two basic categories:

1. languages for assembly systems and
2. languages for translating systems.

Languages for assembly systems are machine-oriented. They are related closely to machine language. When our programming goal is to obtain the most efficient use of a computer's hardware, we write a program in the language of an assembly system. Languages for translating systems are problem-oriented, or user-oriented. When our programming goal is to use a language similar to that normally used to describe the problem, or when our goal is to write a program that is machine independent and can be used on various computers, we write a program in the language of a translating system. Most programs writen by people other than computer specialists are written in the language of a translating system.

Languages for Translating Systems

A translating system has two parts—(1) a processor program, either a compiler or an interpreter, and (2) a language. The language is devised in terms of the problems to be solved and is independent of the computer's hardware. Thus as programmers we can concern ourselves with details of our problem rather than with details of the computer. We write instructions in the language of the translating system, and these are supplied as input to the computer, which in executing the processor program reads and translates them into machine language.

There are numerous translating system languages. Theoretically programs written in any translating system language can be translated and executed on any computer for which there is an appropriate processor program. In practice some program modifications are usually required when programs are switched to a different computer. Although each language has its advantages and disadvantages for a given problem, FORTRAN (on mainframe computers) and BASIC (on microcomputers) are probably the most commonly used languages for scientific applications. There is a variety of excellent manuals available for both FORTRAN (Merchant 1981, Boillot 1981) and BASIC (Didday and Page 1984). The general programming strategy is similar in both languages. For programming simulation models the steps outlined in Figure 3.8 can be followed when using either language. Other languages in common use include PASCAL (Jensen and Wirth 1975), COBOL (Ashley 1974), and ALGOL (McCracken 1962).

Special-Purpose Programming Packages

In addition to machine language software packages and processor programs for various user-oriented languages, most major computing centers also support a variety of special-purpose user-oriented programming packages. Commonly these include statistical packages such as SAS (Barr et al. 1976) or BMDP

(Dixon 1975) that allow us to manipulate and organize data, transform data, and conduct a wide variety of statistical analyses with a minimum amount of programming. Green (1979, p. 142) provides a list of such packages. There also are several general purpose simulation languages that facilitate programming simulation models by taking care of routine bookkeeping automatically (Markowitz et al. 1963, Pritsker 1974, Bobillier et al. 1976). Additional special-purpose simulation languages that are particularly useful for ecological problems also are available at some computing centers (Gustafson and Innis 1973, Sigal and Pritsker 1974, Shaffer 1980).

APPENDIX 5

COMPUTER PROGRAM FOR THE POND SIMULATION MODEL

A5.1. INTRODUCTION

This appendix presents a computer program (POND) intended to provide a hands-on experience in quantifying, validating, and using the pond simulation model described in Chapters 2 through 5. POND is an interactive, user-friendly program written in TRSDOS disk BASIC for use on a TRS80 microcomputer.

The appendix is divided into two sections—(1) a user's guide and (2) a program listing. The user's guide provides instructions necessary to parameterize the pond model as described in Chapter 3, and to run all simulations described in Chapters 4 and 5. The program listing is complete and can be entered into the computer (and saved on an appropriate diskette) exactly as presented.

A5.2. USER'S GUIDE

The following items are required to run POND:

1. TRS80 microcomputer (Models I, III, or IV) or a microcomputer compatible with TRS80
2. 80 column parallel interface printer
3. Diskette containing TRSDOS disk BASIC and POND

To enter or load POND into the computer, turn on the computer and place the diskette in the disk drive. Press the reset button. Type BASIC and press the "enter" key, or

BASIC (ENTER)

the computer will respond with

READY

To use POND type

RUN "POND" (ENTER)

The computer will respond with the POND menu.

```
                POND SIMULATION
1) Conceptual Model          2) Model Quantification
3) Model Validation          4) Model Use
        Select 1-4 or hit <ENTER> to exit?
```

POND is divided into four phases which correspond with the four phases of systems analysis presented in Chapters 2 through 5. Phase 1 (Conceptual Model) contains a brief description of the hypothetical pond system as presented in Chapter 2 (p. 52). Phase 2 (Model Quantification) contains instructions for parameterizing POND based on the information presented in Chapter 3 (Table 3.8). Phase 3 (Model Validation) contains instructions for running the simulations associated with model validation as described in Chapter 4 (p. 151). Phase 4 (Model Use) contains instructions for running the simulations associated with model use as described in Chapter 5 (p. 178).

To enter any of the four phases of POND from the pond simulation menu, type the appropriate number and press "enter." For example, to enter the Conceptual Model phase type

1 (ENTER)

We cannot enter the Model Validation or Model Use phases until the Model Quantification phase has been completed.

To leave any of the four phases of POND press "enter" repeatedly until returned to the pond simulation menu.

Phase 1: Conceptual Model

The Conceptual Model phase of POND contains a brief word description of the hypothetical pond system. It is provided for reference only and requires no user interaction.

Phase 2: Model Quantification

When we enter the Model Quantification phase of POND we see the model quantification menu.

Model Quantification

1) Model Development 2) Display Parameters
3) Edit Parameters 4) Save Parms. (disk)
5) Load Parms. (disk) 6) Run Simulations

Select 1–6 or hit <ENTER> to exit?

To enter any of the six routines of Model Quantification type the appropriate number and press "enter." To leave any of the six routines press "enter" repeatedly until returned to the model quantification menu.

Routine 1: Model Development

This routine prompts us to enter values for model parameters. We enter each value from the keyboard by typing the appropriate number and then pressing "enter." All the information needed to parameterize POND is in Table 3.8. It also is useful to refer to the conceptual model diagram (Figure 2.10) while using this routine, to keep the overall structure of the pond model in mind.

The routine begins with prompts to enter the initial conditions of the model:

Initial Conditions
Aquatic plants (S1) (kg/ha) = 0?

From Table 3.8 we find that $S1 = 40000$, thus we type

40000 (ENTER)

An incorrect entry can be corrected by backspacing over the mistake and retyping before pressing "enter." If we notice the mistake after pressing "enter" it can be corrected using the edit routine (described below) after completing the entry of parameters.

Next we are prompted to enter the initial value for herbivorous fish ($S2$), number of fish stocked ($D2$), and so on. Note that we also are prompted for values of those driving variables that do not change during simulation in the initial conditions section. Next we are prompted to enter values for economic variables.

The routine continues with prompts to enter values of the water temperature driving variable.

Driving variable – Water temperature (Piecewise linear)

Water temp. (D1) (degrees C) = B0 + B1*Julian Date (t)

For 1 < t < 105 B0 = 0?

We enter values for $B0$ and $B1$ for each of the five equations that we used to represent water temperature (Table 3.8). Note that in Table 3.8 we obtain the appropriate values for $B0$ and $B1$ from equations representing the means of the normal distributions from which water temperature is drawn as a random variate. For example we represented water temperature (*D*1) during the first time period ($1 \leq t \leq 105$) as the mean temperature for the period. Thus $B0 = 15.015$ (the mean temperature) and $B1 = 0$ (slope of the regression is zero). Similarly for the second time period ($105 < t < 166$) $B0 = -10.845$ and $B1 = 0.246$. The σ^2 values in Table 3.8 are not used at this point; they will be considered later in the routine.

We enter values for all the equations the same way: Compare the equation that appears on the computer screen with the associated equation in Table 3.8, then type in the appropriate value for each parameter. As another example consider the rate equation for the material transfer from plants to fish ($T_{1,2}$). The prompt on the computer screen is

Transfer – Plant to fish (kg/ha-day)

$$T_{1,2} = \text{B0} + \text{B1} * \text{temperature (D1)} + \text{B2} * \text{weight (A1)} + \text{B3} * \text{temperature (D1)} * \text{weight (A1)}$$

From Table 3.8 we see that for $T_{1,2}$:

$B0 = -1.834$
$B1 = 0.132$
$B2 = 0.422$
$B3 = -0.029$

Note that POND performs many operations automatically, "unseen" by the user. Most unseen operations involve bookkeeping activities which develop internal program structure and handle input and output. However, POND also generates certain equations unseen to the user. The second version of $T_{1,2}$ that operates when $S1 \leq 20000$ (Table 3.8) is one such internally generated equation.

Note that the rate equation for material transfer from plants to natural mortality sink ($T_{1,\text{SI1}}$) is exponential. Type "E" when prompted for the form of this equation.

The model development routine also prompts us to enter variances for random variables if we specify that the model is stochastic. We specify a stochastic model by typing "Y" in response to "is your model stochastic?" For a stochastic model we must enter a daily variance for each of the five water temperature equations. These are the σ^2 values associated with RV3 through RV7 in Table 3.8. For example the variance of the equation for $1 < t < 105$ is 1.457.

We also must specify the variance of the annual water temperature displacement. For all simulations described in the text the variance $= 4$ (σ^2 associated with RV8 in Table 3.8). We also must specify variances associated with the material transfers from source to plants ($T_{\text{SO1},1}$) and from plants to fish ($T_{1,2}$).

For a stochastic model we also must specify a "seed" for the random number generator. Choice of this number is arbitrary. It serves only to initiate a string of internally generated random numbers used to calculate random variables in the model. But note that although choice of the random number seed is arbitrary, replicate stochastic simulations using the same seed will be identical.

Finally this routine prompts us to enter specifications for model output. Output can be directed to either the computer screen (video) or the printer. Output frequency refers to the number of time steps (days) between outputs.

Output Choices

A) Water temperature	B) Aquatic plants
C) Herbivorous fish	D) Fish Harvest
E) Number of fish	F) Weight of fish
G) Transfer–Source to plant	H) Transfer–Plant to fish
I) Transfer–Plant to mortal.	J) Transfer–Fish to mortal.
K) Transfer–Fish to resp.	L) Transfer–Fish to excreta

Enter the letter of each variable to output

Profit will always be output.

(Default = ABCD)?

Output will be displayed left to right in the order specified. For example typing "AHF" will display water temperature, transfer from plants to fish, and weight of fish from left to right. Note that each output item requires approximately 12 columns. Selecting many items will cause the output to "wrap around" and make it difficult to read. Five items will fit on the computer screen and six will fit on an 80-column printer without wrapping around.

The suggested sequence of operations after completion of model development is:

1. display parameters,
2. edit parameters if necessary,
3. save parameters on a diskette, and
4. run simulations.

Routine 2: Display Parameters

This routine allows us to display all of the items that we specified during model development. Parameters can be requested for display in the following groups.

Display Parameters

1) Model parameters 2) Initial conditions
3) Model characteristics 4) Output specifications
5) Profit information

Select 1–5 or hit <ENTER> to exit?

If a mistake is detected when examining any of these displays, it can be corrected using the edit routine (described below) after completing the examination of displays. Note that items displayed on the computer screen can be copied to the printer by pressing "shift," "↓", and "*" simultaneously.

Routine 3: Edit Parameters

This routine allows us to change parameters currently specified in the model. Parameters can be requested for editing in the following groups.

Edit Parameters

1) Model parameters 2) Initial conditions
3) Model characteristics 4) Output specifications
5) Profit information

Select 1–5 or hit <ENTER> to exit?

We are prompted for new values of parameters in the chosen group similar to the way we are prompted for original parameter values in the model development routine. The only difference is that in this routine current parameter values are listed with each prompt. To leave values unchanged press "enter" when prompted. For example within edit group 1 (model parameters) we might be prompted for changes in the rate equation for transfer from fish to respiration sink as follows:

Transfer–Fish to respiration sink (kg/ha-day)

$T_{2,SI2}$ = B0 * (weight (A1) ** B1) exp (B2 * temp. (D1))
* B3 * fish (S2)

B0 = 1.0142 ?
B1 = − 0.2 ?
B2 = 0.07 ?
B3 = 0.0816 ?

Current values for $B1$, $B2$, and $B3$ are correct but the value for $B0$ is incorrect (Table 3.8). To change $B0$ we type in the correct value (0.0142) and then press "enter" when prompted.

Routine 4: Save Parameters on Diskette

This routine allows us to save the model, as we have just parameterized it, on diskette. In TRSDOS disk BASIC a valid filename is a letter followed by seven or fewer symbols. The first symbol must be a letter.

Routine 5: Load Parameters from Diskette

This routine allows us to load the model, as we parameterized it at some previous time, into the computer from diskette. This replaces the entry of model parameters from the keyboard.

If we forget the filename under which we saved model parameters, we can view a list of all files on our diskette by examining the directory. To do this press the reset button and then type "DIR." However, after pressing the reset button we must reload POND, as described at beginning of this appendix.

Routine 6: Run Simulations

This routine allows us to run simulations. Once started, simulations can be stopped temporarily by simultaneously pressing "shift" and "@" and then restarting by pressing "enter." Simulations can be aborted by pressing "X." Do not press "break" to abort simulations.

Phase 3: Model Validation

The Model Validation phase of POND allows us to raise or lower current model parameter values without using the edit routine in the Model Quantification phase. The general procedure is similar to that used in the edit routine but is more efficient for model changes required during model validation.

Phase 4: Model Use

The Model Use phase of POND allows us to change stocking date, harvest date, and number of fish stocked without using the edit routine in the Model Quantification phase. The general procedure is similar to that used in the edit routine but is more efficient for model changes required during model use. Note that increasing the number of fish stocked without also increasing fish biomass (kg/ha) appropriately is equivalent to stocking the pond with more but smaller fish.

A5.3. PROGRAM LISTING

```
10000 ' ******************************************************
10010 ' *                       POND                         *
10020 ' *               Pond Simulation Model                *
10030 ' *                                                    *
10040 ' *                   Lee C. George                    *
10050 ' *           Department of Computer Science           *
10060 ' *               Texas A&M University                 *
10070 ' *            College Station, Texas 77843            *
10080 ' *                                                    *
10090 ' *                   June 18, 1984                    *
10100 ' ******************************************************
10110 '
10120 '
10130 '
10140 ' *************************
10150 ' *   MAIN PROGRAM DRIVER  *
10160 ' *************************
10170 '
10180 CLEAR 100             'Reserve 100 bytes string storage
10190 GOSUB 23310           'Initialize program variables
10200 CLS
10210 GOSUB 22170           'Display main menu
10220 GOSUB 21770           'Command input
10230 IF X>0 THEN GOTO 10260
10240 PRINT "           *** You are currently at the main menu ***";ERASE$;CHR$(27);
10250 GOTO 10220
10260 CLS
10270 '
10280 '          Concept   Quant    Valid    Use
10290 ON X GOSUB  10970,   10380,   11310,   12270
10300 IF X>0 THEN GOSUB 21660
10310 GOTO 10200
10320 '
10330 '
10340 ' *************************
10350 ' *  MODEL QUANTIFICATION  *
10360 ' *************************
10370 '
10380 GOSUB 22260           'Quantification menu
10390 GOSUB 21770           'Command input
10400 IF X=0 RETURN
10410 CLS
10420 '
10430 '          Parm    Disply  Edit    Save    Load    Simul
10440 ON X GOSUB 10560,  18930,  20280,  20450,  20860,  15930
10450 IF X>0 THEN GOSUB 21660
10460 CLS
10470 GOTO 10380
10480 '
10490 '
10500 ' ************************
10510 ' *     DEVELOPMENT      *
10520 ' ************************
10530 '
10540 ' See if existing model is to be re-developed
```

```
10550 '
10560 PRINT TAB(TB-9) "Model Development"
10570 PRINT
10580 GOSUB 22010          'Model developed?
10590 IF Q$="N" THEN RETURN
10600 '
10610 ' Model development
10620 '
10630 GOSUB 13920          'Initial conditions
10640 CLS
10650 GOSUB 10790          'Parameterize model
10660 CLS
10670 GOSUB 14240          'Model specifications
10680 CLS
10690 GOSUB 15450          'Output specifications
10700 '
10710 LET SZ=1             'Set parameterization flag
10720 RETURN
10730 '
10740 '
10750 ' **************************
10760 ' * MODEL PARAMETERIZATION *
10770 ' **************************
10780 '
10790 GOSUB 12770             'Water temperature
10800 CLS
10810 GOSUB 13030             'Sun to plants
10820 CLS
10830 GOSUB 13120             'Plants to fish
10840 GOSUB 13350             'Plants to mortality
10850 GOSUB 21660             'Continue prompt
10860 GOSUB 13620             'Fish to respiration
10870 PRINT
10880 GOSUB 13500             'Fish to excreta
10890 LET X=0
10900 RETURN
10910 '
10920 '
10930 ' **************************
10940 ' *    CONCEPTUAL MODEL    *
10950 ' **************************
10960 '
10970 PRINT TAB(TB-8) "Conceptual Model"
10980 PRINT
10990 PRINT "   The system of interest is a hypothetical pond that is leased"
11000 PRINT "on a monthly basis from the landowner.  Each year the pond"
11010 PRINT "manager (lessee) stocks and harvests herbivorous fish, which"
11020 PRINT "are sold by bulk weight if individual fish weight is heavier"
11030 PRINT "than a legal size limit.  The objective of the pond model is"
11040 PRINT "to examine profit of the operation under various stocking and"
11050 PRINT "harvesting schemes."
11060 PRINT "   We might characterize our conceptual model as a relatively"
11070 PRINT "simple biomass flow model.  The biological portion of the"
11080 PRINT "model is represented by the transfer of biomass bewteen three"
11090 PRINT "state variables, a source, and three sinks.  Biomass enters.
11100 PRINT "the system via plants, which are consumed by herbivorous fish,"
```

```
11110 GOSUB 21660
11120 PRINT "which in turn are harvested.  Biomass leaves the system via"
11130 PRINT "plant mortality, fish respiration, and fish excretion.  The"
11140 PRINT "economic portion of the model is represented by three"
11150 PRINT "auxiliary variables, total cost of lease, gross revenue from"
11160 PRINT "harvest, and profit.  Our description of the system is"
11170 PRINT "multivariate with water temperature (C), number of fish"
11180 PRINT "stocked, assimilation efficiency (%), legal size limit for"
11190 PRINT "selling fish (kg), initial fish size (kg), harvest date,"
11200 PRINT "stocking date, unit cost of lease ($), and unit value of"
11210 PRINT "harvested fish ($) representing system variables."
11220 PRINT "   A complete description of the conceptual model is given in"
11230 PRINT "Chap 2.6., and is diagrammed in Fig. 2.10."
11240 RETURN
11250 '
11260 '
11270 ' **************************
11280 ' *    MODEL VALIDATION    *
11290 ' **************************
11300 '
11310 GOSUB 22360            'Validation menu
11320 GOSUB 21910            'Model parameterized?
11330 IF SZ=0 RETURN
11340 GOSUB 21770            'Command input
11350 IF X=0 THEN RETURN
11360 CLS
11370 '
11380 '           Temp    Sen.An.   Simul
11390 ON X GOSUB 11770,   11880,    11450
11400 IF X>0 THEN GOSUB 21660
11410 CLS
11420 GOTO 11310
11430 '
11440 ' Deterministic or stochastic simulation?
11450 LET SF=STCH
11460 FOR I=0 TO 7
11470   LET SF(I)=STCH(I)
11480 NEXT I
11490 PRINT
11500 LET Q$=""
11510 INPUT "(D)eterministic or (S)tochastic simulation";Q$
11520 IF Q$="S" THEN 11560
11530 IF Q$="D" THEN 11600
11540 PRINT "*** Enter D or S ***"; ERASE$
11550 GOTO 11500
11560 IF STCH<>0 THEN 11640
11570 PRINT "*** Stochastic parameters are not available ***"
11580 PRINT "         Re-develop or edit your model"
11590 RETURN
11600 FOR I=0 TO 7
11610   LET STCH(I)=0
11620 NEXT I
11630 LET STCH=0
11640 PRINT
11650 GOSUB 15930
11660 LET STCH=SF
```

```
11670 FOR I=0 TO 7
11680   LET STCH(I)=SF(I)
11690 NEXT I
11700 RETURN
11710 '
11720 '
11730 ' *************************
11740 ' * TEMPERATURE DISPLACE  *
11750 ' *************************
11760 '
11770 PRINT TAB(TB-12) "Temperature Displacement"
11780 PRINT
11790 PRINT "Enter a temperature displacement (default =";DISP;")";
11800 INPUT DISP
11810 RETURN
11820 '
11830 '
11840 ' *************************
11850 ' *   SENSITIVITY ANALY   *
11860 ' *************************
11870 '
11880 PRINT TAB(TB-10) "Sensitivity Analysis"
11890 PRINT
11900 PRINT "Enter a proportion to vary each parameter"
11910 PRINT "For example, enter -.10 to decrease a parameter value"
11920 PRINT "   by 10%"
11930 PRINT
11940 PRINT "Source to plant transfer (S01,1)      =";SA(1);
11950 INPUT SA(1)
11960 PRINT "Plant to fish transfer (T1,2)         =";SA(2);
11970 INPUT SA(2)
11980 PRINT "Plant to mortality transfer (T1,SI1)  =";SA(3);
11990 INPUT SA(3)
12000 PRINT "Fish to respiration transfer (T2,SI2) =";SA(5);
12010 INPUT SA(5)
12020 PRINT "Fish to excretion transfer (T2,SI3)   =";SA(6);
12030 INPUT SA(6)
12040 PRINT "Plant biomass (S1)                    =";SA(7);
12050 INPUT SA(7)
12060 PRINT "Fish biomass (S2)                     =";SA(8);
12070 INPUT SA(8)
12080 PRINT "Water temperature (D1)                =";SA(14);
12090 INPUT SA(14)
12100 PRINT "Number of fish stocked (D2)           =";SA(9);
12110 INPUT SA(9)
12120 PRINT "Assimilation efficiency (D3)          =";SA(13);
12130 INPUT SA(13)
12140 PRINT "Unit price of fish (D5)               =";SA(10);
12150 INPUT SA(10)
12160 PRINT "Unit cost of lease (D6)               =";SA(11);
12170 INPUT SA(11)
12180 PRINT "Min. harvest size limit (D7)          =";SA(12);
12190 INPUT SA(12)
12200 RETURN
12210 '
12220 '
```

```
12230 ' **************************
12240 ' *       MODEL USE        *
12250 ' **************************
12260 '
12270 GOSUB 22450          'Use menu
12280 GOSUB 21910          'Model parameterized?
12290 IF SZ=0 RETURN
12300 GOSUB 21770          'Command input
12310 IF X=0 RETURN
12320 CLS
12330 '
12340 '          Stock  Harv   Dens   Simul
12350 ON X GOSUB 12450, 12550, 12650, 11450
12360 IF X>0 THEN GOSUB 21660
12370 CLS
12380 GOTO 12270
12390 '
12400 '
12410 ' **************************
12420 ' *    STOCKING DATE       *
12430 ' **************************
12440 '
12450 PRINT TAB(TB-7) "Stocking date"
12460 PRINT
12470 GOSUB 21440
12480 RETURN
12490 '
12500 '
12510 ' **************************
12520 ' *      HARVEST DATE      *
12530 ' **************************
12540 '
12550 PRINT TAB(TB-6) "Harvest date"
12560 PRINT
12570 GOSUB 21550
12580 RETURN
12590 '
12600 '
12610 ' **************************
12620 ' *    STOCKING DENSITY    *
12630 ' **************************
12640 '
12650 PRINT TAB(TB-8) "Stocking density"
12660 PRINT
12670 GOSUB 21280
12680 RETURN
12690 '
12700 '
12710 ' **************************
12720 ' *     POND PARAMETERS    *
12730 ' **************************
12740 '
12750 ' Water temperature driving variable
12760 '
12770 PRINT "Driving variable - Water temperature  (Piecewise linear)"
12780 PRINT "   Water temp. (D1) (degrees C) = B0 + B1 * Julian day (t)"
```

```
12790 PRINT
12800 PRINT "   For   1 <= t <= 105   B0 =";Y(25);
12810 INPUT Y(25)
12820 PRINT "                          B1 =";Y(26);
12830 INPUT Y(26)
12840 PRINT "   For 105 <  t <  166   B0 =";Y(27);
12850 INPUT Y(27)
12860 PRINT "                          B1 =";Y(28);
12870 INPUT Y(28)
12880 PRINT "   For 166 <= t <= 258   B0 =";Y(29);
12890 INPUT Y(29)
12900 PRINT "                          B1 =";Y(30);
12910 INPUT Y(30)
12920 PRINT "   For 258 <  t <  319   B0 =";Y(31);
12930 INPUT Y(31)
12940 PRINT "                          B1 =";Y(32);
12950 INPUT Y(32)
12960 PRINT "   For 319 <= t <= 365   B0 =";Y(33);
12970 INPUT Y(33)
12980 PRINT "                          B1 =";Y(34);
12990 INPUT Y(34)
13000 RETURN
13010 '
13020 '
13030 PRINT "Transfer - Source to plant (kg/ha-day)"
13040 PRINT "   TS01,1 = (B0 + B1 * temperature (D1)) * plant (S1) / 1000"
13050 PRINT "    B0 =";Y(9);
13060 INPUT Y(9)
13070 PRINT "    B1 =";Y(10);
13080 INPUT Y(10)
13090 RETURN
13100 '
13110 '
13120 PRINT "Transfer - Plant to fish (kg/ha-day)"
13130 PRINT "   T1,2 = B0 + B1 * temperature (D1) + B2 * weight (A1) +"
13140 PRINT "          B3 * temperature (D1)* weight (A1)"
13150 PRINT "    B0 =";Y(11);
13160 INPUT Y(11)
13170 PRINT "    B1 =";Y(12);
13180 INPUT Y(12)
13190 PRINT "    B2 =";Y(13);
13200 INPUT Y(13)
13210 PRINT "    B3 =";Y(14);
13220 INPUT Y(14)
13230 PRINT
13240 '
13250 ' Plant limiting density
13260 '
13270 LET Y(35)=20000
13280 '
13290 ' Plant to fish limiting rate    f2=f2 exp(-Y(37)(Y(35)-s1))
13300 '
13310 LET Y(37)=-0.00013863
13320 RETURN
13330 '
13340 '
```

```
13350 PRINT "Transfer - Plant to mortality sink (kg/ha-day)"
13360 PRINT "   Linear      : T1,SI1 = (B0 + B1 * plant (S1)) * plant (S1)"
13370 PRINT "   Exponential : T1,SI1 = (exp(B0 * plant (S1)) - 1)"
13380 PRINT "                               * plant (S1)"
13390 GOSUB 13800
13400 LET PM=2
13410 PRINT "      B0 =";Y(15);
13420 INPUT Y(15)
13430 IF Q$="E" THEN RETURN
13440 PRINT "      B1 =";Y(16);
13450 INPUT Y(16)
13460 LET PM=1
13470 RETURN
13480 '
13490 '
13500 PRINT "Transfer - Fish to excreta (kg/ha-day)"
13510 PRINT "   T2,SI3 = (1- assimilation efficiency (D3)) * (T1,2)"
13520 PRINT
13530 PRINT "   Enter an assimilation efficiency between 0 and 1"
13540 PRINT "      (default =";X8;")";
13550 INPUT X8
13560 IF X8>=0 AND X8<=1 THEN 13590
13570 PRINT "*** Enter a number between 0 and 1 ***"; ERASE$
13580 GOTO 13540
13590 GOSUB 21660
13600 RETURN
13610 '
13620 PRINT "Transfer - Fish to respiration sink (kg/ha-day)"
13630 PRINT "   T2,SI2 = B0 * (weight (A1) ** B1) * exp(B2 * temp. (D1))"
13640 PRINT "             * B3 * fish (S2)"
13650 PRINT "      B0 =";Y(19);
13660 INPUT Y(19)
13670 PRINT "      B1 =";Y(20);
13680 INPUT Y(20)
13690 PRINT "      B2 =";Y(21);
13700 INPUT Y(21)
13710 PRINT "      B3 =";Y(22);
13720 INPUT Y(22)
13730 RETURN
13740 '
13750 '
13760 ' **************************
13770 ' *  LINEAR OR EXPONENTIAL *
13780 ' **************************
13790 '
13800 PRINT
13810 LET Q$=""
13820 INPUT "Would you like a (L)inear or (E)xponential model";Q$
13830 IF Q$="L" OR Q$="E" THEN RETURN
13840 PRINT "*** Enter L or E ***"; ERASE$
13850 GOTO 13810
13860 '
13870 '
13880 ' **************************
13890 ' *   POND INITIAL COND.   *
13900 ' **************************
```

```
13910 '
13920 PRINT TAB(TB-9) "Initial Conditions"
13930 PRINT
13940 '
13950 ' Plant biomass
13960 '
13970 PRINT "Aquatic plants (S1) (kg/ha) =";X1;
13980 INPUT X1
13990 IF X1>=0 THEN 14050
14000 PRINT "*** Enter a positive value ***"; ERASE$
14010 GOTO 13970
14020 '
14030 ' Fish biomass and number of fish
14040 '
14050 GOSUB 21280
14060 '
14070 ' Weight of individual fish
14080 '
14090 LET X4=X2/X3
14100 PRINT "Initial weight of individual fish (D9) (kg) =";X4
14110 PRINT
14120 PRINT "Simulation time (Julian date):"
14130 GOSUB 21440
14140 GOSUB 21550
14150 CLS
14160 GOSUB 15170
14170 RETURN
14180 '
14190 '
14200 ' *************************
14210 ' * MODEL CHARACTERISTICS *
14220 ' *************************
14230 '
14240 PRINT TAB(TB-10) "Model Characteristics"
14250 PRINT
14260 LET Q$=""
14270 INPUT "Is your model stochastic (Y-N)";Q$
14280 IF Q$="Y" OR Q$="N" THEN 14310
14290 PRINT "*** Enter Y or N ***"; ERASE$
14300 GOTO 14260
14310 PRINT
14320 IF Q$="Y" THEN 14450
14330 '
14340 LET STCH=0          'Reset stochastic model flag
14350 '
14360 ' Reset stochatic variable flags
14370 ' Set standard deviations to zero
14380 '
14390 FOR I=0 TO 7
14400    LET STCH(I)=0
14410    LET VR(I)=0
14420 NEXT I
14430 RETURN
14440 '
14450 LET STCH=-1         'Set stochastic model flag
14460 PRINT "Stochastic water temperature model"
```

```
14470 PRINT "   Enter the daily variance of the water temperature equations"
14480 PRINT
14490 '
14500 PRINT "For   1 <= t <= 105   var =";VR(1);
14510 INPUT VR(1)
14520 PRINT "For 105 <  t <  166   var =";VR(2);
14530 INPUT VR(2)
14540 PRINT "For 166 <= t <= 258   var =";VR(3);
14550 INPUT VR(3)
14560 PRINT "For 258 <  t <  319   var =";VR(4);
14570 INPUT VR(4)
14580 PRINT "For 319 <= t <= 365   var =";VR(5);
14590 INPUT VR(5)
14600 CLS
14610 PRINT "Enter an annual water temperature variance."
14620 PRINT "Note: Annual water temperature is used in model validation."
14630 PRINT "   (default =";VR(0);")";
14640 INPUT VR(0)
14650 PRINT
14660 PRINT "Enter variance for SOURCE to PLANT transfer (TSO1,1)"
14670 PRINT "   (default =";VR(6);")";
14680 INPUT VR(6)
14690 PRINT
14700 PRINT "Enter variance for PLANT to FISH transfer (T1,2)"
14710 PRINT "   (default =";VR(7);")";
14720 INPUT VR(7)
14730 CLS
14740 '
14750 ' Set stochastic variable flags
14760 '
14770 FOR I=0 TO 7
14780    IF VR(I)=0 THEN STCH(I)=0
14790    IF VR(I)>0 THEN STCH(I)=-1
14800 NEXT I
14810 '
14820 GOSUB 14910        'Random number generator seed
14830 '
14840 RETURN
14850 '
14860 '
14870 ' ************************
14880 ' * RANDOM NUMBER GENERTR *
14890 ' ************************
14900 '
14910 LET R4=PEEK(&H40AA)
14920 LET R5=PEEK(&H40AB)
14930 LET R6=PEEK(&H40AC)
14940 '
14950 PRINT USING "   Current random number generator seed is  ### ### ###";R4,R5,R6
14960 PRINT "   Enter a new seed (e.g. 134,95,12)"
14970 LET R1=R4
14980 LET R2=R5
14990 LET R3=R6
15000 INPUT "      or hit <ENTER> for current seed";R1,R2,R3
15010 IF R1>=0 AND R1<256 AND R2>=0 AND R2<256 AND        R3>=0 AND R3<256 THEN 15070
15020 PRINT "*** Seed values must lie between 0 and 255 ***"; ERASE$
```

```
15030 GOTO 14970
15040 '
15050 ' Poke random number generator seed into memory
15060 '
15070 POKE &H40AA,R1
15080 POKE &H40AB,R2
15090 POKE &H40AC,R3
15100 RETURN
15110 '
15120 '
15130 ' ************************
15140 ' *   PROFIT INFORMATION  *
15150 ' ************************
15160 '
15170 PRINT TAB(TB-9) "Profit Information"
15180 PRINT
15190 PRINT "Enter the unit price of harvested fish (D5) ($/kg) =";X5;
15200 INPUT X5
15210 IF X5>=0 THEN 15240
15220 PRINT "*** Enter a positive amount ***"; ERASE$
15230 GOTO 15190
15240 PRINT
15250 PRINT "Enter the unit cost of the lease (D6) ($/day) =";X6;
15260 INPUT X6
15270 IF X6>=0 THEN 15300
15280 PRINT "*** Enter a positive amount ***"; ERASE$
15290 GOTO 15250
15300 PRINT
15310 PRINT "   Tctal cost of lease (A3) ($) = ";(Y(2)-Y(1)+1)*X6
15320 PRINT
15330 PRINT "Enter the minumum size limit for selling fish (D7) (kg) =";X7;
15340 INPUT X7
15350 IF X7>=0 THEN 15380
15360 PRINT "*** Enter a positive value ***"; ERASE$
15370 GOTO 15330
15380 RETURN
15390 '
15400 '
15410 ' **************************
15420 ' *  OUTPUT SPECIFICATIONS *
15430 ' **************************
15440 '
15450 PRINT TAB(TB-10) "Output Specifications"
15460 PRINT
15470 LET V$="V"
15480 IF KPW=0 THEN 15540          'Is printer available?
15490 INPUT "Output to (P)rinter or (V)ideo (default=video)";V$
15500 IF V$="V" OR V$="P" THEN 15540
15510 PRINT "*** Enter a P or V ***"; ERASE$
15520 GOTO 15470
15530 '
15540 LET PR=1
15550 INPUT "Enter the output frequency (default = 1 = each time step)";PR
15560 IF PR>0 THEN 15620
15570 PRINT "*** Enter a positive value greater than 1 ***"; ERASE$
15580 GOTO 15540
```

```
15590 '
15600 ' Output Options
15610 '
15620 CLS
15630 PRINT TAB(TB-7) "Output Choices"
15640 FOR I=1 TO OPT
15650    IF I/2=INT(I/2) THEN 15690
15660    PRINT
15670    PRINT CHR$(I+64)+") "+OPT$(I);
15680    GOTO 15700
15690    PRINT TAB(TB) CHR$(I+64)+") "+OPT$(I);
15700 NEXT I
15710 PRINT
15720 PRINT
15730 LET OPT$=LEFT$(ALPHA$,OPT)
15740 PRINT "Enter the letter of each variable to output."
15750 PRINT "   Profit will always be output."
15760 LET Q$="ABCD"
15770 PRINT "   (Default = ";Q$;")";
15780 INPUT Q$
15790 LET NOPT=LEN(Q$)
15800 FOR I=1 TO NOPT
15810    LET PR(I)=INSTR(OPT$,MID$(Q$,I,1))
15820    IF PR(I)>0 THEN 15850
15830    PRINT "*** Enter only capital letters A - ";RIGHT$(OPT$,1);" ***"; ERASE$
15840    GOTO 15760
15850 NEXT I
15860 RETURN
15870 '
15880 '
15890 ' *************************
15900 ' *    POND SIMULATION    *
15910 ' *************************
15920 '
15930 PRINT TAB(TB-7) "Run Simulations"
15940 GOSUB 21910              'Model parameterized?
15950 IF SZ=0 RETURN
15960 '
15970 PRINT
15980 LET IREP=1
15990 IF STCH=0 THEN 16010
16000 INPUT "Enter the number of replicate simulations (default = 1)";IREP
16010 GOSUB 18450              'Start simulation prompt
16020 IF Q$="" THEN RETURN
16030 GOSUB 22760              'Model information heading
16040 '
16050 LET ADISP=0              'Annual water temp displacement
16060 FOR I=0 TO 7
16070    LET SD(I)=SQR(VR(I))
16080 NEXT I
16090 '
16100 ' Begin simulation replications
16110 '
16120 FOR JREP=1 TO IREP
16130    GOSUB 23040           'Replication headings
16140 '
```

```
16150     LET S1=X1           'Iniital plants
16160     LET S2=X2           'Initial fish
16170     LET S3=0            'Harvest
16180     LET D2=X3           'Initial number of fish
16190     LET D5=X5           'Unit price
16200     LET D6=X6           'Unit cost of lease
16210     LET D7=X7           'Size limit
16220     LET D3=X8           'Assimilation
16230     LET A4=0            'Profit
16240     LET SO=0            'Source
16250     LET D1=0            'Water temperature
16260     LET Z1=0            'Natural mortality
16270     LET Z2=0            'Respiration
16280     LET Z3=0            'Excretion
16290     LET F1=0            'Source to plants
16300     LET F2=0            'Plants to fish
16310     LET F3=0            'Plants to mortality
16320     LET F4=0            'Fish to mortality
16330     LET F5=0            'Fish to respiration
16340     LET F6=0            'Fish to excreta
16350     LET KL=2            'Output line count
16360 '
16370 ' Sensitivity Analysis
16380 '
16390     LET S1=S1+S1*SA(7)
16400     LET S2=S2+S2*SA(8)
16410     LET D2=D2+D2*SA(9)
16420        LET A1=S2/D2   'Initial weight of fish
16430     LET D5=D5+D5*SA(10)
16440     LET D6=D6+D6*SA(11)
16450     LET D7=D7+D7*SA(12)
16460     LET D3=D3+D3*SA(13)
16470 '
16480     GOSUB 18630        'Output labels
16490     LET T=0
16500     GOSUB 17930        'Output initial state
16510 '
16520     LET DELTAT=0
16530 '
16540 ' Begin simulation run
16550 '
16560     FOR T=Y(1) TO Y(2)
16570        IF INKEY$="X" THEN RETURN
16580        LET DELTAT=DELTAT+1
16590 '
16600 ' Stochastic water temperature model.(3 piecewise linear)
16610 '
16620        IF T>105 THEN 16670
16630        LET D1=ADISP+DISP+Y(25)+Y(26)*T
16640        IF STCH(1) THEN
                LET D1=FN RN(D1,SD(1))
16650        GOTO 16920
16660 '
16670        IF T>=166 THEN 16720
16680        LET D1=ADISP+DISP+Y(27)+Y(28)*T
16690        IF STCH(2) THEN
```

```
            LET D1=FN RN(D1,SD(2))
16700       GOTO 16920
16710 '
16720       IF T>258 THEN 16770
16730       LET D1=ADISP+DISP+Y(29)+Y(30)*T
16740       IF STCH(3) THEN
               LET D1=FN RN(D1,SD(3))
16750       GOTO 16920
16760 '
16770       IF T>=319 THEN 16820
16780       LET D1=ADISP+DISP+Y(31)+Y(32)*T
16790       IF STCH(4) THEN
               LET D1=FN RN(D1,SD(4))
16800       GOTO 16920
16810 '
16820       LET D1=ADISP+DISP+Y(33)+Y(34)*T
16830       IF STCH(5) THEN
              LET D1=FN RN(D1,SD(5))
16840 '
16850 ' Temp sensitivity analysis
16860       LET D1=D1+D1*SA(14)
16870 '
16880 '
16890 ' Compute transfers
16900 '
16910 ' Source to plants
16920       LET F1=(Y(9)+Y(10)*D1)*S1
16930       LET F1=F1/1000
16940       IF STCH(6) THEN
              F1=FN RN(F1, SD(6))
16950       LET F1=F1+F1*SA(1)
16960       IF F1<0 THEN F1=0
16970 '
16980 ' Plants to fish
16990       LET F2=Y(11)+Y(12)*D1+Y(13)*A1+Y(14)*D1*A1
17000       LET F2=F2*S2
17010       IF S1<Y(35) THEN
              F2=F2*EXP(Y(37)*(Y(35)-S1))
17020       IF STCH(7) THEN
              F2=FN RN(F2, SD(7))
17030       LET F2=F2+F2*SA(2)
17040       IF F2<0 THEN F2=0
17050 '
17060 ' Plants to mortality
17070       IF PM=2 THEN 17100
17080       LET F3=(Y(15)+Y(16)*S1)*S1
17090       GOTO 17110
17100       LET F3=(EXP(Y(15)*S1)-1)*S1
17110       LET F3=F3+F3*SA(3)
17120       IF F3<0 THEN F3=0
17130 '
17140 ' Fish to respiration
17150       LET Y(0)=0
17160       IF A1<>0 THEN Y(0)=A1[Y(20)
17170       LET F5=S2*Y(19)*Y(0)*EXP(Y(21)*D1)*Y(22)/Y(23)
```

```
17180       LET F5=F5+F5*SA(5)
17190       IF F5<0 THEN F5=0
17200 '
17210 ' Fish to excreta
17220       LET F6=(1-D3)*F2
17230       LET F6=F6+F6*SA(6)
17240       IF F6<0 THEN F6=0
17250 '
17260 '
17270 ' Update system components
17280 '
17290 ' Source
17300       LET SO=SO-F1
17310 '
17320 ' Respiration sink
17330       LET Z2=Z2+F5
17340 '
17350 ' Mortality sink
17360       LET Z1=Z1+F3
17370 '
17380 ' Excretrion sink
17390       LET Z3=Z3+F6
17400 '
17410 ' Plant biomass
17420       LET S1=S1+F1-F2-F3
17430       IF S1<0 THEN S1=0
17440 '
17450 ' Fish biomass
17460       LET S2=S2+F2-F5-F6
17470       IF S2<0 THEN S2=0
17480       LET S3=S2
17490 '
17500 ' Weight of individual fish
17510       LET A1=S2/D2
17520 '
17530 ' Number of fish
17540 '   LET D2=0
17550 '   IF A1>0 THEN D2=S2/A1
17560 '   IF D2<0 THEN D2=0
17570 '
17580 '
17590 ' Profit
17600       LET A2=0
17610       IF A1>=D7 THEN A2=S2*D5
17620       LET A3=D6*DELTAT
17630       LET A4=A2-A3
17640 '
17650 ' Output state of system
17660 '
17670       GOSUB 17920
17680       IF KL<KVL OR V$="P" THEN 17720
17690       LET KL=3
17700       PRINT
17710       GOSUB 18630
17720    NEXT T
17730 '
```

```
17740 ' Output last day of simulation and harvest
17750 '
17760     LET T=T-1
17770     GOSUB 17930
17780     IF V$="P" THEN 17820
17790     PRINT "Fish harvest on day";Y(2);"is";S2;"kg/ha"
17800     PRINT
17810     GOTO 17840
17820     LPRINT
17830     LPRINT "Fish harvest on day";Y(2);"is";S2;"kg/ha"
17840 NEXT JREP
17850 RETURN
17860 '
17870 '
17880 ' **************************
17890 ' *    OUTPUT RESULTS      *
17900 ' **************************
17910 '
17920 IF INT((T-Y(1))/PR)<>(T-Y(1))/PR THEN 18380
17930 IF V$="P" THEN 17990
17940 IF T>0 THEN 17970
17950 PRINT "    ";
17960 GOTO 18030
17970 PRINT USING "####";T;
17980 GOTO 18030
17990 IF T>0 THEN 18020
18000 LPRINT "    ";
18010 GOTO 18030
18020 LPRINT USING "####";T;
18030 FOR I=1 TO NOPT
18040     ON PR(I) GOTO 18050, 18070, 18090, 18110, 18130,           18150, 18170, 18190,
18210, 18230, 18250, 18270
18050     LET VAR=D1
18060     GOTO 18280
18070     LET VAR=S1
18080     GOTO 18280
18090     LET VAR=S2
18100     GOTO 18280
18110     LET VAR=S3
18120     GOTO 18280
18130     LET VAR=D2
18140     GOTO 18280
18150     LET VAR=A1
18160     GOTO 18280
18170     LET VAR=F1
18180     GOTO 18280
18190     LET VAR=F2
18200     GOTO 18280
18210     LET VAR=F3
18220     GOTO 18280
18230     LET VAR=F4
18240     GOTO 18280
18250     LET VAR=F5
18260     GOTO 18280
18270     LET VAR=F6
18280     IF V$="P" THEN 18310
```

```
18290    PRINT USING "  ##.#####[[[[";VAR;
18300    GOTO 18320
18310    LPRINT USING "  ##.#####[[[[";VAR;
18320 NEXT I
18330 IF V$="P" THEN 18360
18340 PRINT USING "  $##,###.##-";A4
18350 GOTO 18370
18360 LPRINT USING "  $##,###.##-";A4
18370 LET KL=KL+1
18380 RETURN
18390 '
18400 '
18410 ' **************************
18420 ' *   SIMULATION PROMPT    *
18430 ' **************************
18440 '
18450 PRINT
18460 PRINT "Type START to begin simulation or hit <ENTER> to exit."
18470 LET Q$=""
18480 PRINT "   Hit the 'X' key to abort during simulation";Q$
18490 PRINT "   or hit <SHIFT> @ to pause and <ENTER> to resume";
18500 INPUT Q$
18510 IF Q$="" THEN RETURN
18520 IF Q$="START" THEN 18550
18530 PRINT"*** Type START or hit <ENTER> ***"; ERASE$
18540 GOTO 18470
18550 PRINT
18560 RETURN
18570 '
18580 '
18590 ' *************************
18600 ' *     OUTPUT LABELS     *
18610 ' *************************
18620 '
18630 IF V$="P" THEN 18760
18640 PRINT "TIME";
18650 FOR I=1 TO NOPT
18660    PRINT LEFT$("  "+LB$(PR(I))+"            ",14);
18670 NEXT I
18680 PRINT "   Profit ($)"
18690 PRINT "====";
18700 FOR I=1 TO NOPT+1
18710    PRINT "  ===========";
18720 NEXT I
18730 PRINT
18740 RETURN
18750 '
18760 LPRINT "TIME";
18770 FOR I=1 TO NOPT
18780    LPRINT LEFT$("  "+LB$(PR(I))+"            ",14);
18790 NEXT I
18800 LPRINT "   Profit ($)"
18810 LPRINT "====";
18820 FOR I=1 TO NOPT+1
18830    LPRINT "  ===========";
18840 NEXT I
```

```
18850 LPRINT
18860 RETURN
18870 '
18880 '
18890 ' *************************
18900 ' *   DISPLAY PARAMETERS   *
18910 ' *************************
18920 '
18930 GOSUB 22520        'Parm menu
18940 GOSUB 21910        'Model parameterized?
18950 IF SZ=0 THEN RETURN
18960 GOSUB 21770        'Command input
18970 IF X=0 RETURN
18980 CLS
18990 '           Parms  Initl  Char   Output Profit
19000 ON X GOSUB 19100, 19520, 19690, 20100, 19960
19010 IF X>0 THEN GOSUB 21660
19020 CLS
19030 GOTO 18930
19040 '
19050 '
19060 ' *************************
19070 ' *   MODEL PARAMETERS    *
19080 ' *************************
19090 '
19100 PRINT TAB(TB-8) "Model Parameters"
19110 PRINT
19120 PRINT "Driving variable - Water temperature (D1)"
19130 PRINT "   D1 =";Y(25);"+";Y(26);"* t"; TAB(43); "For   1 <= t <= 105"
19140 PRINT "   D1 =";Y(27);"+";Y(28);"* t"; TAB(43); "For 105 <  t <  166"
19150 PRINT "   D1 =";Y(29);"+";Y(30);"* t"; TAB(43); "For 166 <= t <= 258"
19160 PRINT "   D1 =";Y(31);"+";Y(32);"* t"; TAB(43); "For 258 <  t <  319"
19170 PRINT "   D1 =";Y(33);"+";Y(34);"* t"; TAB(43); "For 319 <= t <= 365"
19180 PRINT
19190 PRINT "Transfer - Source to plant (kg/ha-day)"
19200 PRINT "   Limiting plant density =";Y(35)
19210 PRINT "   TSO1,1 = (";Y(9);"+";Y(10);"* temp (D1)) * plant (S1) / 1000"
19220 GOSUB 21660
19230 PRINT "Transfer - Plant to fish (kg/ha-day)"
19240 PRINT "If plant density >=";Y(35);"then"
19250 PRINT "   T1,2 =";Y(11);"+";Y(12);"* temp (D1) +";Y(13);"* weight (A1) +"
19260 PRINT "         ";Y(14);"* temp (D1) * weight (A1)"
19270 PRINT
19280 PRINT "If plant density  <";Y(35);"then"
19290 PRINT "   Limiting rate = T1,2 * exp(";Y(37);"* (";Y(35);" -"
19300 PRINT "                        plant (S1)))"
19310 GOSUB 21660
19320 PRINT "Transfer - Plant to mortality sink (kg/ha-day)"
19330 IF PM=2 THEN 19360
19340 PRINT "   T1,SI1 = (";Y(15);"+";Y(16);"* plant (S1)) * plant (S1)"
19350 GOTO 19370
19360 PRINT "   T1,SI1 = exp(";Y(15);"* plant (S1)) * plant (S1)"
19370 PRINT
19380 PRINT "Transfer - Fish to excreta (kg/ha-day)"
19390 PRINT "   T2,SI3 = (plant to fish) * 1 -";X8
19400 PRINT "   Assimilation efficiency =";X8
```

```
19410 PRINT
19420 PRINT "Transfer - Fish to respiration sink (kg/ha-day)"
19430 PRINT "   T2,SI2 = ";Y(19);"* (weight (A1) **";Y(20);") *"
19440 PRINT "          exp(";Y(21);"* temp (D1)) *";Y(22);"* fish (S2)"
19450 RETURN
19460 '
19470 '
19480 ' *************************
19490 ' *  INITIAL CONDITIONS   *
19500 ' *************************
19510 '
19520 PRINT TAB(TB-9) "Initial Conditions"
19530 PRINT
19540 PRINT "Aquatic plants (S1) (kg/ha) =";X1
19550 PRINT "Herbviorous fish (S2) (kg/ha) =";X2
19560 PRINT "Number of fish stocked (D2) =";X3
19570 PRINT "Initial weight of individual fish (A1) (kg) =";X4
19580 PRINT
19590 PRINT "Simulation time (Julian date):"
19600 PRINT "   Stocking time (t begin) =";Y(1)
19610 PRINT "   Harvest time (t end)    =";Y(2)
19620 RETURN
19630 '
19640 '
19650 ' *************************
19660 ' * MODEL CHARACTERISTICS *
19670 ' *************************
19680 '
19690 PRINT TAB(TB-7) "Model Characteristics"
19700 PRINT
19710 IF STCH=0 THEN 19870
19720 PRINT "Stochastic water temperature model"
19730 PRINT "   Daily variance of water temperature equations"
19740 PRINT "      For   1 <= t <= 105   var =";VR(1)
19750 PRINT "      For 105 <  t <  166   var =";VR(2)
19760 PRINT "      For 166 <= t <= 258   var =";VR(3)
19770 PRINT "      For 258 <  t <  319   var =";VR(4)
19780 PRINT "      For 319 <= t <= 365   var =";VR(5)
19790 PRINT "   Annual water temperature variance =";VR(0)
19800 PRINT
19810 PRINT "Variance of SOURCE to PLANT transfer (TS01,1) =";VR(6)
19820 PRINT
19830 PRINT "Variance of PLANT to FISH transfer (T1,2) =";VR(7)
19840 GOSUB 21660
19850 PRINT "Random number generator seed =";R1;R2;R3
19860 RETURN
19870 PRINT
19880 PRINT "Model is not stochastic."
19890 RETURN
19900 '
19910 '
19920 ' *************************
19930 ' *  PROFIT INFORMATION   *
19940 ' *************************
19950 '
19960 PRINT TAB(TB-9) "Profit Information"
```

```
19970 PRINT
19980 PRINT "Unit cost of lease (D6) ($) = ";X6
19990 PRINT "Total cost of lease (A3) ($) = ";(Y(2)-Y(1)+1)*X6
20000 PRINT
20010 PRINT "Unit price of harvested fish (D5) ($/kg) = ";X5
20020 PRINT "Minimum size limit for selling fish (D7) (kg) =";X7
20030 RETURN
20040 '
20050 '
20060 ' *************************
20070 ' *    OUTPUT OPTIONS     *
20080 ' *************************
20090 '
20100 PRINT TAB(TB-7) "Output Options"
20110 PRINT
20120 PRINT "Printer option = ";
20130 IF V$="P" THEN PRINT "ON"
20140 IF V$<>"P" THEN PRINT "OFF"
20150 PRINT "Output frequency =";PR
20160 PRINT
20170 PRINT "Output choices:"
20180 FOR I=1 TO NOPT
20190    PRINT USING "% % - %                                    %";MID$(ALPHA$,PR(I),1),OPT$(PR(I))
20200 NEXT I
20210 RETURN
20220 '
20230 '
20240 ' **************************
20250 ' *     EDIT PARAMETERS    *
20260 ' **************************
20270 '
20280 GOSUB 22600            'Edit menu
20290 GOSUB 21910            'Model parameterized?
20300 IF SZ=0 RETURN
20310 GOSUB 21770            'Command input
20320 IF X=0 RETURN
20330 CLS
20340 '           Parms  Initl  Char   Output Profit
20350 ON X GOSUB 10790, 13920, 14240, 15450, 15170
20360 IF X>0 THEN GOSUB 21660
20370 CLS
20380 GOTO 20280
20390 '
20400 '
20410 ' *************************
20420 ' *  SAVE MODEL (DISK)    *
20430 ' *************************
20440 '
20450 PRINT TAB(TB-11) "Save Model (diskette)"
20460 GOSUB 21910            'Model parameterized?
20470 IF SZ=0 RETURN
20480 PRINT
20490 PRINT "Enter a valid file name to store your model"
20500 PRINT "   or hit <ENTER> to exit"
20510 PRINT "   For example:   SIMULATE"
20520 PRINT
```

```
20530 LET Q$=""
20540 LINE INPUT "File name :";Q$
20550 IF Q$="" THEN RETURN
20560 LET L=LEN(Q$)
20570 IF L=1 THEN 20590
20580 IF INSTR(Q$,":")=L-1 THEN 20600
20590 LET Q$=Q$+":0"
20600 PRINT
20610 PRINT "Hit <ENTER> when diskette is mounted on drive ";RIGHT$(Q$,1);
20620 INPUT P$
20630 OPEN "O",1,Q$
20640 FOR I=1 TO 40
20650    PRINT #1,Y(I);
20660 NEXT I
20670 PRINT #1, PM; FM; X1; X2; X3; X4; X5; X6; X7; X8; STCH;
20680 FOR I=0 TO 7
20690    PRINT #1, STCH(I); VR(I);
20700 NEXT I
20710 PRINT #1,DISP; R1; R2; R3; V$
20720 PRINT #1, PR; NOPT;
20730 FOR I=1 TO NOPT
20740    PRINT #1,PR(I);
20750 NEXT I
20760 CLOSE 1
20770 PRINT
20780 PRINT "Model ";Q$;" saved."
20790 RETURN
20800 '
20810 '
20820 ' ************************
20830 ' *  LOAD MODEL (DISK)   *
20840 ' ************************
20850 '
20860 PRINT TAB(TB-11) "Load Model (diskette)"
20870 PRINT
20880 IF SZ=0 THEN 20980
20890 PRINT TAB(TB-14) "*** Model already exists ***"
20900 LET Q$=""
20910 INPUT "Replace (Y-N)";Q$
20920 IF Q$="N" THEN RETURN
20930 IF Q$="Y" THEN 20970
20940 PRINT "*** Enter Y or N ***";
20950 PRINT ERASE$
20960 GOTO 20900
20970 PRINT
20980 LINE INPUT "Enter the file name of the model to load? ";Q$
20990 PRINT
21000 INPUT "Hit <ENTER> when diskette is mounted";P$
21010 OPEN "I",1,Q$
21020 FOR I=1 TO 40
21030    INPUT #1,Y(I)
21040 NEXT I
21050 INPUT #1, PM, FM, X1, X2, X3, X4, X5, X6, X7, X8, STCH
21060 FOR I=0 TO 7
21070    INPUT #1, STCH(I), VR(I)
```

```
21080 NEXT I
21090 INPUT #1, DISP, R1, R2, R3, V$
21100 INPUT #1, PR, NOPT
21110 FOR I=1 TO NOPT
21120    INPUT #1,PR(I)
21130 NEXT I
21140 LET SZ=1
21150 POKE &H40AA, R1
21160 POKE &H40AB, R2
21170 POKE &H40AC, R3
21180 CLOSE 1
21190 PRINT
21200 PRINT "Model ";Q$;" loaded."
21210 RETURN
21220 '
21230 '
21240 ' *************************
21250 ' *     INITIAL FISH      *
21260 ' *************************
21270 '
21280 PRINT "Herbivorous fish (S2) (kg/ha) =";X2;
21290 INPUT X2
21300 IF X2>0 THEN 21330
21310 PRINT "*** Enter a positive value ***"; ERASE$
21320 GOTO 21280
21330 PRINT "Number of fish stocked (D2) =";X3;
21340 INPUT X3
21350 IF X3>0 THEN RETURN
21360 PRINT "*** Enter a value greater than 0 ***"; ERASE$
21370 GOTO 21340
21380 '
21390 '
21400 ' *************************
21410 ' *     STOCKING TIME     *
21420 ' *************************
21430 '
21440 PRINT "   Stocking time (t begin) =";Y(1);
21450 INPUT Y(1)
21460 IF Y(1)>0 THEN RETURN
21470 PRINT "*** Enter a positive integer value ***"; ERASE$
21480 GOTO 21440
21490 '
21500 '
21510 ' *************************
21520 ' *     HARVEST TIME      *
21530 ' *************************
21540 '
21550 PRINT "   Harvest time (t end)    =";Y(2);
21560 INPUT Y(2)
21570 IF Y(2)>Y(1) THEN RETURN
21580 PRINT "*** Enter a positive integer greater than";Y(1);"***"; ERASE$
21590 GOTO 21550
21600 '
21610 '
21620 ' **************************
21630 ' *    CONTINUE PROMPT     *
```

```
21640 ' **************************
21650 '
21660 PRINT
21670 PRINT TAB(TB-12) "Hit <ENTER> to continue";
21680 INPUT Q$
21690 CLS
21700 RETURN
21710 '
21720 '
21730 ' **************************
21740 ' *      COMMAND INPUT     *
21750 ' **************************
21760 '
21770 PRINT
21780 PRINT
21790 LET X=0
21800 PRINT TAB(TB-18) "Select 1 -";N"or hit <ENTER> to exit";
21810 INPUT X
21820 IF X>=0 AND X<=N THEN RETURN
21830 PRINT TAB(TB-14) "*** Enter a number 1 -";N;"***"; ERASE$
21840 GOTO 21790
21850 '
21860 '
21870 ' **************************
21880 ' * IS MODEL PARAMETERIZED *
21890 ' **************************
21900 '
21910 IF SZ=1 THEN RETURN
21920 PRINT
21930 PRINT TAB(TB-20) "*** Model has not been parameterized ***"
21940 RETURN
21950 '
21960 '
21970 ' *************************
21980 ' * MODEL DEVELOPED PROMPT*
21990 ' *************************
22000 '
22010 IF SZ<>1 THEN 22070
22020 LET Q$=""
22030 INPUT "The model is already developed.  Re-develop (Y-N)";Q$
22040 IF Q$="Y" OR Q$="N" THEN RETURN
22050 PRINT "*** Enter Y or N ***"; ERASE$
22060 GOTO 22020
22070 LET Q$="Y"
22080 RETURN
22090 '
22100 '
22110 ' *************************
22120 ' *        MENUS          *
22130 ' *************************
22140 '
22150 ' Main menu
22160 '
22170 PRINT TAB(TB-15) "P O N D   S I M U L A T I O N"
22180 PRINT
22190 PRINT TAB(10) "1) Conceptual Model" TAB(35) "2) Model Quantification"
```

```
22200 PRINT TAB(10) "3) Model Validation" TAB(35) "4) Model Use"
22210 LET N=4
22220 RETURN
22230 '
22240 ' Quantification menu
22250 '
22260 PRINT TAB(TB-10) "Model Quantification"
22270 PRINT
22280 PRINT TAB(10) "1) Model Development" TAB(35) "2) Display Parameters"
22290 PRINT TAB(10) "3) Edit parameters" TAB(35) "4) Save Parms (diskette)"
22300 PRINT TAB(10) "5) Load Parms (diskette)" TAB(35) "6) Run Simulations"
22310 LET N=6
22320 RETURN
22330 '
22340 ' Validation menu
22350 '
22360 PRINT TAB(TB-8) "Model Validation"
22370 PRINT
22380 PRINT TAB(10) "1) Displace Temperature" TAB(35) "2) Sensitivity Analysis"
22390 PRINT TAB(10) "3) Run Simulations"
22400 LET N=3
22410 RETURN
22420 '
22430 ' Model use menu
22440 '
22450 PRINT TAB(TB-5) "Model Use"
22460 PRINT
22470 PRINT TAB(10) "1) Stocking date" TAB(35) "2) Harvest date"
22480 PRINT TAB(10) "3) Stocking density" TAB(35) "4) Run Simulations"
22490 LET N=4
22500 RETURN
22510 '
22520 ' Display Param menu
22530 '
22540 PRINT TAB(TB-9) "Display Parameters"
22550 GOSUB 22640
22560 RETURN
22570 '
22580 ' Edit parameters
22590 '
22600 PRINT TAB(TB-8) "Edit Parameters"
22610 GOSUB 22640
22620 RETURN
22630 '
22640 PRINT
22650 PRINT TAB(10) "1) Model parameters" TAB(35) "2) Initial conditions"
22660 PRINT TAB(10) "3) Model characteristics" TAB(35) "4) Output Specifications"
22670 PRINT TAB(10) "5) Profit information"
22680 LET N=5
22690 RETURN
22700 '
22710 '
22720 ' *************************
22730 ' *    OUTPUT HEADINGS    *
22740 ' *************************
22750 '
```

```
22760 IF V$="P" THEN 22880
22770 PRINT "P O N D   S I M U L A T I O N"
22780 PRINT
22790 IF STCH=0 THEN PRINT "Deterministic"
22800 IF STCH=-1 THEN PRINT "Stochastic"
22810 PRINT "Stocking time =";Y(1); TAB(35) "Harvest time =";Y(2)
22820 PRINT "Unit cost of lease (D6) ($) = ";X6;
      TAB(35) "Total cost of lease (A3) ($) = ";(Y(2)-Y(1)+1)*X6
22830 PRINT "Number of fish stocked (D2) =";X3
22840 PRINT "Price per kg of fish ($) = ";X5; TAB(35) "Minimum size =";X7;"kg"
22850 PRINT "Daily water temperature displacement ="; DISP
22860 RETURN
22870 '
22880 LPRINT "P O N D   S I M U L A T I O N"
22890 LPRINT
22900 IF STCH=0 THEN LPRINT "Deterministic"
22910 IF STCH=-1 THEN LPRINT "Stochastic"
22920 LPRINT "Stocking time =";Y(1); TAB(35) "Harvest time =";Y(2)
22930 LPRINT "Unit cost of lease (D6) ($) = ";X6;
      TAB(35) "Total cost of lease (A3) ($) = "; (Y(2)-Y(1)+1)*X6
22940 LPRINT "Number of fish stocked (D2) =";X3
22950 LPRINT "Price per kg of fish ($) = ";X5; TAB(35) "Minimum size =";X7;"kg"
22960 LPRINT "Daily water temperature displacement =";DISP
22970 RETURN
22980 '
22990 '
23000 ' *************************
23010 ' * REPLICATION HEADINGS  *
23020 ' *************************
23030 '
23040 IF V$="P" THEN 23090
23050 PRINT
23060 PRINT "Rep";JREP;"out of";IREP
23070 PRINT
23080 GOTO 23120
23090 LPRINT
23100 LPRINT "Rep";JREP;"out of";IREP
23110 LPRINT
23120 IF STCH=0 THEN RETURN
23130 IF V$="P" THEN 23200
23140 PRINT "Seed =";PEEK(&H40AA); PEEK(&H40AB); PEEK(&H40AC)
23150 '
23160 LET ADISP=FN RN(0,SD(0))
23170 PRINT "Annual water temperature displacement =";ADISP
23180 PRINT
23190 RETURN
23200 LPRINT "Seed =";PEEK(&H40AA); PEEK(&H40AB); PEEK(&H40AC)
23210 LET ADISP=FN RN(0,SD(0))
23220 LPRINT "Annual water temperature displacement =";ADISP
23230 LPRINT
23240 RETURN
23250 '
23260 '
23270 ' *************************
23280 ' *     INITITALIZE       *
```

```
23290 ' **************************
23300 '
23310 GOSUB 24150             'System specifications
23320 RANDOM                  'Random number seeding
23330 '
23340 ' Random NORMAL variate with mean M and standard dev. SD
23350 '
23360 DEF FN RN(XM,SD)=XM+SD*SQR(-2.0*LOG(RND(0)))*          COS(6.282*RND(0))
23370 '
23380 DIM Y(40)                             'Model parameters
23390 DIM OPT$(20)                          'Output options
23400 DIM LB$(20)                           'Output labels
23410 DIM PR(20)                            'Output choices
23420 DIM STCH(7)                           'Stochastic models
23430 DIM SF(7)                             'Save stoch model
23440 DIM VR(7)                             'Variances of stoch mod
23450 DIM SD(7)                             'Stand Dev of stoch mod
23460 DIM SA(14)                            'Sensitivity Analy
23470 '
23480 ' Model coefficients
23490 '
23500 FOR I=1 TO 40
23510    LET Y(I)=0
23520 NEXT I
23530 '
23540 ' Stochastic variable flags
23550 '
23560 FOR I=0 TO 7
23570    LET STCH(I)=0
23580    LET VR(I)=0
23590    LET SD(I)=0
23600 NEXT I
23610 '
23620 FOR I=0 TO 14
23630    LET SA(I)=0
23640 NEXT I
23650 '
23660 LET ERASE$=CHR$(27)+CHR$(27)          'Two reverse line feed
23670 LET TB=IVW/2                          'Video center
23680 LET SZ=0                              'Parameterization flag
23690 LET ALPHA$="ABCDEFGHIJKLMNOPQRSTUVWXYZ"
23700 LET STCH=0                            'Stochastic model flag
23710 LET X1=0                              'Initial plants
23720 LET X2=0                              'Initial fish
23730 LET X3=0                              'Initial number of fish
23740 LET D3=0                              'Assimilation
23750 LET D5=0                              'Initial unit price of fish
23760 LET D6=0                              'Initial unit cost of lease
23770 LET D7=0                              'Initial size limit for sell
23780 '
23790 '
23800 ' Read Output Options, Output Labels, Var/Label Conversion
23810 '
23820 LET OPT=0
23830 LET OPT=OPT+1
23840 READ OPT$(OPT)
```

```
23850 IF OPT$(OPT)="END" THEN 23880
23860 READ LB$(OPT)
23870 GOTO 23830
23880 LET OPT=OPT-1
23890 RETURN
23900 '
23910 '
23920 ' **************************
23930 ' *  OUTPUT OPTIONS/LABELS *
23940 ' **************************
23950 '
23960 DATA Water temperature, Water Temp.
23970 DATA Aquatic plants, Plant Bio.
23980 DATA Herbivorous fish, Fish Bio.
23990 DATA Fish Harvest, Harvest
24000 DATA Number of fish, Number Fish
24010 DATA Weight of fish, Weight Fish
24020 DATA Transfer - Source to plant, Source-Plant
24030 DATA Transfer - Plant to fish, Plant-Fish
24040 DATA Transfer - Plant to mortal., Plant-Mortal
24050 DATA Transfer - Fish to mortal., Fish-Mortal
24060 DATA Transfer - Fish to resp., Fish-Resp.
24070 DATA Transfer - Fish to excreta, Fish-Excreta
24080 DATA END
24090 '
24100 '
24110 ' **************************
24120 ' *     SPECIFICATIONS     *
24130 ' **************************
24140 '
24150 DEFINT I-N
24160 '
24170 ' Precision - Single=DEFSNG, Double=DEFDBL
24180 DEFSNG A-H, O-Z
24190 '
24200 ' Video Width and Tab Setting
24210 LET IVW=63
24220 '
24230 ' Number of video lines
24240 LET KVL=16
24250 '
24260 ' Printer Width (KPW=0 for no printer available)
24270 LET KPW=130
24280 '
24290 'Output Format
24300 LET PAT$="##.#####[[[["
24310 RETURN
```

APPENDIX 6

ANSWERS TO EXERCISES

A6.1. ANSWERS TO CHAPTER 2 EXERCISES

Exercise 2.7.1

A conceptual model diagram of the plant–herbivore system described in Chapter 2 is presented in Figure A6.1. The basic process to be modeled is biomass flow through a terrestrial ecosystem. Thus biomass is the quantity of primary interest. State variables representing points of accumulation of biomass within the system include live plants ($S1$), herbivores ($S2$), plant litter ($S3$), herbivore excreta ($S4$), and harvested herbivores ($S5$). (We have included herbivore excreta as a state variable based on the assumption that we are interested in the amount of plant "fertilizer" returned to the litter layer by herbivores. If we were not explicitly interested in the accumulation of herbivore excreta, this state variable would be omitted.) Biomass enters the system by transfer from a biomass source (SO1) to live plants. Biomass leaves the system by transfer from herbivores to a respiration sink (SI1) and by transfers from plant litter and herbivore excreta to a decomposition sink (SI2). Within the system biomass is transferred from live plants to herbivores and plant litter, and from herbivores to herbivore excreta and harvested herbivores.

An environmental temperature driving variable ($D1$) influences biomass transfers from the source to live plants ($T_{\mathrm{SO1},1}$), from live plants to herbivores ($T_{1,2}$), from herbivores to the respiration sink ($T_{2,\mathrm{SI1}}$), and from herbivores to herbivore excreta ($T_{2,4}$) (Figure A6.1). A soil water tension driving variable ($D2$) influences biomass transfers from the source to live plants ($T_{\mathrm{SO1},1}$), from plant litter to the decomposition sink ($T_{3,\mathrm{SI2}}$), and from herbivore excreta to the decomposition sink ($T_{4,\mathrm{SI2}}$). A management scheme driving variable ($D3$) influences the biomass transfer from herbivores to harvested herbivores ($T_{2,5}$). The ratio of live plant biomass to herbivore biomass, which is an auxiliary

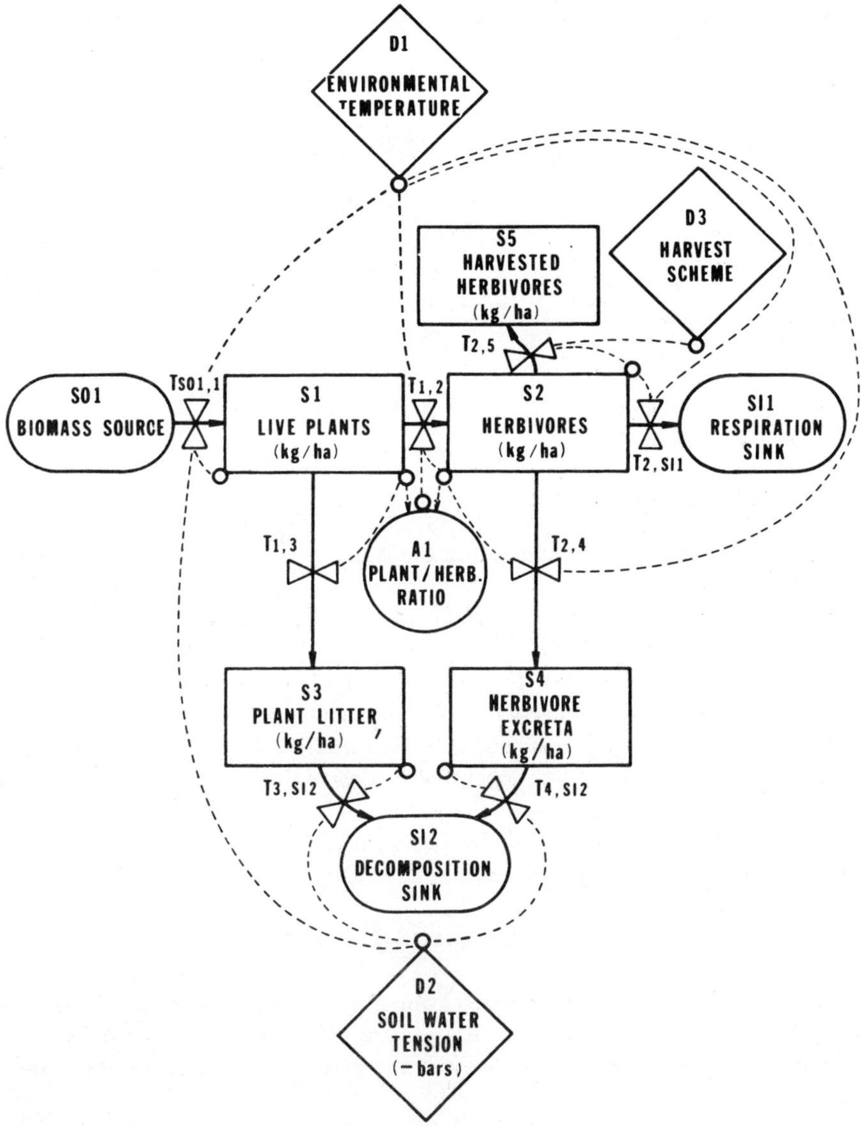

FIGURE A6.1. Conceptual model diagram representing biomass dynamics of a hypothetical plant–herbivore system.

variable ($A1$), influences the biomass transfer from live plants to herbivores ($T_{1,2}$).

We might characterize our conceptual model of the hypothetical plant–herbivore system as a relatively simple biomass flow model. We represent the transfer of biomass between five state variables, a source, and two sinks. The control of biomass transfers can be represented as:

$$\begin{aligned}
T_{SO1,1} &= f(D1, D2, S1) \\
T_{1,2} &= f(D1, A1, S2) \\
T_{1,3} &= f(S1) \\
T_{2,4} &= f(D1, S2) \\
T_{2,SI1} &= f(D1, S2) \\
T_{2,5} &= f(D3, S2) \\
T_{3,SI2} &= f(D2, S3) \\
T_{4,SI2} &= f(D2, S4)
\end{aligned}$$

There are no submodels representing the transfer of material other than biomass. Our description of the system is multivariate with state variables measured in units such as kg/ha, material transfer rates measured in units such as kg/ha-day, the environmental temperature driving variable measured in units such as °C, the soil water tension driving variable measured in units such as –bars, and the management scheme driving variable and the ratio of live plant biomass to herbivore biomass auxiliary variable being unitless. (Note that there certainly are alternative conceptual models of the plant–herbivore system that would serve our purposes equally well. In most circumstances it is naive to suppose we know enough about the system of interest that there is only one "correct" model for our purposes.)

Exercise 2.7.2

A matrix of material transfers representing our conceptual model of the plant–herbivore system is presented in Figure A6.2. The one-way, essentially nonbranching route of biomass flow through the system is obvious from this representation of the conceptual model.

A6.2. ANSWERS TO CHAPTER 3 EXERCISES

Exercise 3.7.1

Choice of the basic time unit for simulation of the plant–herbivore model involves reconciling the difference in the level of resolution needed to address our questions of interest and level of resolution of available data. Data on the sys-

MATERIAL TRANSFER FROM EACH COMPONENT	MATERIAL TRANSFER TO EACH COMPONENT						
	S1	S2	S3	S4	S5	S11	S12
SO1	X						
S1		X	X				
S2				X	X	X	
S3							X
S4							X
S5							

FIGURE A6.2. Matrix of material transfers representing the conceptual model of the plant-herbivore system (Figure A6.1).

tem have been collected on a daily basis, which provides sufficient resolution to represent appropriately temporal changes in transfer rates within the system. But it seems reasonable that weekly, or even monthly, time steps would give sufficient resolution to address our questions of interest. We will choose 1 day as the time step for simulation because this allows us to maintain maximum resolution consistent with available data, facilitates parameterization of model equations, and given the relatively small size of the model should not be too costly in computing time or money. Note that this choice is arbitrary and models developed using time steps anywhere between 1 day and 1 month probably would work equally well.

Choice of the functional form and parameterization of the rate equation representing biomass transfer from the biomass source to live plants ($T_{SO1,1}$) are based on data in Table 3.11 and expert opinion concerning constraints on primary production due to competition to high plant densities (p. 127). Plots of data relating primary production to environmental temperature and relating primary production to soil water tension suggest linear relationships. Simple linear regression of net primary production (g produced/g live plant biomass-day) on temperature (°C) yields the equation

$$\text{production} = 0.029 - 0.0006 \text{ temperature}$$

Regression is significant ($P < 0.0001$) with $r^2 = 0.913$. Simple linear regression of net primary production on soil water tension (–bars) yields the equation

$$\text{production} = 0.029 - 0.008 \text{ soil water tension}$$

Regression is significant ($P < 0.0001$) with $r^2 = 0.898$. Multiple linear regression of net primary production on temperature and soil water tension yields the equation

$$\text{production} = 0.029 - 0.002 \text{ temperature} + 0.019 \text{ soil water tension}$$

Regression is significant ($P < 0.0001$) with $r^2 = 0.928$. Multiple linear regression of net primary production on temperature, soil water tension, and their interaction yields the equation

$$\text{production} = 0.025 - 0.002 \text{ temperature} + 0.026 \text{ soil water tension} - 0.00009 \text{ temperature} \times \text{soil water tension}$$

Regression is significant ($P < 0.0006$) with $r^2 = 0.934$.

Each of the four regressions above predicts net primary production well. From a strictly empirical point of view we might use any of these equations in the model. A consideration in terms of selecting the best regression equation is that temperature and soil water tension are highly correlated. Once one of them is in the regression model, addition of the other does not improve the model much. The best regression equation is

$$\text{production} = 0.029 - 0.0006 \text{ temperature}$$

because it contains only one independent variable and predicts net primary production very well (slightly better than the simple linear regression using soil water tension). However, let's assume that we are interested in a more mechanistic representation of net primary production that includes both temperature and soil water tension. Then our choice is the equation

$$\text{production} = 0.029 - 0.002 \text{ temperature} + 0.019 \text{ soil water tension}$$

(We do not include the interaction term since it adds little predictive capability.) Thus when live plant biomass is relatively low and competition is not important $T_{SO1,1}$ (kg/ha-day) can be represented as

$$T_{SO1,1} = (0.029 - 0.002D1 + 0.019D2)S1 \text{ if } S1 \leq 50{,}000$$

However, when live plant biomass becomes relatively high and competition slows production

$$T_{SO1,1} = (0.029 - 0.002D1 + 0.019D2)S1\left(\frac{10{,}000}{S1}\right) \text{ if } S1 > 50{,}000$$

The rate equation representing biomass transfer from live plants to herbivores ($T_{1,2}$) is based on data in Table 3.12 and expert opinion concerning constraints on consumption of live plants by herbivores due to competition when plants become relatively scarce (p. 127). A plot of data relating consumption to environmental temperature suggests a linear relationship. Simple linear regression of consumption (g consumed/g herbivore biomass-day) on temperature (°C) yields the equation

$$\text{consumption} = -0.322 + 0.032 \text{ temperature}$$

Regression is significant ($P < 0.0001$) with $r^2 = 0.998$. Thus when plants are relatively abundant $T_{1,2}$ (kg/ha-day) can be represented as

$$T_{1,2} = (-0.322 + 0.032D1)S2 \text{ if } A1 \geq 1500$$

However, when plants become relatively scarce

$$T_{1,2} = (-0.322 + 0.032D1)S2\left(\frac{A1}{1500}\right) \text{ if } A1 < 1500$$

Rate equations representing biomass transfers from herbivores to the respiration sink ($T_{2,\mathrm{SI1}}$) and from herbivores to herbivore excreta ($T_{2,4}$) are based on data in Table 3.12. Plots of data relating respiration to environmental temperature and excretion to environmental temperature suggest that both relationships are linear. Simple linear regression of respiration (g respired/g herbivore biomass-day) on temperature (°C) yields the equation

$$\text{respiration} = -0.061 + 0.0065 \text{ temperature}$$

Regression is significant ($P < 0.001$) with $r^2 = 0.995$. Thus $T_{2,\mathrm{SI1}}$ (kg/ha-day) can be represented as

$$T_{2,\mathrm{SI1}} = (-0.061 + 0.0065D1)S2$$

Simple linear regression of excretion (g excreted/g herbivore biomass-day) on temperature (°C) yields the equation

$$\text{excretion} = -0.255 + 0.025 \text{ temperature}$$

Regression is significant ($P < 0.0001$) with $r^2 = 0.999$. Thus $T_{2,4}$ (kg/ha-day) can be represented as

$$T_{2,4} = (-0.255 + 0.025D1)S2$$

Rate equations representing biomass transfers from plant litter to the decomposition sink ($T_{3,\mathrm{SI2}}$) and from herbivore excreta to the decomposition sink ($T_{4,\mathrm{SI2}}$) are based on data in Table 3.13. Plots of data relating plant litter decomposition to soil water tension and herbivore excreta decomposition to soil water tension suggest that both relationships are exponential. However, plots relating the natural log of plant litter decomposition to soil water tension and the natural log of herbivore excreta decomposition to soil water tension are linear. Simple linear regression of ln (plant litter decomposition (g decomposed/g plant litter-day)) on soil water tension (–bars) yields the equation

$$\ln(\text{decomposition}) = -2.448 - 0.458 \text{ soil water tension}$$

Regression is significant ($P < 0.0001$) with $r^2 = 0.994$. Detransforming the equation yields

$$\text{decomposition} = 0.0865 e^{-0.458 \text{ soil water tension}}$$

Thus $T_{3,SI2}$ (kg/ha-day) can be represented as

$$T_{3,SI2} = (0.0865 e^{-0.458 D2}) S3$$

Simple linear regression of ln (herbivore excreta decomposition (g decomposed/g herbivore excreta-day)) on soil water tension (–bars) yields the equation.

$$\ln(\text{decomposition}) = -1.644 - 0.502 \text{ soil water tension}$$

Regression is significant ($P < 0.0001$) with $r^2 = 0.996$. Detransforming the equation yields

$$\text{decomposition} = 0.1932 e^{-0.502 \text{ soil water tension}}$$

Thus $T_{4,SI2}$ (kg/ha-day) can be represented as

$$T_{4,SI2} = (0.1932 e^{-0.502 D2}) S4$$

The rate equation representing biomass transfer from live plants to plant litter ($T_{1,3}$) is based on information from literature (p. 124). Since 0.01 g of plant litter is produced for every 1 g of live plant biomass in the system every day,

$$T_{1,3} = 0.01 S1$$

The rate equation representing biomass transfer from herbivores to harvested herbivores ($T_{2,5}$) is based on the particular harvest schemes ($D3$) that we wish to simulate. Since all harvest will be on the last day of the year (p. 59),

$$T_{2,5} = D3\, S2$$

where $D3 = 0$ if $t < 365$
and $0 \leq D3 \leq 1.0$ (depending upon our choice of harvest scheme) if $t = 365$

Choice of the functional forms and parameterization of driving variable equations representing seasonal dynamics of environmental temperature and soil water tension both are based on expert opinion (Figure 3.20). Both can be represented as half sine waves peaking during the middle of the year (for convenience let's say when $t = 180$), with temperature peaking at 40 (°C) and soil water tension peaking at 3(−bars). Use of the functional form suggested earlier (p. 124) yields equations for temperature ($D1$)

$$D1 = 40 \sin\left(\frac{\pi t}{365}\right)$$

and soil water tension ($D2$)

$$D2 = 3 \sin\left(\frac{\pi t}{365}\right)$$

if the sine is calculated based on radians.

Or

$$D1 = 40 \sin\left(\frac{t}{2}\right)$$

$$D2 = 3 \sin\left(\frac{t}{2}\right)$$

if the sine is calculated based on degrees.

The single auxiliary variable ($A1$) represents the ratio of live plant biomass to herbivore biomass. The equation is

$$A1 = \frac{S2}{S3}$$

State variable equations for live plants ($S1$), herbivores ($S2$), plant litter ($S3$), herbivore excreta ($S4$), and harvested herbivores ($S5$) are

$$S1_{t+1} = S1_t + T_{SO1,1} - T_{1,2} - T_{1,3}$$

$$S2_{t+1} = S2_t + T_{1,2} - T_{2,SI1} - T_{2,4} - T_{2,5}$$

$$S3_{t+1} = S3_t + T_{1,3} - T_{3,SI2}$$

$$S4_{t+1} = S4_t + T_{2,4} - T_{4,SI2}$$

$$S5_{t+1} = S5_t + T_{2,5}$$

Exercise 3.7.2

Plant–herbivore model equations are presented formally by computing sequence of equation types in Table A6.1. A BASIC program to solve the model on a TRS80 microcomputer is presented in Table A6.2.

TABLE A6.1

Formal presentation of plant-herbivore model equations organized by computing sequence of equation types. Variables are as identified in Figure A6.1

Initial Conditions of Model

$S1 = 50{,}000$
$S2 = 1.0$
$S3 = 6600$
$S4 = 0.1$
$S5 = 0.0$
$t_{begin} = 1$ (January 1)
$t_{end} = 365$ (December 31)

Driving Variable Equations

$D1 = 40 \sin(\pi t/365)$
$D2 = 3 \sin(\pi t/365)$
$D3 = 0$

Auxiliary Variable Equations

$A1 = S1/S2$

Rate Equations

$T_{SO1,1} = (0.029 - 0.002D1 + 0.019D2)S1$ if $S1 \leq 50{,}000$
$T_{SO1,1} = (0.029 - 0.002D1 + 0.019D2)S1(10{,}000/S1)$ if $S1 > 50{,}000$
$T_{1,2} = (-0.322 + 0.032D1)S2$ if $A1 \geq 1500$
$T_{1,2} = (-0.322 + 0.032D1)S2(A1/1500)$ if $A1 < 1500$
$T_{1,3} = 0.01S1$
$T_{2,SI1} = (-0.061 + 0.0065D1)S2$
$T_{2,4} = (-0.255 + 0.025D1)S2$
$T_{2,5} = D3 \times S2$
$T_{3,SI2} = (0.0865e^{-0.458D2})S3$
$T_{4,SI2} = (0.1932e^{-0.502D2})S4$

State Variable Equations

$S1_{t+1} = S1_t + T_{SO1,1} - T_{1,2} - T_{1,3}$
$S2_{t+1} = S2_t + T_{1,2} - T_{2,SI1} - T_{2,4} - T_{2,5}$
$S3_{t+1} = S3_t + T_{1,3} - T_{3,SI2}$
$S4_{t+1} = S4_t + T_{2,4} - T_{4,SI2}$
$S5_{t+1} = S5_t + T_{2,5}$

TABLE A6.2

Listing of a BASIC computer program for the plant-herbivore model

```
10 LPRINT "DETERMINISTIC MODEL OF THE PLANT-HERBIVORE SYSTEM"
40 DIM S(4), S1(366), S2(366), S3(366), S4(366)
50 DIM Y1(366),Y2(366),Y3(366),HV(366)
60 S(1)= 50000: S(2)=1: S(3)=6600: S(4)=0: REM INITIAL CONDITIONS
70 FOR I=0 TO 365: REM BEGIN DAILY DO LOOP
80 IF I=0 GOTO 310
90 D1=(40*SIN((3.1416*I)/365.0))
100 D2=(3*SIN((3.1416*I)/365.0))
110 A1=S(1)/S(2):REM AUXILIARY VARIABLE
120 C=1
130 IF A1<1500 THEN C=A1/1500
140 TI=(0.029 + (0.0192 * D2)-(0.00204*D1))*S(1): REM BEGIN RATE EQUATIONS
150 IF S(1)>50000 THEN T1=T1*(10000/S(1))
170 T2=0.01*S(1)
190 T3=((-0.32226+(0.0326*D1))*S(2))*C
230 T4=((-0.2548+(0.02525*D1))*S(2))
250 T5=((-0.06119+(0.00652*D1))*S(2))
270 T6=((0.0865*2.718[(-0.458*D2))*S(3))
290 T7=((0.1933*2.718[(-0.5021*D2))*S(4))
310 D3=0
320 T8=0
330 IF I=365 THEN T8=S(2)*D3
340 PRINT T8
350 S1(I)=S(1): REM STORES BEGINNING CONDITION FOR DAY I
360 S2(I)=S(2)
370 S3(I)=S(3)
380 S4(I)=S(4)
390 Y1(I)=D1
400 Y2(I)=A1
410 HV(I)=T8
420 Y3(I)=D2
430 IF I <>0 THEN GOSUB 550
440 NEXT I
450 LPRINT:LPRINT:LPRINT "DAY";TAB(5);"TEMP"; TAB(12);"P/H";TAB(20);
"LIVE PL";
460 LPRINT TAB(30);"HERB";TAB(39);"PLANT LITTER";TAB(55);"HERB EXCRETA";
TAB(70);"HARVEST"
470 FOR J= 0 TO 365 STEP 30
480 LPRINT J;TAB(5);INT(Y1(J));TAB(10);Y2(J);TAB(20);S1(J);TAB(30);S2(J);
490 LPRINT TAB(39);S3(J);TAB(55);S4(J);TAB(70);HV(J)
500 NEXT J
510 LPRINT " 365"; TAB(5);INT(Y1(365));TAB(10);Y2(365);TAB(20);S1(365);
TAB(30);
520 LPRINT S2(365);TAB(39);S3(365);TAB(55);S4(365);TAB(70);HV(365)
540 END
550 S(1)=S(1)+T1-T3-T2-T8: REM BEGIN STATE VARIABLE EQUATIONS
560 S(2)=S(2)+T3-T4-T5
570 S(3)=S(3)+T2-T6
580 S(4)=S(4)+T4-T7
590 RETURN
```

Exercise 3.7.3

Results of the initial 1-year baseline simulation are summarized in Figure A6.3.

A6.3. ANSWERS TO CHAPTER 4 EXERCISES

Exercise 4.8.1

Baseline simulation results indicate that plant–herbivore model behavior generally is reasonable, however herbivore excreta biomass is negative both early and late in the year (Figure A6.3). Examination of daily material transfers early and late in the year indicates that herbivore consumption ($T_{1,2}$), respiration ($T_{2,SI1}$), and excretion ($T_{2,4}$) all are negative when environmental temperature is below 10°C. One way to correct such unreasonable model behavior is to add statements in the computer program that set negative transfers equal to zero before state variable equations are updated each time step. Addition of such statements implies that our understanding of the processes occurring within the system is incomplete and that rate equations representing the processes do not perform well outside the range of data used to parameterize them. Temperature data used to parameterize equations for herbivore consumption, respiration, and excretion range from 10°C to 40°C. Simulated temperatures under baseline conditions range from 0°C to 40°C. We do not know how herbivore consumption, respiration, and excretion are related to temperature below 10°C, but we do know that these rates cannot be negative.

Herbivore excreta decomposition ($T_{4,SI2}$) also is negative early and late in the year. However, this occurs because herbivore excreta biomass ($S4$) becomes negative and not because the regression equation relating excreta decomposition (g decomposed/g excreta-day) to soil water tension predicts negative numbers within the range of simulated soil water tensions.

Exercise 4.8.2

A BASIC program to solve the model on a TRS80 microcomputer that includes additional program statements to prevent material transfers from becoming negative is presented in Table A6.3. Results of the new baseline simulation using the modified computer program are essentially the same as those of the initial baseline (Figure A6.3) except that herbivore excreta biomass no longer becomes negative and no material transfers become negative. Seasonal dynamics of live plant biomass and herbivore biomass look reasonable in comparison to data from the real system (Table 4.9) with regard to the timing of

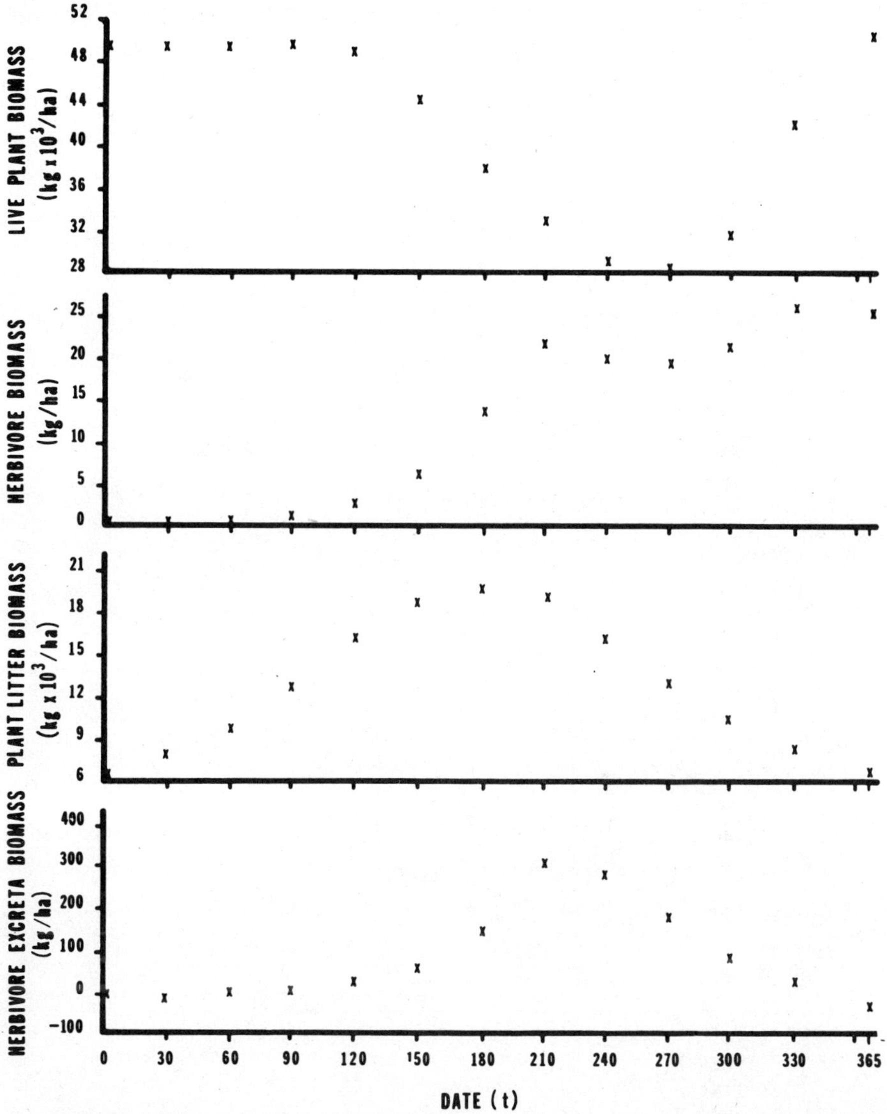

FIGURE A6.3. Results of the initial baseline simulation of the plant–herbivore model.

TABLE A6.3

Listing of a BASIC computer program for the plant-herbivore model, which includes additional program statements (not included in Table A6.2) to prevent material transfers from becoming negative. New statements are flagged (*)

```
 10 LPRINT "DETERMINISTIC MODEL OF THE PLANT-HERBIVORE SYSTEM"
 40 DIM S(4), S1(366), S2(366), S3(366), S4(366)
 50 DIM Y1(366),Y2(366),Y3(366),HV(366)
 60 S(1)= 50000: S(2)=1: S(3)=6600: S(4)=0: REM INITIAL CONDITIONS
 70 FOR I=0 To 365: REM BEGIN DAILY DO LOOP
 80 IF I=0 GOTO 310
 90 D1=(40*SIN((3.1416*I)/365.0))
 100 D2=(3*SIN((3.1416*I)/365.0))
 110 A1=S(1)/S(2):REM AUXILIARY VARIABLE
 120 C=1
 130 IF A1<1500 THEN C=A1/1500
 140 TI=(0.029 + (0.0192 * D2)-(0.00204*D1))*S(1): REM BEGIN
RATE EQUATIONS
 150 IF S(1)>50000 THEN T1=T1*(10000/S(1))
*160 IF T1<0 THEN T1=0
 170 T2=0.01*S(1)
*180 IF T2<0 THEN T2=0
 190 T3=((-0.32226+(0.0326*D1))*S(2))*C
*200 IF T3<0 THEN T3=0
*210 TTAL=T2+T3
*220 IF S(1)<TTAL THEN T3=S(1)-T2
 230 T4=((-0.2548+(0.02525*D1))*S(2))
*240 IF T4<0 THEN T4=0
 250 T5=((-0.06119+(0.00652*D1))*S(2))
*260 IF T5<0 THEN T5=0
 270 T6=((0.0865*2.718[(-0.458*D2))*S(3))
*280 IF T6<0 THEN T6=0
 290 T7=((0.1933*2.718[(-0.5021*D2))*S(4))
*300 IF T7<0 THEN T7=0
 310 D3=0
 320 T8=0
 330 IF I=365 THEN T8=S(2)*D3
 340 PRINT T8
 350 S1(I)=S(1): REM STORES BEGINNING CONDITION FOR DAY I
 360 S2(I)=S(2)
 370 S3(I)=S(3)
 380 S4(I)=S(4)
 390 Y1(I)=D1
 400 Y2(I)=A1
 410 HV(I)=T8
 420 Y3(I)=D2
 430 IF I <>0 THEN GOSUB 550
 440 NEXT I
 450 LPRINT:LPRINT:LPRINT "DAY";TAB(5);"TEMP"; TAB(12);"P/H";TAB(20);
"LIVE PL";
```

TABLE A6.3 (*Continued*)

```
 460 LPRINT TAB(30);"HERB";TAB(39);"PLANT LITTER";TAB(55);"HERB EXCRETA";
TAB(70);"HARVEST"
 470 FOR J= 0 TO 365 STEP 30
 480 LPRINT J;TAB(5);INT(Y1(J));TAB(10);Y2(J);TAB(20);S1(J);TAB(30);S2(J);
 490 LPRINT TAB(39);S3(J);TAB(55);S4(J);TAB(70);HV(J)
 500 NEXT J
 510 LPRINT " 365";TAB(5);INT(Y1(365));TAB(10);Y2(365);TAB(20);S1(365);
TAB(30);
 520 LPRINT S2(365);TAB(39);S3(365);TAB(55);S4(365);TAB(70);HV(365)
 540 END
 550 S(1)=S(1)+T1-T3-T2-T8: REM BEGIN STATE VARIABLE EQUATIONS
 560 S(2)=S(2)+T3-T4-T5
 570 S(3)=S(3)+T2-T6
 580 S(4)=S(4)+T4-T7
 590 RETURN
```

maximum and minimum values and the relative amplitude and periodicities of fluctuations.

Four additional simulations appropriate at this point include those in which environmental temperatures are (1) increased and (2) decreased and soil water tensions are (1) increased and (2) decreased by 10 percent relative to baseline values. These simulations examine model behavior over a reasonably wide range of input conditions (range of values for driving variables) to attempt to expose additional inadequacies in the model. Results indicate that changes in both temperature (Figure A6.4) and soil water tension (Figure A6.5) alter seasonal biomass dynamics of live plants, herbivores, plant litter, and herbivore excreta. However, the model exhibits no unreasonable behavior in view of what we know about the real system. (The fact that plants grow faster when environmental conditions are cold and dry may be counter-intuitive based on our personal experience, but it is consistent with data on the hypothetical plant–herbivore system (Table 3.11).

Exercise 4.8.3

Modifications to the BASIC computer program in Table A6.3 that will represent environmental temperature ($D1$) as a random variable drawn from a normal distribution with

$$\mu = 40 \sin\left(\frac{\pi t}{365}\right), \ \sigma^2 = 2$$

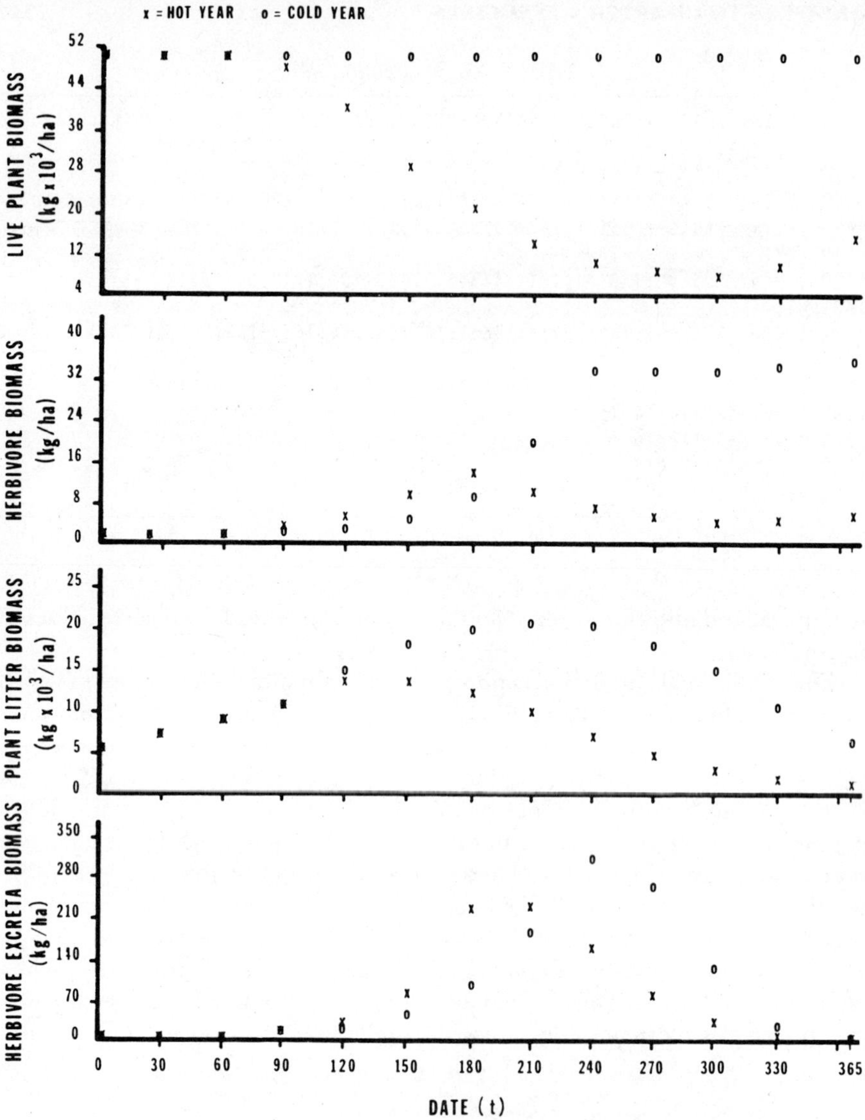

FIGURE A6.4. Results of plant–herbivore model simulations in which environmental temperatures were increased (hot year) and decreased (cold year) by 10 percent relative to baseline.

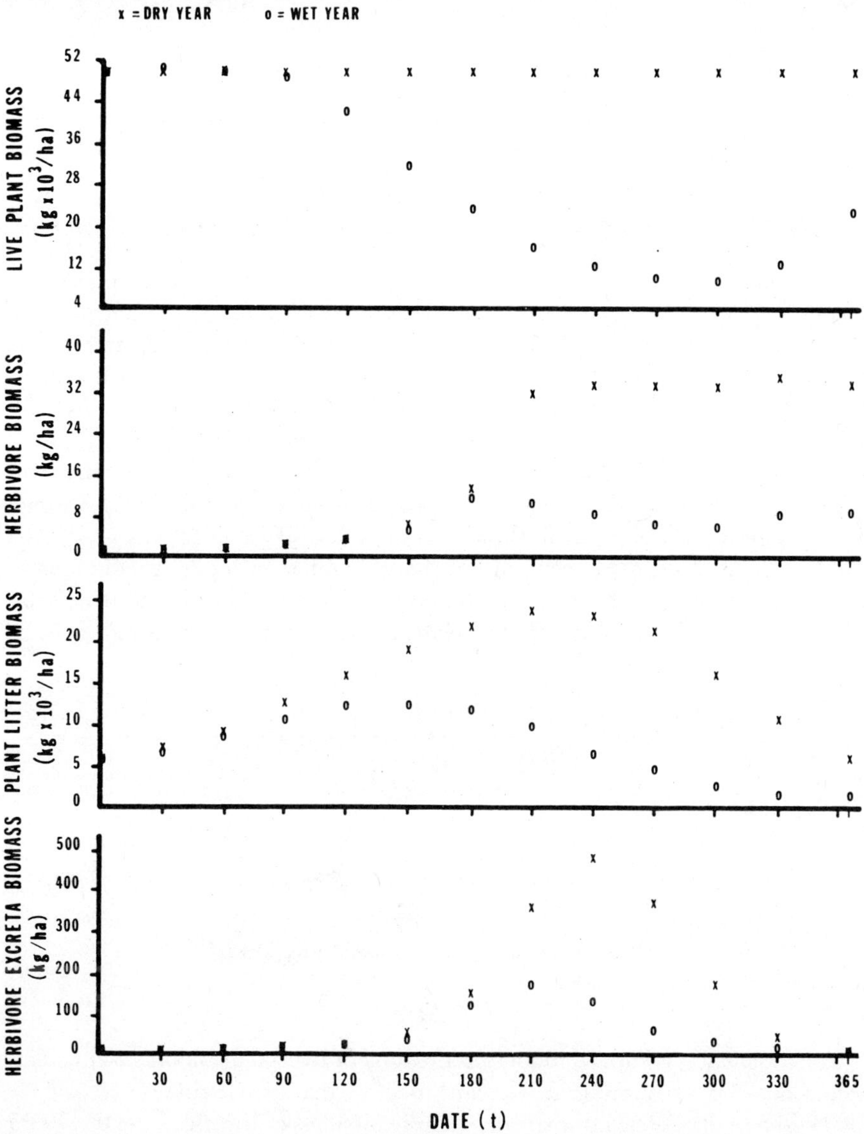

FIGURE A6.5. Results of plant–herbivore model simulations in which soil water tensions were increased (dry year) and decreased (wet year) by 10 percent relative to baseline.

include the addition of two statements. The statement

```
30 RANDOM
```

is added between statement numbers 10 and 40. The statement

```
95 D1 = D1 + 2*SQR( - 2.0*LOG(RND(0)))*COS(6.282*RND(0))
```

is added between statement numbers 90 and 100. To statistically compare stochastic model predictions of herbivore biomass with real-system data (Table 4.9) we first must establish the conditions of the comparison and determine the number of stochastic baseline simulations to run. Let's assume that we want to show that a true difference of $\delta = 2.5$ kg/ha in herbivore biomass on day 365 between model predictions and real-system data is statistically significant at a significance level $\alpha = 0.05$ with a probability $P = 0.80$ that the difference will be detected if it exists. Real-system sample size is $n_2 = 5$ and our best estimate of real-system variability in herbivore biomass on day 365 (σ_2) is $s_2 = 1.42$ (Table 4.9). A reasonable estimate of the variability in model predictions of herbivore biomass on day 365 (σ_1) based on 50 preliminary baseline simulations is $s_1 = 1.24$. The number of stochastic simulations (n_1) required is calculated as

$$n_1 \geq \frac{\sigma_1^2}{\delta^2/(t_{\alpha,\gamma} + t_{2(1-P),\gamma})^2 - (\sigma_2^2/n_2)}$$

if we assume that $n_1 = 7$, then

$$\gamma = n_1 + n_2 - 2 = 7 + 5 - 2 = 10$$

$$7 \geq \frac{(1.24)^2}{(2.5)^2/(t_{0.05,10} + t_{2(1-0.80),10})^2 - ((1.42)^2/5)}$$

$$7 \geq 6.30$$

To statistically compare model predictions of live plant biomass with real-system data let's assume that we want to show that a true difference of $\delta = 5000$ kg/ha on day 365 is statistically significant at a significance level $\alpha = 0.05$ with probability $P = 0.80$ that the difference will be detected if it exists. A reasonable estimate of the variability in model predictions of live plant biomass on day 365 based on 50 preliminary baseline simulations is 341. Calculating sample size as above we find that fewer than three samples are required. Since we must run seven replicate simulations to test for differences in herbivore biomass we also will use results from those seven simulations to test for differences in plant biomass (even though the experiment is slightly overdesigned).

Model predictions of biomass of live plants and herbivores on day 365 based on seven replicate stochastic baseline simulations (Table A6.4) can be compared to real-system data (Table 4.9) using two two-sample t-tests. Results of the t-tests indicate no statistically significant differences between model predictions and real-system data for live plant biomass ($P > 0.49$) or herbivore biomass ($P > 0.52$).

Exercise 4.8.4

Results of sensitivity analysis of the deterministic version of the plant–herbivore model are presented in Table A6.5. Each material transfer rate was varied by + 10 percent and − 10 percent relative to baseline and effects on biomass of live plants, herbivores, plant litter, and herbivore excreta on last day of the year ($t = 365$) were noted. Biomass of live plants and plant litter are relatively insensitive to changes in any of the material transfer rates. Biomass of both herbivores and herbivore excreta are relatively sensitive to changes in material transfer rates representing herbivore consumption ($T_{1,2}$), respiration ($T_{2,\mathrm{SI1}}$), and excretion ($T_{2,4}$). Biomass of herbivore excreta is also relatively sensitive to changes in material transfer rates representing excreta decomposition ($T_{4,\mathrm{SI2}}$). Seasonal dynamics of state variables in response to changes in material transfer rates are qualitatively similar to baseline in most cases. However, herbivore biomass remains low and exhibits essentially no seasonal dynamics when consumption ($T_{1,2}$) is decreased by 10 percent and when excretion ($T_{2,4}$) is increased by 10 percent. When respiration ($T_{2,\mathrm{SI1}}$) is increased by 10 percent herbivore biomass increases much more slowly than baseline.

TABLE A6.4

Plant-herbivore model predictions of biomass of live plants and herbivores on day 365 based on seven replicate stochastic baseline simulations

	Biomass (kg/ha)	
Replication	Live Plants	Herbivores
1	50,521	25.78
2	50,597	28.50
3	50,087	25.18
4	50,893	23.73
5	50,985	27.16
6	50,567	25.45
7	50,547	24.90

TABLE A6.5

Results of sensitivity analysis of the deterministic version of the plant-herbivore model, indicating changes relative to baseline in biomass of live plants, herbivores, plant litter, and herbivore excreta on the last day of the year (t = 365) resulting from changes of + 10 percent and − 10 percent in the indicated material transfer rate

Material Transfer Rate and (Percent Change Relative to Baseline)		Live Plants (kg/ha)	Live Plants Percent Change	Herbivores (kg/ha)	Herbivores Percent Change	Plant Litter (kg/ha)	Plant Litter Percent Change	Herbivore Excreta (kg/ha)	Herbivore Excreta Percent Change
$T_{SO1,1}$	(+ 10)	50,194	− <1	31	+ 19	6709	+ 3	0.0848	+ 18
	(− 10)	50,047	− <1	21	− 19	5833	− 10	0.0603	− 16
$T_{1,2}$	(+ 10)	49,955	− 1	28	+ 8	6441	− 1	0.0763	+ 6
	(− 10)	50,026	− <1	<1	− 99.9	6604	+ 1	<0.0001	− 99.9
$T_{1,3}$	(+ 10)	50,690	+ 1	21	− 19	6474	− 1	0.0582	− 19
	(− 10)	50,043	− <1	32	+ 23	6049	− 7	0.0877	+ 22
$T_{2,SI1}$	(+ 10)	50,438	+ <1	5	− 81	6590	+ 1	0.0135	− 81
	(− 10)	50,158	+ <1	27	+ 4	6477	− 1	0.0725	+ 1
$T_{2,4}$	(+ 10)	50,008	− <1	<1	− 99.9	6604	+ 1	<0.0001	− 99.9
	(− 10)	50,326	+ <1	28	+ 8	6449	− 1	0.0679	− 6
$T_{3,SI2}$	(+ 10)	50,219	0	26	0	5873	− 10	0.0718	0
	(− 10)	50,219	0	26	0	7319	+ 12	0.0718	0
$T_{4,SI2}$	(+ 10)	50,219	0	26	0	6516	0	0.0321	− 55
	(− 10)	50,219	0	26	0	6516	0	0.1628	+ 127

A6.4. ANSWERS TO CHAPTER 5 EXERCISES

Exercise 5.7.1

To statistically compare model predictions of live plant biomass with and without herbivores in the system let's assume that we want to show that a true difference of δ = 5000 kg/ha on day 365 of year 5 is statistically significant at a significance level α = 0.05 with probability P = 0.80 that the difference will be detected if it exists. Under these circumstances we already have calculated

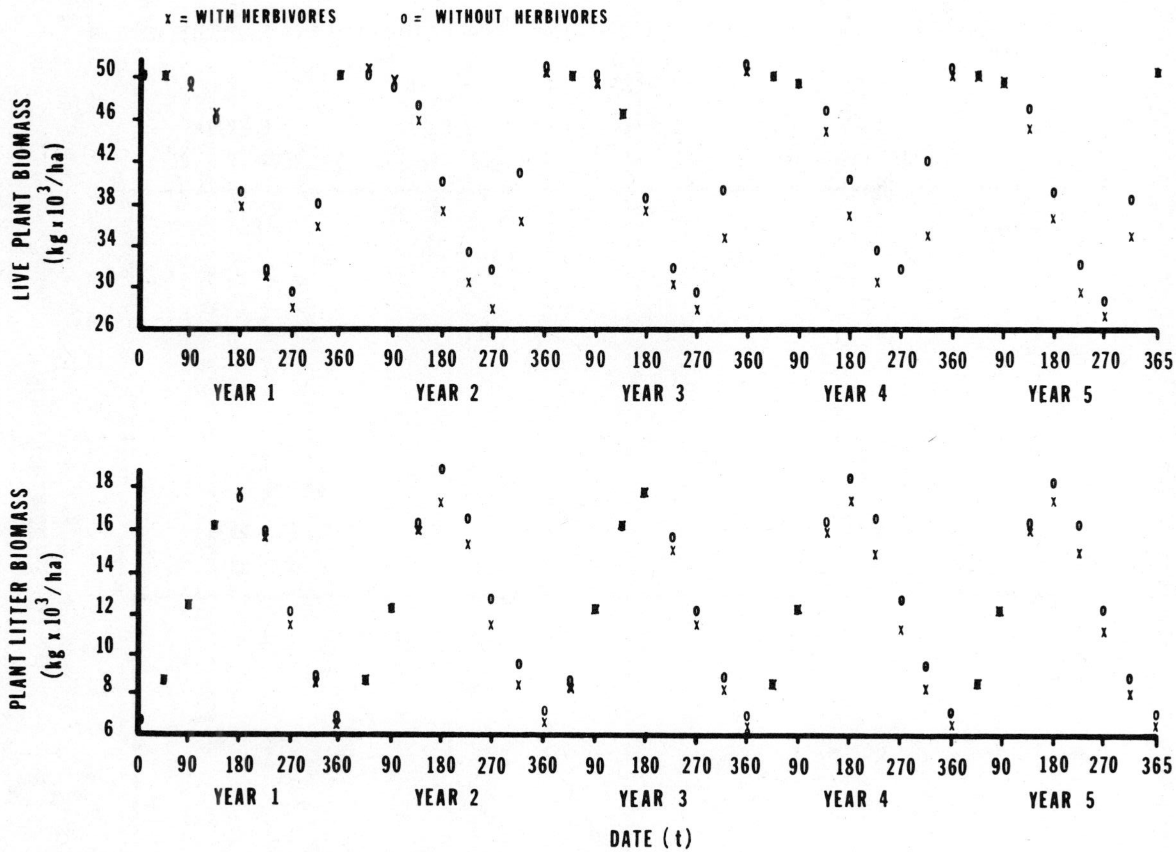

FIGURE A6.6. Results of three replicate stochastic simulations of the plant–herbivore model with and without herbivores.

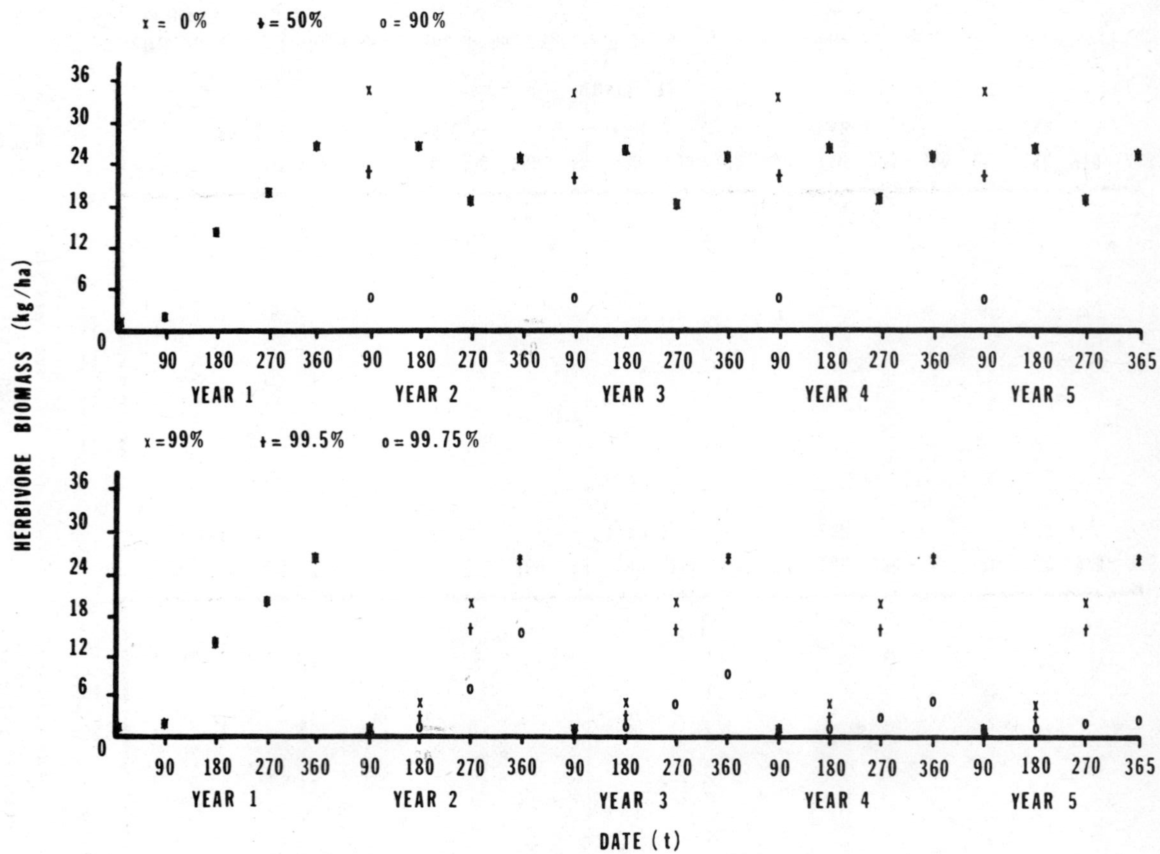

FIGURE A6.7. Results of deterministic plant–herbivore model simulations under several harvest schemes.

that fewer than three samples are required (p.320). To statistically compare model predictions of plant litter biomass with and without herbivores let's assume that we want to show that a true difference of δ = 650 kg/ha on day 365 of year 5 is statistically significant at α = 0.05 and P = 0.80. A reasonable estimate of the variability in plant litter biomass on day 365 based on 50 preliminary baseline simulations is 149. Calculating sample size in the usual way we find that three samples are required. Thus we will run three replicate simulations to test for differences in both live plant biomass and plant litter biomass with and without herbivores.

Model predictions of live plant biomass with and without herbivores and plant litter biomass with and without herbivores on day 365 of year 5 are not significantly different ($P > 0.08$ and $P > 0.26$, respectively). Biomass dynamics of both live plants and plant litter over a 5-year period also are essentially the same with and without herbivores (Figure A6.6). We conclude that introducing herbivores will have little effect on the island plant community.

Exercise 5.7.2

To determine the proportion of herbivore biomass that could be harvested on a sustained yield basis assuming that all harvest occurs on the last day of the year we run a series of deterministic 5-year simulations in which we vary the proportion harvested and observe herbivore biomass dynamics.

Results of deterministic simulations in which 0, 50, 90, 99, 99.5, and 99.75 percent of the herbivore population is harvested on the last day of the year indicate that the population can recover within one year even when 99.5 percent is harvested (Figure A6.7). In fact the population recovers within 180 days at harvest levels up to 90 percent. However the population fails to recover within a year at a 99.75 percent harvest level. Thus the proportion that can be harvested on a sustained yield basis is about 0.995.

REFERENCES

Ashley, R. 1974. *ANS COBOL*. Wiley, New York.

Barr, A. J., J. H. Goodnight, J. P. Sall, and J. T. Helwig. 1976. *A User's Guide to SAS 76*. SAS Institute Inc., Raleigh, N.C.

Berryman, A. A. 1981. *Population Systems: A General Introduction*. Plenum, New York.

Bertalanffy, L. von. 1964. *General Systems Theory: Foundations, Development, Applications*. George Braziller, New York.

Beverton, R. J. H. and S. J. Holt. 1957. On the dynamics of exploited fish populations. Min. Agric. Fish. Food, H.M.S.O., London.

Bledsoe, L. J. and G. M. Van Dyne. 1971. A compartment model simulation of secondary succession. In B. C. Patten (ed.), *Systems Analysis and Simulation in Ecology (Vol. I)*. Academic, New York.

Bliss, L. C., O. W. Heal, and J. J. Moore. 1981. *Tundra Ecosystems: A Comparative Analysis*. Cambridge University Press, Cambridge, Mass.

Bobillier, P. A., B. C. Kahan, and A. R. Probst. 1976. *Simulation with GPSS and GPSS V*. Prentice-Hall, Englewood Cliffs, N.J.

Boillot, M. 1981. *Understanding FORTRAN*. West, St. Paul, Minn.

Breymeyer, A. L. and G. M. Van Dyne. 1980. *Grasslands, Systems Analysis and Man*. Cambridge University Press, Cambridge, Mass.

Brown, G. M., J. Hammack, and M. F. Tillman. 1976. Mallard population dynamics and management models. *J. Wildl. Manage.* **40**:542–555.

Caswell, H. 1976. The validation problem. In B. C. Patten (ed.), *Systems Analysis and Simulation in Ecology (Vol. IV)*. Academic, New York.

Cochran, W. G. and G. M. Cox. 1957. *Experimental Design*. Wiley, New York.

Daniel, C. and F. S. Wood. 1971. *Fitting Equations to Data*. Wiley, New York.

Didday, R. and R. Page. 1984. *Using BASIC*. West, St. Paul, Minn.

Dixon, W. J. 1975. *BMDP: Biomedical Computer Programs*. University of California, Berkeley.

Draper, N. R. and H. Smith. 1966. *Applied Regression Analysis*. Wiley, New York.

Forrester, J. W. 1961. *Industrial Dynamics*. M.I.T. Press, Cambridge, Mass.

Giordano, F. R. and M. D. Weir. 1985. *A First Course in Mathematical Modeling*. Brooks/Cole Publishing, Monterey, Calif.

Glass, N. R. 1971. Computer analysis of predation energetics in the largemouth bass. In B. C. Patten (ed.), *Systems Analysis and Simulation in Ecology (Vol. I)*. Academic, New York.

Gold, H. J. 1977. *Mathematical Modeling of Biological Systems: An Introductory Guidebook.* Wiley, New York.

Grant, W. E., S. O. Fraser, and K. G. Isakson. 1984. Effect of vertebrate pesticides on non-target wildlife populations: Evaluation through modelling. *Ecol. Modelling* 21:85–108.

Grant, W. E. and W. L. Griffin. 1979. A bioeconomic model of the Gulf of Mexico shrimp fishery. *Trans. Am. Fish. Soc.* **108:**1–13.

Graybeal, W. J. and U. W. Pooch. 1980. *Simulation Principles and Methods.* Winthrop, Cambridge.

Green, R. H. 1979. *Sampling Design and Statistical Methods for Environmental Biologists.* Wiley, New York.

Gustafson, J. D. and G. S. Innis. 1973. SIMCOMP version 2.1 user's manual and maintenance documentation. Technical Report No. 217, Natural Resource Ecology Laboratory, Colorado State University, Fort Collins.

Halfon, E. 1979. *Theoretical Systems Ecology: Advances and Case Studies.* Academic, New York.

Hall, C. A. S. and J. W. Day, Jr. 1977. *Ecosystem Modeling in Theory and Practice: An Introduction with Case Histories.* Wiley, New York.

Hastings, N. A. J. and J. B. Peacock. 1975. *Statistical Distributions.* Butterworth, London.

Hilborn, R. 1976. Optimal exploitation of multiple stocks by a common fishery: A new methodology. *J. Fish. Res. Board Can.* **33:**1–5.

Holling, C. S. 1978. *Adaptive Environmental Assessment and Management.* Wiley, New York.

Horst, T. J. 1977. Use of the Leslie matrix for assessing environmental impact with an example for a fish population. *Trans. Am. Fish. Soc.* **106:**253–257.

Innis, G. S. 1978. *Grassland Simulation Model.* Springer-Verlag, New York.

Innis, G. S. 1979. A spiral approach to ecosystem simulation, I. In G. S. Innis and R. V. O'Neill (eds.), *Systems Analysis of Ecosystems.* International Cooperative Publishing House, Burtonsville, Md.

Innis, G. S. and R. V. O'Neill. 1979. *Systems Analysis of Ecosystems.* International Cooperative Publishing House, Burtonsville, Md.

Jeffers, J. N. R. 1972. *Mathematical Models in Ecology.* Blackwell Scientific Publishers, Oxford, England.

Jeffers, J. N. R. 1978. *An Introduction to Systems Analysis: With Ecological Applications.* University Park Press, Baltimore, Md.

Jensen, K. and N. Wirth. 1975. *PASCAL: User Manual and Report (2nd Ed.)*,Springer-Verlag, New York.

Johnson, N. I. and S. Kotz. 1969. *Discrete Distributions.* Houghton Mifflin, Boston.

Jones, D. D. and C. J. Walters. 1976. Catastrophy theory and fisheries regulation. *J. Fish. Res. Board Can.* **33:**2829–2833.

Kitching, R. L. 1983. *Systems Ecology: An Introduction to Ecological Modelling.* University of Queensland Press, St. Lucia, Queensland.

Kowal, N. E. 1971. A rationale for modeling dynamic ecological systems. In B. C. Patten (ed.), *Systems Analysis and Simulation in Ecology (Vol. I).* Academic, New York.

Larkin, P. A. 1971. Simulation studies of the Adams River sockeye salmon (*Oncorhynchus narka*). *J. Fish. Res. Board Can.* **28:**1493–1502.

Lassiter, R. R. and D. W. Hayne. 1971. A finite difference model for simulation of dynamic processes in ecosystems. In B. C. Patten (ed.), *Systems Analysis and Simulation in Ecology (Vol. I).* Academic, New York.

Law, A. M. and W. D. Kelton. 1982. *Simulation and Analysis*. McGraw-Hill, New York.

Leslie, P. H. 1945. On the use of matrices in certain population mathematics. *Biometrika* **33**:183–212.

Leslie, P. H. 1948. Some further notes on the use of matrices in population mathematics. *Biometrika* **35**:213–245.

Mankin, J. B., R. V. O'Neill, H. H. Shugart, and B. W. Rust. 1977. The importance of validation in ecosystem analysis. In G. S. Innis (ed.), *New Directions in the Analysis of Ecological Systems. Part 1*. Simulation Council Proceedings Series 5(1), La Jolla, Calif.

Markowitz, N. M., H. N. Karr, and B. Havsner. 1963. *SIMSCRIPT: A Simulation Programming Language*. Prentice-Hall, Englewood Cliffs, N.J.

Matis, J. H., B. C. Patten, and G. C. White. 1979. *Compartmental Analysis of Ecosystem Models. Statistical Ecology, Vol IX*. International Cooperative Publishing House, Burtonsville, Md.

Matis, J. H. and T. E. Wehrly. 1979. An approach to a compartmental model with multiple sources of stochasticity for modeling ecological systems. *In* J. H. Matis, B. C. Patten, and G. C. White (eds.),*Compartmental Analysis of Ecosystem Models*, pp. 195–222, International Cooperative Publishing House, Burtonsville, Md.

McCracken, D. D. 1962. *A Guide to ALGOL Programming*. Wiley, New York.

Merchant, M. J. 1981. *FORTRAN 77 Language and Style: A Structured Guide to FORTRAN 77*. Wadsworth, Belmont, Calif.

Odum, H. T. 1972. An energy circuit language for ecological and social systems: Its physical basis. In B. C. Patten (ed.), *Systems Analysis and Simulation in Ecology (Vol. II)*. Academic, New York.

Odum, H. T. 1983. *Systems Ecology: An Introduction*. Wiley, New York.

Olinick, M. 1978. *An Introduction to Mathematical Models in Social and Life Sciences*. Addison-Wesley, Reading, Mass.

Olson, J. S. 1960. *Health Physics Annual Reports*. Oak Ridge National Laboratory, Oak Ridge, Tenn.

O'Neill, R. V. 1979. A review of linear compartmental analysis in ecosystem science. In J. H. Matis, B. C. Patten, and G. C. White (eds.), *Compartmental Analysis of Ecosystem Models. Statistical Ecology, Vol. IX*, pp. 3–28, International Cooperative Publishing House, Burtonsville, Md.

Ott, L. 1984. *An Introduction to Statistical Methods and Data Analysis*. Duxbury, Boston.

Patten, B. C. 1969. Ecological systems analysis and fisheries science. *Trans. Am. Fish. Soc.* **98**:570–581.

Patten, B. C. 1971a. *Systems Analysis and Simulation in Ecology (Vol. I)*. Academic, New York.

Patten, B. C. 1971b. A primer for ecological modeling and simulation with analog and digital computers. In B. C. Patten (ed.), *Systems Analysis and Simulation in Ecology (Vol. I)*. Academic, New York.

Patten, B. C. 1972. *Systems Analysis and Simulation in Ecology (Vol. II)*. Academic, New York.

Patten, B. C. 1975. *Systems Analysis and Simulation in Ecology (Vol. III)*. Academic, New York.

Patten, B. C. 1976. *Systems Analysis and Simulation in Ecology (Vol. IV)*. Academic, New York.

Paulik, G. J. 1969. Computer simulation models for fisheries research, management and teaching. *Trans. Am. Fish. Soc.* **98**:551–559.

Paulik, G. J. 1972. Digital simulation modeling in resource management and the training of applied ecologists. In B. C. Patten (ed.), *Systems Analysis and Simulation in Ecology (Vol. II)*. Academic, New York.

Paulik, G. J. and J. W. Greenough, Jr. 1966. Management analysis for a salmon resource system. In K. E. F. Watt (ed.), *Systems Analysis in Ecology*. Academic, New York.

Pielou, E. C. 1981. The usefulness of ecological models: A stock-taking. *Quart. Rev. Biol.* **56**:17–31.

Pritsker, A. A. B. 1974. *The GASP IV Simulation Language*. Wiley, New York.

Reichle, D. E. 1981. *Dynamic Properties of Forest Ecosystems*. Cambridge University Press, Cambridge, Mass.

Rohlf, F. J. and R. R. Sokal. 1969. *Statistical Tables*. Freeman & Co., San Francisco.

Roseberry, J. L. 1979. Bobwhite population responses to exploitation: real and simulated. *J. Wildl. Manage.* **43**:285–305.

Rykiel, E. J., Jr. and N. T. Kuenzel. 1971. Analog computer models of "The wolves of Isle Royale." In B. C. Patten (ed.), *Systems Analysis and Simulation in Ecology (Vol. I)*. Academic, New York.

Saila, S. B. 1972. Systems analysis applied to some fisheries problems. In B. C. Patten (ed.), *Systems Analysis and Simulation in Ecology (Vol. II)*. Academic, New York.

Salas, S. L. and E. Hille. 1982. *Calculus: One and Several Variables*. Wiley, New York.

Searle, S. R. 1966. *Matrix Algebra for the Biological Sciences*. Wiley, New York.

Shaffer, W. A. 1980. DYNAMO. *Simulation* **34**:134–136.

Shugart, H. H. and R. V. O'Neill. 1979. *Systems Ecology*. Dowden, Hutchinson, & Ross, Stroudsburg, Penn.

Sigal, C. E. and A. A. B. Pritsker. 1974. SMOOTH: A combined continuous-discrete network simulation language. *Simulation* **22**:65–73.

Silliman, R. P. 1969. Analog computer simulation and catch forecasting in commercially fished populations. *Trans. Am. Fish. Soc.* **98**:560–569.

Sinclair, A. R. E. 1973. Regulation and population models for a tropical ruminant. *E. Afr. Wildl. J.* **11**:307–316.

Smith, F. E. 1973. Analysis of ecosystems. In D. E. Reichle (ed.), *Analysis of Temperate Forest Ecosystems*. Springer-Verlag, New York.

Smith, R. L. 1980. *Ecology and Field Biology*. Harper & Row, New York.

Snedecor, G. W. and W. G. Cochran. 1967. *Statistical Methods*. Iowa State University Press, Ames.

Sokal, R. R. and F. J. Rohlf. 1969. *Biometry*. Freeman & Co., San Francisco.

Spain, J. D. 1982. *BASIC Microcomputer Models in Biology*. Addison-Wesley, Reading, Mass.

Steinhorst, R. K. 1979. Parameter identifiability, validation, and sensitivity analysis of large system models. In G. S. Innis and R. V. O'Neill (eds.), *Systems Analysis of Ecosystems*. International Cooperative Publishing House, Burtonsville, Md.

Tillman, M. F. and D. Stadelman. 1976. Development and example-application of a simulation model to the northern anchovy fishery. *Fish. Bull.* **74**:118–130.

Tipton, A. R. 1980. Mathematical modeling in wildlife management. In S. D. Schemnitz (ed.), *Wildlife Management Techniques Manual (4th Ed.)*. Wildlife Society, Washington, D.C.

Usher, M. B. 1972. Developments in the Leslie matrix model. In J. N. R. Jeffers (ed.), *Mathematical Models in Ecology*. Blackwell Scientific Publishers, Oxford, England.

Walters, C. J. 1969. A generalized computer simulation model for fish population studies. *Trans. Am. Fish. Soc.* **98**:505–512.

Walters, C. J. 1980. Systems principles in fisheries management. In R. T. Lackey and L. A. Nielsen (eds.), *Fisheries Management*. Halsted, New York.

Walters, C. J. and J. Brandy. 1972. Periodic harvest as a method for increasing big game yields. *J. Wildl. Manage.* **36**:128–134.

Walters, C. J. and R. Hilborn. 1976. Adaptive control of fishing systems. *J. Fish. Res. Board Can.* **33**:145–159.

Walters, C. J., R. Hilborn, and R. Peterman. 1975. Computer simulation of barren-ground caribou dynamics. *Ecol. Modelling* **1**:303–315.

Watt, K. E. F. 1966. *Systems Analysis in Ecology*. Academic, New York.

Watt, K. E. F. 1968. *Ecology and Resource Management*. McGraw-Hill, New York.

Weinberg, G. M. 1975. *An Introduction to General Systems Thinking*. Wiley, New York.

Zeigler, B. P. 1976. *Theory of Modelling and Simulation*. Wiley, New York.

Ziesenis, J. S. and L. Adams. 1980. Computer applications in wildlife management. In S. D. Schemnitz (ed.), *Wildlife Management Techniques Manual (4th Ed.)*. Wildlife Society, Washington D.C.

INDEX

Adams, L., 266
ALGOL, 269
American Fisheries Society, Transactions of the, 36
Analog computer(s), 267
  diagrams, 51
Anchovy, 3
Animal and Plant Health Inspection Service, 23, 27
Animal Damage Control, 25
Ashley, R., 269
Auxiliary variables, 17
  description of, 44–45
  equations for, 62–63
  symbol for, 41

Barr, A. J., 269
BASIC, 80, 82, 84, 269
Bass, largemouth, 35
Berryman, A. A., 34
Bertalanffy, L. von, 6
Bertalanffy growth equation, 265
Beverton, R. J. H., 3
Bledsoe, L. J., 35, 61, 223
Bliss, L. C., 2
BMDP, 269
Bobillier, P. A., 270
Bobwhite quail, 3
Boillot, M., 269
Box-and-arrow diagram, 48
Brandy, J., 3
Breymeyer, A. L., 2
Brown, G. M., 3
Buffalo, African, 3

Canadian Journal of Fisheries and Aquatic Sciences, 36
Caribou, barren-ground, 3
Caswell, H., 19, 35, 129
Catastrophy theory, 3
COBOL, 269
Cochran, W. G., 78, 140, 170, 172, 188
Compound 1080, 23, 25, 26, 27
Computers, 266
  analog, 267
  digital, 267–268
  hardware, 268
  mainframe, 267
  micro, 267
  mini, 267
  software, 268
  see also Programming, computer
Control valves:
  description of, 45–46
  symbol for, 41
Cox, G. M., 170
Coyote control, 23, 26
Cuttlefish, 34

Daniel, C., 71
Day, J. W., 35
Deterministic models, *see* Models, deterministic
Didday, R., 269
Difference equations, 17, 62, 240–241, 244
  numerical approximation of, 245–247
  representing population growth, 249–251
  simulation models based on, 13
Differential equations, 17, 61, 240–241
  degree of, 243–244
  linear, 244
  order of, 243–244

Differential equations (*Continued*)
representing population growth, 248–249
terminology of, 242–244
type of, 243
Digital computers, 267–268
Distributions:
F-, 257, 258
frequency, 73, 74–75
normal, 76–79
Poisson, 77
statistical, 73, 76–79
uniform, 77
Dixon, W. J., 270
Draper, N. R., 252, 258, 259, 263
Driving variables, 17, 46
description of, 43–44
equations for, 62–63
symbol for, 41
Duncan's Multiple Range Test, 172, 174, 175

Eigenvalues, 234
Eigenvectors, 234
Energy circuit language, 51
Equations:
auxiliary variable, 62–63
computer coding of, 80–81
computing sequence for model, 62–63, 80
driving variable, 62–63
fitted to data, 71
formal presentation of, 80–83
functional forms of, 18, 65–69
parameterization of, 18, 70–79
rate, 62–63
state variable, 62–63
Euler method of numerical approximation, 246–247

Fisheries and Aquatic Sciences, Canadian Journal of, 36
Fisheries Research Board of Canada, Journal of the, 36
Fishery Conservation and Management Act, 27
Fishery Conservation Zone, 28
Forrester, J. W., 7, 35, 41, 48
FORTRAN, 80, 82, 84, 269

Giordano, F. R., 246
Glass, N. R., 35
Gold, H. J., 7, 14, 34
Grant, W. E., 3, 23, 24, 27, 29
Graybeal, W. J., 35
Green, R. H., 270
Greenough, J. W., 35
Griffin, W. L., 3, 27, 29
Gulf of Mexico Fisheries Management Council, 27, 28, 34
Gustafson, J. D., 270

Halfon, E., 35
Hall, C. A. S., 35
Hastings, N. A. J., 77, 78
Hayne, D. W., 35
Hilborn, R., 3
Hille, E., 241
Holling, C. S., 2, 19, 20, 35, 129
Holt, S. J., 3
Horst, T. J., 3

Information network, 48
Information transfers:
description of, 45–46
symbol for, 41
Innis, G. S., 2, 14, 35, 41, 48, 270
International Biological Program, 2, 35
Isle Royale, 35

Jeffers, J. N. R., 7, 14, 34, 35
Jensen, K., 269
Johnson, N. I., 78
Jones, D. D., 3
Journal of the Fisheries Research Board of Canada, 36
Journal of Wildlife Management, 36

Kelton, W. D., 35, 74, 77
Kitching, R. L., 2, 14, 17, 34, 266
Kolmogorov-Smirnov test, 140
Kotz, S., 78
Kowal, N. E., 35
Kuenzel, N. T., 35

Larkin, P. A., 3
Lassiter, R. R., 35
Latent roots, 234
Latent vectors, 234
Law, A. M., 35, 74, 77

Leslie matrix, 3, 35, 220, 223
  analytical treatment of, 234–235
  simulation of population dynamics using, 231–234
  with time-varying rates, 235
Leslie, P. H., 3, 223
Lindeman's trophic dynamic model, 242
Lotka-Volterra equations, 241–242

Mallards, 3
Mankin, J. B., 19, 129
Markowitz, N. M., 270
Material transfers:
  description of, 45–46
  matrix of, 49–51
  symbol for, 41
Mathematical modeling, *see* Modeling
Mathematical models:
  compartmental structure of, 17
  definition of, 1
  description of, 8
  in ecology, 35
  in wildlife and fisheries, 3, 35
  see also Models; Simulation models
Matis, J. H., 2, 214
Matrix algebra, 17, 61, 223
  addition, 225
  adjugate matrix, 230, 231
  basic calculations in, 225–231
  cofactors, 228–229
  determinants, 227–228
  identity matrix, 229
  inversion, 226, 229
  multiplication, 225–226
  subtraction, 225
  terminology of, 224
  transposition, 226
  use in wildlife and fisheries, 231–240. *See* also Leslie matrix
McCracken, D. D., 269
Merchant, M. J., 269
Modeling:
  communication of results of, 21
  compartment, 2
  ecological, 2
  in natural resource management, 35
  population dynamics, 3, 35
  see also Simulation modeling
Models:
  abstract, 8
  analytical, 10
  classification of, 7–11
  compartment, 35
  compartmental structure of, 61, 62
  components of, 17
  computer coding of, 19, 80–82
  conceptual diagrams for, 48–49
  conceptual formulation of, 16–17, 23–25, 28–30, 37–58
  correlative, 8, 19, 132, 137
  definition of, 7
  deterministic, 9, 18, 137, 138–139, 150, 170, 172, 220
  dynamic, 8
  ecosystem, 2
  in education, 4
  empirical, 8, 19, 132, 137
  explanatory, 8, 19, 132, 137
  formal representation of, 48–51, 80–83
  Leslie matrix, 35. *See also* Leslie matrix
  mechanistic, 8, 19, 132, 137
  microcomputer, 35
  physical, 8
  predictions of compared to real-system data, 140–149
  quantitative specification of, 17–19, 25, 30, 60–123
  quantitative structure for, 61–65
  random variables represented in, 73–79
  in research, 4
  sensitivity analysis of, 20, 149–151
  simulation, 11. *See also* Simulation modeling; Modeling
  static, 8
  stochastic, 9, 18, 19, 21, 73–79, 85–88, 137, 138–139, 150, 171, 172, 220
  tuning of, 134–137
  use of, 20–21, 26–27, 31–34, 168–199
  validation of, 19–20, 25, 30–31, 35, 129–165. *See also* Validation, model
  verbal, 7, 8
  see also Mathematical models; Simulation models

National Marine Fisheries Service, 31, 34

Octopus, 34
Odum, H. T., 35, 51
Olinick, M., 246
Olson, J. S., 2

O'Neill, R. V., 2, 35
Optimization, 3
Ott, L., 172
Owl, horned, 26

Page, R., 269
PASCAL, 269
Patten, B. C., 2, 3, 12, 35, 51, 266, 267
Paulik, G. J., 3, 35
Peacock, J. B., 77, 78
Pielou, E. C., 3
Pond model example:
  computer program for, 271–303
  conceptual model formulation for, 52–58
  quantitative specification of, 88–123
  use of, 178–199
  validation of, 151–165
Pooch, U. W., 35
Population dynamics:
  Leslie matrix, representation of, 231–235
  simulation of, 223
Pritsker, A. A. B., 270
Probability transition matrix, 223
  analytical treatment of, 238–240
  simulation of ecological succession using, 235–237
  with time-varying probabilities, 240
Programming, computer:
  analog, 267
  coding model equations, 80–81
  constructing flow-diagrams, 81–82, 83
  debugging code, 82
  digital, 268–270
  languages for, 268–269
  special-purpose packages for, 269–270
  verifying code, 82
  writing code, 82
  see also Computers

Rate equations, 62–63
Rectangular method of numerical approximation, 246–247
Regression:
  assumptions of, 259
  calculations associated with, 254–258, 259–261
  correlation coefficient of, 253, 255
  error mean square about, 253, 255
  F-value associated with, 257, 258
  independent variables selected for, 262–263
  linear transformations associated with, 263–265
  matrix format for, 259–261
  multiple linear, 259–265
  simple linear, 253
  slope of, 253, 255
  statistical distributions generated from, 78–79
  terminology of, 253–254
  uses in wildlife and fisheries, 252–253, 265
Reichle, D. E., 2
Rohlf, F. J., 74, 86, 140, 219, 258
Roseberry, J. L., 3
Rykiel, E. J., 35

Saila, S. B., 35
Salas, S. L., 241
Salmon, 35
  sockeye, 3
SAS, 269
Scalar, 224
Scientific method, 7
Searle, S. R., 223, 234
Sensitivity analysis, 20, 149
  interpretation of, 150–151
Shaffer, W. A., 270
Shrimp fishery, 3, 27–34
Shugart, H. H., 2, 35
Sigal, C. E., 270
Silliman, R. P., 3
Simulation modeling:
  analyzing and interpreting results of, 172–178
  based on difference equations, 13–14
  baseline simulation as part of, 18–19, 82–88
  basic concepts of, 11–12
  basic time unit for, 18, 70
  case history:
    effect of vertebrate pesticides, 22–27
    Gulf of Mexico shrimp fishery, 27–34
  dangers of, 4–5, 200–221
  definition of, 11
  in engineering and business, 35

evaluation of management policies or environmental situations through, 169–178
example of, 12–14
experimental designs for, 170–171
"gaming", 4
overviews of, 34–35
in wildlife and fisheries, 3, 4, 221
see also Modeling
Simulation models:
basic time unit for, 18, 70
compared to analytical models, 11
description of, 11
modification of to represent alternative situations, 169
relationship to statistical models, 78
in wildlife and fisheries, 3
see also Models; Mathematical models
Sinclair, A. R. E., 3
Sinks, 46
description of, 45
symbol for, 41
Smith, F. E., 49, 149
Smith, H., 252, 258, 259, 263
Smith, R. L., 241, 242
Snedecor, G. W., 78, 140, 172, 188
Sokal, R. R., 74, 86, 140, 219, 258
Sources, 46
description of, 45
symbol for, 41
Spain, J. D., 35, 71
Squid, 34
Stadelman, D., 3
State variables, 17, 45, 46
description of, 42–43
equations for, 62–63
symbol for, 41
Statistical distributions, 73, 76–79
Statistical significance *vs.* practical significance, 146–148
Steinhorst, R. K., 19, 129, 149
Stochastic dynamic programming, 3
Stochastic models, *see* Models, stochastic
Submodels, description of, 46–48
Succession, ecological, 35
probability transition matrix representation of, 235–237
simulation of, 223
System of interest:
bounding of, 17, 38–42
categorization of components within, 42–45
concept of, 7
identification of relationships among components within, 45–48
mathematical formulations for, 17–18
Systems:
definition of, 1, 5–6
ecological, 2, 6
general theory of, 6
levels of resolution of, 6
nesting of, 6
overlapping of, 6
Systems analysis:
approaches to, 14
benefits of, 3
case history:
effect of vertebrate pesticides, 22–27
Gulf of Mexico shrimp fishery, 27–34
concepts of, 5–14
considerations before application of, 14–16
dangers of, 4–5, 200–221
definition of, 1, 7, 14
in ecology, 2, 34–35
in engineering and business, 35
in environmental management, 35
goals of, 7
iteration of phases of, 22
in natural resource management, 2
origin of, 1
overviews of, 34–35
phase I of, 16–17, 23–25, 28–30, 37–58
phase II of, 17–19, 25, 30, 60–123
phase III of, 19–20, 25, 30–31, 129–165
phase IV of, 20–21, 26–27, 31–34, 168–199
phases of, 14–22
terminology of, 5–14
in wildlife and fisheries, 2, 3, 4, 35, 221
Systems ecologists, 2, 3
Systems ecology, 2, 35

Texas A&M University, 23
Sea Grant College Program at, 27
Tillman, M. F., 3
Tipton, A. R., 3, 35
Transactions of the American Fisheries Society, 35

t-test, 140, 141, 142, 143

United Nations, Food and Agriculture Organization of, 34
United States Department of Agriculture, 23
United States Fish and Wildlife Service, 25
Usher, M. B., 35, 61, 223

Validation, model:
  based on:
    model mechanisms, 132–133
    model structure, 132–133
    qualitative model behavior, 133–137
    quantitative model behavior, 137–149
  compared to hypothesis testing, 132
  steps in, 129–130
  using quantitative versus qualitative tests, 148–149
  see also Models, validation of
Van Dyne, G. M., 2, 35, 61, 223
Variables:
  auxiliary, 17
  driving, 17
  random, 73–79
  state, 17
Vector, 224

Walters, C. J., 3, 35
Watt, K. E. F., 2, 35
Wehrly, T. E., 214
Weinberg, G. M., 35
Weir, M. D., 246
Wilcoxon signed rank test, 140
Wildlife Management, Journal of, 36
Wirth, N., 269
Wolves, 35
Wood, F. S., 71

Zeigler, B. P., 35
Ziesenis, J. S., 266